DUAL RESONANCE MODELS AND SUPERSTRINGS

DUAL RESONANCE MODELS AND SUPERSTRINGS

Paul H Frampton
University of North Carolina at Chapel Hill

Published by

World Scientific Publishing Co Pte Ltd.
P. O. Box 128, Farrer Road, Singapore 9128.

Thanks are due to the authors, the American Physical Society and North-Holland Publishing Co. for permission to reproduce the articles reprinted in this volume.

Library of Congress Cataloging-in-Publication data is available.

DUAL RESONANCE MODELS AND SUPERSTRINGS

ISBN 9971-50-080-9
9971-50-081-7 (pbk)

Printed in Singapore by Kyodo-Shing Loong Printing Industries Pte Ltd.

PREFACE TO REPRINTED AND EXPANDED EDITION

The development of string and superstring theories has an interesting history mainly because the periods of intense research activity 1968-74 and 1984- are separated by some ten years during which only a small number of papers were published. "Dual Resonance Models" was the only book on dual models and strings published during the first period. Shortly after its publication, the bottom dropped out of the field. The book went out of print in 1979, by which time interest in the subject had waned.

The main reason for reprinting the 1974 text of "Dual Resonance Models" is the widespread resurgence of interest in string theories. The book provides a comprehensive set of references for the period 1968-74 and, beyond that, many of the ideas and techniques are still germane in present superstring theories. A Supplement on superstrings, together with a small set of reprinted papers, has been added; hence the new title, "Dual Resonance Models and Superstrings." In the reprinted text from "Dual Resonance Models," typographical errors have been corrected and references updated.

As a digression, my first intimation that something unexpected had occurred in string theory (after ten years) came while visiting CERN, Switzerland, in August 1984, when Lars Brink told me from Michael Green and John Schwarz, who were then at the Aspen Center for

Physics, Colorado, that O(32) is anomaly free for open superstrings. This message was conveyed to me because just four months earlier I had stated at the ICTP in Trieste, Italy, that all open superstrings were potentially anomalous; Green and Schwarz had been in the audience. Now they had discovered the elegant mechanism by which O(32) uniquely escapes the fatal difficulty of chiral anomalies.

The construction of the E(8) x E(8) heterotic string by David Gross and his collaborators at Princeton University gave a more geometrical scheme for introduction of Yang-Mills symmetry than the unsatisfying multiplicative Chan-Paton factor of old. These developments compelled the scientific community to take seriously the idea of a superstring as a "theory of everything." Firstly, it is a real candidate for a consistent model of quantum gravity; the theory not only can describe gravity but actually requires that gravity must exist. Second, by compactification to four spacetime dimensions one can hope to make contact with the gauge field theories which successfully describe strong and electroweak interactions at experimentally accessible energies.

Of course, there is room for skepticism. Superstrings have no basis in, or even motivation from, experiment. Further, superstrings seem to have been discovered by historical accident and therefore may be just one of many possible consistent frameworks for quantum gravity. Certainly experiment, and not just mathematical consistency, is the ultimate test of any physical theory. Nevertheless, it is when one makes detailed calculations in superstring theory that one is constantly impressed by the remarkable properties which emerge. Superstrings could provide at least an existence theorem for a finite quantum gravity unified with gauge field theory.

The abrupt 1974 demise of dual resonance models for strong interactions was foreseeable because an alternative approach, gauge field theory, had been potentially available for almost twenty years. An equally abrupt demise of superstrings would require an alternative, better approach to quantum gravity. At present, to my knowledge, a plausible competitor to the superstring is nowhere in sight. At this time, therefore, the superstring is expected to remain of interest to physicists for several years; it seems likely that it will take that long to establish any meaningful confrontation with experiment. If and when any agreement between superstrings and experiment is found, it will certainly justify an entirely new book on the subject.

I wish to thank Professor J. C. Pati for encouragement to reprint and expand "Dual Resonance Models." The Benjamin/Cummings Publishing Company generously assigned and transferred its 1974 copyright. I also thank Professor T. Kobayashi for the opportunity to give lectures on the superstrings Supplement at Tokyo Metropolitan University in April 1986. Finally, I thank Jody Gollan for typing the new camera-ready material.

Chapel Hill
May 1986

Paul H. Frampton

ORIGINAL PREFACE*

Despite the fact that local field theory is successful in describing electromagnetic, and to a lesser extent, weak interactions, it seems to be plagued with difficulties when applied to strong interactions. Firstly, the available methods in field theory are largely tied to perturbation expansions in the coupling constant and the coupling constant for strong interactions is large; secondly, because of the proliferation of strongly-interacting particles it seems impracticable to associate a new field to each new particle.

Because of these problems, the S-matrix approach was developed which rejects the field as the important concept and attempts to study directly the transition matrix elements. This work is based on the fundamental postulates of (1) Poincaré invariance, (2) Cluster decomposition, (3) Analyticity, (4) Unitarity and (5) Crossing symmetry. The main difficulty here lies in disentangling the complicated non-linear relations implied by unitarity; in general, rather drastic simplifying assumptions must be made.

*To "Dual Resonance Models" (Benjamin, 1974).

During the last few years, however, some new light has been shed on the S-matrix approach through the introduction of the duality idea, and that is our present subject.

As is well known, the concept of duality in strong interactions was first arrived at in 1967 by Dolen, Horn and Schmid from the study of the constraints imposed by analyticity and crossing symmetry through the technique of finite-energy sum rules. It was found that the direct-channel resonances and the cross-channel Regge poles provided, in an average sense, equivalent descriptions of the same phenomena.

The notion of duality received its first precise formulation with the advent of dual resonance models, starting from Veneziano's proposal in 1968 of the Euler B function model for the four-point function. This model demonstrated how the direct-channel and cross-channel descriptions can be precisely equivalent for a sum over an infinite number of resonances. The presentation of this model clarified this question and precipitated a great deal of significant progress. First it was found that the model could be straightforwardly extended to a multiparticle amplitude embodying the same principles, and, more remarkably, that the resultant amplitude was completely factorisable on a finite degeneracy. Secondly, towards the end of 1969 Virasoro pointed out that for a particular intercept value there were sufficient gauge relations to allow the possibility of eliminating indefinite-metric ghosts, although the complete proof that this indeed happens was not arrived at until 1972.

The model thus exhibits enormous mathematical consistency: of course this simplest model is far from describing Nature--for example, the mass spectrum is quite unrealistic and there are no fermions.

There followed a search for more complicated and improved models. In 1971, Ramond, Neveu, and Schwarz developed a dual theory with several advantages over the earlier one. It added additional degrees of freedom, and enlarged the gauge algebra, in such a way that ghosts were still absent, and fermions could be included in a consistent way. Subsequently, a variety of methods have been used to look for an even better model but although proposals exist no one has yet demonstrated that his particular model satisfies all the required postulates.

Despite the lack of realism, the construction of these models represents a significant advance in the S-matrix approach to strong interactions. It teaches us that we may try to satisfy the basic postulates in a resonance approximation, with the advantage that the implications of unitarity are greatly simplified for resonance exchange.

What follows has been developed out of lectures given at the Nordic School in Spötind, Norway (January 1972), at Bielefeld University, West Germany (Autumn 1972), at Syracuse University, New York (Spring 1973) and at the Ettore Majorana School in Erice, Sicily (July 1973). The material has been up-dated, approximately to April 1974.

In Part I we introduce the phenomenological concept of duality after giving some elementary discussion of Regge poles and resonances. This explains the motivation for constructing the dual resonance models. Part II deals with the Veneziano function, its multiparticle generalisation and derives the exponential degeneracy of states. Here the very important projective group O(2,1) is first introduced. In Part III the operator formulation of the model is analysed, making extensive use of the projective group. The no-ghost theorem is proved here.

The treatment of internal symmetry, particularly isospin, is made in Part IV. The difficulty of introducing broken symmetries (such as SU_3) is pointed out, and then the rubber string derivation of the Veneziano model is given. The main subject of Part V is the introduction of fermions, and the principal properties of the Neveu-Schwarz-Ramond theory are worked through in some detail. In Part VI the symmetric group approach is used to classify the earlier models, and to lead the way towards improved ones. To correct, at least partially, for our theoretical bias we outline in Part VII some of the phenomenological applications of the generalised Euler B function formula. Finally, in an Appendix, we show how, in the limit of small Regge slope, dual resonance models reduce to lagrangian field theories.

Throughout, we give fairly complete derivations for all the algebraic and group theoretic results and only in a few less important cases are results cited without proof.

Over the last several years I have benefited in my knowledge of dual resonance models from interactions with many other theorists. A partial list includes: D. Amati, L. Brink, R. C. Brower, P. G. O. Freund, S. Fubini, M. Gell-Mann, P. Goddard, M. Jacob, Z. Koba (deceased), C. Lovelace, S. Mandelstam, Y. Nambu, A. Neveu, H. B. Nielsen, D. I. Olive, P. Ramond, R. J. Rivers, J. Scherk, C. Schmid, J. H. Schwarz, G. Veneziano and M. A. Virasoro. I am grateful to these, and others, for enlightenment.

Finally it is a pleasure to thank Mrs. Joyce McManus, Mrs. Betty Osborne, Frau Irmela Schmidts and Mrs. Marjorie Warner for typing the manuscript.

April 1974 P. H. Frampton

TABLE OF CONTENTS

DUAL RESONANCE MODELS AND SUPERSTRINGS

1

DUALITY

1.1 INTRODUCTION

Here we introduce the background material necessary to understand the motivation for constructing dual resonance models. We discuss in turn kinematic definitions, unitarity, resonances, Regge poles, superconvergence and finite energy sum rules. After outlining the concepts of global and local duality we go on to the related questions of the Harari-Freund ansatz, absence of exotics, exchange degeneracy, and duality diagrams. The role of unitarity-violating approximations such as the use of zero-width resonances and real linearly rising trajectories is emphasised. After a discussion of the general situation of mid-1968 we end with a brief account of multiparticle production.

The main purpose here is to introduce the vocabulary of duality, and to emphasise various significant points which we will use later. It is not intended to make an exhaustive account of the topics covered, and therefore rather copious references are given both to the original papers and to some review articles.

Broadly speaking we shall be concerned here with the *real* world while later on we shall be concerned almost entirely with a model world (the Veneziano model world). There, however, we shall often identify and discuss important features of the model world in terms of the vocabulary introduced here, and thus the present discussions will constitute an invaluable dictionary of duality to have available when it is needed.

1.2 DEFINITIONS AND KINEMATICS

We begin by defining momentum and energy variables for the two into two particles scattering amplitude. All

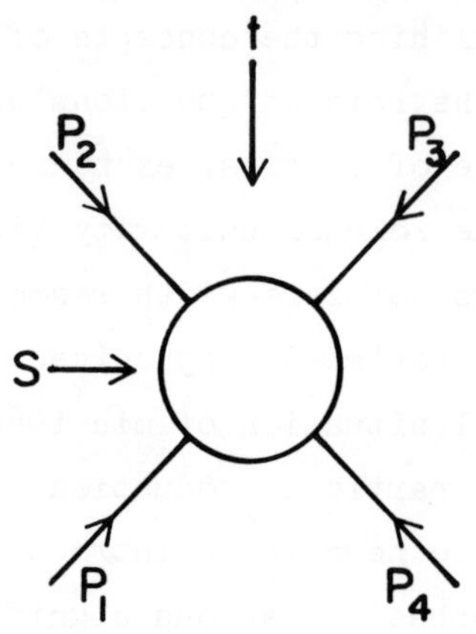

Figure 1.1
Energy and Momentum Variables.

momenta are taken as incoming and they are labelled as indicated in Figure 1.1. Momentum conservation then reads

$$\sum_{i=1}^{4} p_{i\mu} = 0 \tag{1.1}$$

and we can define the scalar energy variables

$$s = (p_1 + p_2)^2 = (p_3 + p_4)^2 \tag{1.2}$$

$$t = (p_2 + p_3)^2 = (p_4 + p_1)^2 \tag{1.3}$$

$$u = (p_1 + p_3)^2 = (p_2 + p_4)^2 \tag{1.4}$$

subject to the constraint

$$s + t + u = \sigma = \sum_{i=1}^{4} m_i^2 \tag{1.5}$$

with $p_i^2 = m_i^2$ (we use the timelike metric + ---).

When we consider the N-point function, the momenta will again be labelled p_i (i = 1, 2, ..., N) all incoming and corresponding to a particular cyclic ordering. The channel energy variables are then defined by

$$s_{ij} = (p_i + p_{i+1} + \cdots + p_j)^2 =$$
$$(p_{j+1} + p_{j+2} + \cdots + p_N + p_1 + \cdots + p_{i-1})^2 \tag{1.6}$$

The number of such planar variables is clearly $\frac{1}{2}N(N-3)$. Unless we are considering a space-time of sufficiently high dimensionality $d \geq (N-1)$ not all of these are independent. The number which is independent is otherwise given by

$$\sum_{r=1}^{d-1} r + (N-d)\,d - N = N(d-1) - \frac{1}{2}d(d+1) \tag{1.7}$$

for $d < (N-1)$. To derive this,note that we can go to a frame where particle 1 is at rest, particle 2 is moving along the z-axis, particle 3 is moving in the y-z plane, and so on, and then at the end impose momentum conservation and N mass shell conditions $p_i^2 = m_i^2$. For d = 4 the number becomes 3N-10 for $N \geq 4$; for an N = 3 vertex the correct answer is of course zero for all d.

Returning to the case N = 4 we define the s-channel centre-of-mass scattering angle θ_s by

$$z_s = \cos\theta_s = \frac{s^2 + s(2t - \sigma) + (m_1^2 - m_2^2)(m_3^2 - m_4^2)}{\sqrt{s_{12}^+ s_{12}^- s_{34}^+ s_{34}^-}} \tag{1.8}$$

in which

$$s_{ij}^{\pm} = s - (m_i \pm m_j)^2 \tag{1.9}$$

and similarly we define z_t, z_u.

The physical regions are defined by the inequality

$$\begin{Vmatrix} m_1^2 & p_1 \cdot p_2 & p_1 \cdot p_3 \\ p_2 \cdot p_1 & m_2^2 & p_2 \cdot p_3 \\ p_3 \cdot p_1 & p_3 \cdot p_2 & m_3^2 \end{Vmatrix} \geq 0 \tag{1.10}$$

Next we give our normalizations of states and S-matrix elements, and thence the connection to experimentally measured quantities. We adopt the covariant state normalization for single particle states

$$\langle p' \, \tau' \, \lambda' \mid p \, \tau \, \lambda \rangle = (2\pi)^3 \, 2p_0 \, \delta^3(\underline{p} - \underline{p}')\delta_{\lambda\lambda'}\delta_{\tau\tau'} \tag{1.11}$$

where τ, λ are particle-type, helicity labels respectively. States with N-particles are normalized similarly:

$$\langle \{p_i' \, \tau_i' \, \lambda_i'\}_{i=1,2,\ldots,N} \mid \{p_i, \tau_i, \lambda_i\} \rangle =$$

$$= \prod_{i=1}^{N} [(2\pi)^3\, 2p_{i0}\, \delta^3(\underline{p}_i' - \underline{p}_i)\delta_{\lambda_i \lambda_i'} \delta_{\tau_i \tau_i'}]. \quad (1.12)$$

To relate to experimental observables (cross-sections and decay widths) we introduce the S-matrix element connecting an initial state i to a final state f by

$$S_{fi} = \delta_{fi} + i(2\pi)^4\, \delta^4(p_f - p_i)\, \langle f|T|i\rangle \quad (1.13)$$

so that the transition probability per unit time, P_{fi}, is given by

$$P_{fi} = (2\pi)^4\, \delta^4(p_f - p_i)\, |\langle f|T|i\rangle|^2 \quad . \quad (1.14)$$

In these equations p_i, p_f are the total momenta of the initial, final states.

For an initial state with two spinless particles of momenta p_1,p_2 and a final state of (N-2) spinless particles of momenta $-p_3$, $-p_4$, ..., $-p_N$ this becomes

$$P_{fi} = (2\pi)^4 \int \prod_{i=3}^{N} \frac{d^3p_i}{2p_{0i}\,(2\pi)^3}$$

$$|\langle -p_3, -p_4, \ldots, -p_N|T|p_1\, p_2\rangle|^2\, \delta^4(p_1 + p_2 - p_f) \quad (1.15)$$

The incident flux is just $2\sqrt{s^+_{12}\, s^-_{12}}$ so that the cross-section for 2 → (N-2) particles is

$$\sigma = \frac{1}{2\sqrt{s^+_{12}\, s^-_{12}}}\, P_{fi} \quad (1.16)$$

In particular, for N = 4, we introduce the notation for the S-matrix element

$$<-p_3, -p_4 \, |T| \, p_1, p_2> = A(s,t) \tag{1.17}$$

and, doing the phase space integrals, we find for N = 4

$$\sigma = \frac{1}{2\sqrt{s^+_{12} s^-_{12}}} \int d\Omega \int \frac{p^2 dp}{4E_3 E_4} \delta(\sqrt{s} - E_3 - E_4) \, |A(s,t)|^2 \tag{1.18}$$

with $E_{3,4} = (p^2 + (m_{3,4})^2)^{1/2}$ whence

$$\sigma = \frac{1}{64\pi^2} \quad \frac{s^+_{34} s^-_{34}}{s^+_{12} s^-_{12}} \int_{-1}^{+1} d(\cos\theta_s) \, |A(s,t)|^2 \tag{1.19}$$

or

$$\frac{d\sigma}{dt} = \frac{1}{16\pi s^+_{12} s^-_{12}} \, |A(s,t)|^2 = \frac{1}{64\pi k^2 s} \, |A(s,t)|^2 \tag{1.20}$$

with $k^2 = (s^+_{12} s^-_{12})/(4s)$ as the square of the incident 3-momentum in the centre-of-mass.

It will be useful to have the formula for decay of a spin J resonance into two spin 0 particles of momenta $-p_1$, $-p_2$. For this we write the amplitude in momentum space as

$$<-p_1 \, -p_2 \, |T| \, P, J, \Lambda> =$$

$$= ig_{Joo} \, \varepsilon^{(\lambda)}_{\mu_1 \, \dots \, \mu_J} (P) \, p_1^{\mu_1} \, p_1^{\mu_2} \dots p_1^{\mu_J} \tag{1.21}$$

corresponding to the interaction Lagrangian

$$L_{int} = g_{Joo} \, \phi_{\mu_1 \, .. \, \mu_J} \, \partial^{\mu_1} \, .. \, \partial^{\mu_J} \, \phi_1 \phi_2 \tag{1.22}$$

and the decay width is then given by

$$\Gamma(J \to OO) = \frac{1}{2m_J} \int \frac{d^3p_1\, d^3p_2}{(2\pi)^6 4E_1E_2} (2\pi)^4\, \delta^4(p_f - p_i) \cdot$$

$$\cdot \frac{1}{(2J+1)} \sum_\lambda \; |\langle -p_1\, -p_2\, |T|\, P\, J\, \lambda\rangle|^2 \tag{1.23}$$

$$= \frac{g^2_{Joo}\; k^{2J+1}}{8\pi(2J+1)\, m_J^{\;2}\, C_J} \tag{1.24}$$

with

$$C_J = \frac{(2J)!}{2^J\, (J!)^2} \tag{1.25}$$

Finally we discuss unitarity of the S-matrix which is most conveniently expressed in terms of the partial wave amplitudes. Unitarity requires that

$$i(T^+_{\beta\alpha} - T_{\beta\alpha}) = (2\pi)^4 \sum_\gamma \int d\gamma\; T^+_{\beta\gamma}\; T_{\gamma\alpha}\;\, \delta^4(p_f - p_i) \tag{1.26}$$

Putting $|\alpha\rangle = |\beta\rangle = |p_1p_2\rangle$ one arrives at the optical theorem

$$2\; \mathrm{Im}\, \langle p_1p_2\, |T|\, p_1p_2\rangle = 2\; \mathrm{Im}\; A(s,o) \tag{1.27}$$

$$= (2\pi)^4 \sum_\gamma \int d\gamma\; |\langle\gamma|T|p_1p_2\rangle|^2\; \delta^4(p_1 + p_2 - p_\gamma) \tag{1.28}$$

$$= 2\, \sqrt{s^+_{12}s^-_{12}}\; \sigma_{tot} \tag{1.29}$$

Hence the total cross-section σ_{tot} is given by

$$\sigma_{tot} = \frac{1}{\sqrt{s^+_{12}s^-_{12}}}\; \mathrm{Im}\; A(s,o) \tag{1.30}$$

Take now elastic scattering of two spinless particles and define partial wave amplitudes $a_\ell(s)$ through

$$<-p_3\ -p_4\ |T|\ p_1\ p_2> = A(s,t)$$

$$= \sum_{\ell=0}^{\infty} (2\ell+1)\ a_\ell(s)\ P_\ell(z_s) \quad (1.31)$$

We will now find the conditions imposed by unitarity on $a_\ell(s)$, which take a particularly simple and convenient form. Full unitarity reads

$$2\ \text{Im}\ A(s,t) = (2\pi)^4 \sum_\gamma \int d\gamma\ <-p_3\ -p_4\ |T|\ \gamma>$$

$$\cdot <\gamma\ |T|\ p_1\ p_2>\ \delta^4(p_1 + p_2 - p_\gamma)$$

$$(1.32)$$

If we consider a kinematic region where only the elastic intermediate state is accessible $|\gamma> = |p_x p_y>$. Then we find the requirement that

$$2 \sum_{\ell=0}^{\infty} (2\ell + 1)\ \text{Im}\ a_\ell(s)\ P_\ell(z_s) =$$

$$= (2\pi)^4 \int \frac{d^3p_x\ d^3p_y}{(2\pi)^6\ 4E_x E_y} \sum_{\ell',\ell''=0}^{\infty} (2\ell'+1)(2\ell''+1)\ \cdot$$

$$\cdot\ a^*_{\ell'}(s) a_{\ell''}(s)\ P_{\ell'}(\cos\theta'_s)\ P_{\ell''}(\cos\theta''_s)\ \cdot$$

$$\cdot\ \delta^4(p_x + p_y - p_1 - p_2) \quad (1.33)$$

$$= \frac{k}{4} \int_{-1}^{+1} d(\cos\theta') \int_{0}^{2\pi} d\phi' \sum_{\ell',\ell''=0}^{\infty} (2\ell'+1)(2\ell''+1)$$

$$a^{*}_{\ell'}(s)\ a_{\ell''}(s)\ P_{\ell'}(\cos\theta'_{s})\ P_{\ell''}(\cos\theta''_{s}) \qquad (1.34)$$

In these equations θ' is the angle between $\underline{p}_1$ and $\underline{p}_x$, θ'' is the angle between $\underline{p}_x$ and $\underline{p}_4$ while ϕ' is the azimuthal angle of $\underline{p}_x$ relative to the plane of $\underline{p}_1$, $\underline{p}_3$. Hence

$$\cos\theta'' = \cos\theta\ \cos\theta' + \sin\theta\ \sin\theta'\ \cos\phi' \qquad (1.35)$$

Now we use the addition theorem for Legendre polynomials to find

$$P_{\ell''}(\cos\theta'') = P_{\ell''}(\cos\theta)\ P_{\ell''}(\cos\theta') + \sum_{m=1}^{\infty} \frac{(\ell''-m)!}{(\ell''+m)!} P^{m}_{\ell''}(\cos\theta)\ P^{m}_{\ell''}(\cos\theta') \cdot \cos m\phi' \qquad (1.36)$$

to deduce (using orthogonality) that elastic unitarity requires

$$\mathrm{Im}\ a\ (s) = \frac{k}{8\pi\sqrt{s}} |a_{\ell}(s)|^2 \qquad (1.37)$$

Hence defining

$$\hat{a}_{\ell}(s) = \frac{k}{8\pi\sqrt{s}} a_{\ell}(s) \qquad (1.38)$$

one has

$$\mathrm{Im}\ \hat{a}_{\ell}(s) = |\hat{a}_{\ell}(s)|^2.$$

(Note: For two identical particles, the phase space integration carries an additional factor 1/2! So in such a case we define $\hat{a}_\ell(s) = \frac{k}{16\pi\sqrt{s}} a_\ell(s)$ (identical particles).)

Full unitarity implies the possible addition of manifestly positive terms on the right hand side of the unitarity equation. Thus full unitarity requires

$$\text{Im } \hat{a}_\ell(s) \leq |\hat{a}_\ell(s)|^2 \tag{1.39}$$

which means that $\hat{a}(s)$ must lie in or on the elastic unitarity circle (EUC) with centre $+ \frac{1}{2} i$ and radius 1/2. This will be useful for checking when our approximations are manifestly unitarity-violating.

Another consequence of unitarity is that $\Gamma(J \to 00) \geq 0$ if the decay is kinematically allowed ($k \geq 0$). Hence to avoid negative lifetimes we must have $g^2_{J00} \geq 0$. More generally unitarity of the S-matrix implies that our interaction Lagrangian L_{int} be hermitian since one writes formally the time-ordered operator.

$$S = T\,[\exp\,(i \int dx\, L_{int})] \tag{1.40}$$

which again implies that g_{J00} is real and hence $g^2_{J00} \geq 0$.

1.3 SINGLE VARIABLE DISPERSION RELATIONS

The amplitude $A(s,t)$ is expected to be a real analytic function of the three complex variables s, t, u subject to the constraint $s + t + u = \sigma$. Let us define

$\nu = \frac{1}{2}(s - u)$ and write a dispersion relation in ν at fixed real t. There will be (say) a pole at $\nu = \nu_s$ corresponding to a bound state of squared mass $\nu_s + \frac{1}{2}(\sigma - t)$ in the s-channel and then a branch cut starting at $\nu = \nu_R$ extending to the right. Similarly on the left-hand-side of the ν-plane we may have a pole at $\nu = \nu_u$ (a u-channel bound state) and a cut running from $\nu = \nu_L$ to the left. Now we use Cauchy's theorem

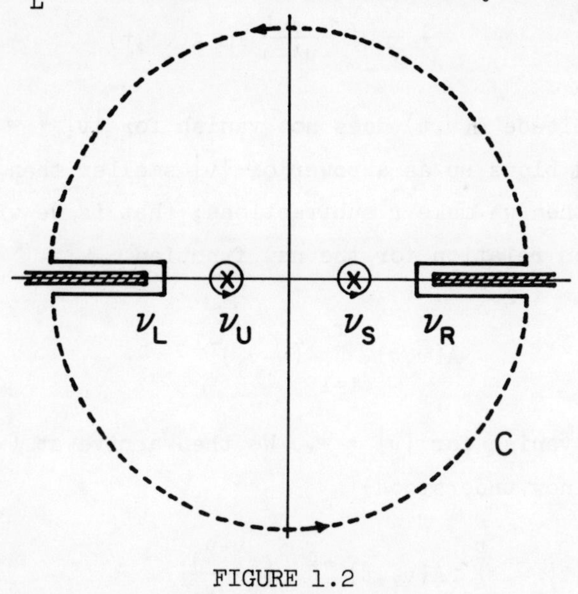

FIGURE 1.2
Complex ν-Plane

for the contour indicated in Figure 2 to arrive at

$$A(\nu,t) = \frac{g_s}{\nu_s - \nu} + \frac{g_u}{\nu_u - \nu} + \frac{1}{2\pi i} \int_c \frac{d\nu'}{\nu' - \nu} \, \mathrm{Im}\, A(\nu',t) \quad (1.41)$$

If now $A(\nu,t)$ vanishes as $|\nu| \to \infty$ we can drop the contour at infinity. Further since A is a real function we may write for the discontinuity

$$A(\nu + i\varepsilon, t) - A(\nu - i\varepsilon, t) = A(\nu + i\varepsilon, t) - A^*(\nu + i\varepsilon, t)$$
$$= 2 \, \mathrm{Im} \, A(\nu, t) \qquad (1.42)$$

Whence

$$A(\nu,t) = \frac{g_s}{\nu-\nu_s} + \frac{g_u}{\nu-\nu_u} + \frac{1}{\pi}\int_{\nu_R}^{\infty} \frac{d\nu'}{\nu'-\nu} \, \mathrm{Im} \, a(\nu',t)$$
$$+ \frac{1}{\nu}\int_{-\infty}^{\nu_L} \frac{d\nu'}{\nu'-\nu} \, \mathrm{Im} \, A(\nu',t) \qquad (1.43)$$

If the amplitude $A(\nu,t)$ does not vanish for $|\nu| \to \infty$ but instead blows up as a power of $|\nu|$ smaller than the integer r then we make r subtractions; that is, we write a dispersion relation for the new function

$$A(\nu,t) \prod_{i=1}^{r} (\nu-\nu_i)^{-1} \qquad (1.44)$$

which does vanish for $|\nu| \to \infty$. We then arrive at (with pole terms now understood)

$$A(\nu,t) = \sum_{i=1}^{r} A(\nu_i,t) \prod_{\substack{j=1 \\ j\neq i}}^{r} \left(\frac{\nu-\nu_j}{\nu_i-\nu_j}\right)$$

$$+ \frac{1}{\pi} \prod_{j=1}^{r} (\nu-\nu_j) \int_{-\infty}^{\infty} d\nu' \prod_{k=1}^{r} (\nu'-\nu_k) \times$$
$$\times \frac{\mathrm{Im} \, A(\nu',t)}{\nu'-\nu} \qquad (1.45)$$

thus introducing r subtraction constants $A(\nu_i,t)$.

1.4 RESONANCES AND REGGE POLES

We now introduce two general features observed in hadronic cross-sections, taken as a function of the direct channel energy s

(i) At low energies certain (nonexotic) two-body channels show bumps in the cross-section as a function of s - both for the total cross-section and for specific final states. Other (exotic) two-body channels do not show this structure, and have a smooth s dependence at low energies.

(ii) At high energies total cross-sections tend to approximately constant values while for specific final states the cross-section has a smooth power-law-like dependence (now for all channels, exotic and nonexotic) with the power being correlated to the quantum numbers exchanged in the crossed channel. Further the differential cross-section $d\sigma/dt$ generally shows a forward peaking and there is often marked fixed-t structure.

As is well-known these features are conveniently parametrised by (i) direct channel resonances [1] and (ii) Regge poles [original papers 2-14, Reviews 15-20] respectively, and we now briefly discuss these in turn.

For simplicity let us take again elastic spinless scattering and a direct-channel resonance (R) of mass m_R, total width Γ_R and spin J. Then its contribution to A(s,t) may be characterised by a pole on the second sheet at $s = m_R{}^2 - im_R\,\Gamma_R$ by

$$A(s,t) = \frac{G^2_{J00}\;P_J(z_s)}{m_R{}^2 - im_R\,\Gamma_R - s} + \;..... \qquad (1.46)$$

with z_s evaluated at $s = m_R{}^2$ and

$$G_{J00}^2 = g_{J00}^2 \; k^{2J} \; (C_J)^{-1} \tag{1.47}$$

so that the partial elastic width is

$$\Gamma(J \rightarrow 00) = \frac{G_{J00}^2 \; k}{8\pi \; m_R{}^2(2J + 1)} \tag{1.48}$$

As we already mentioned to avoid difficulties with unitarity we need $g_{J00}^2 \geq 0$. In general this requires also $(G_{J00})^2 \geq 0$ except for the special circumstance that there is a bound state with $k^2 < 0$ and odd J in which case we need $(G_{J00})^2 \leq 0$. A state in a theory which has $g_{J00}^2 < 0$ is referred to as a ghost.

It is a very fruitful approach in dispersion integrals to make a narrow resonance approximation $(\Gamma_R \rightarrow 0)$ to estimate the imaginary part at low energies. One then uses the well-known principal part formula

$$\frac{1}{x - i\varepsilon} = PP(\frac{1}{x}) + i\pi \; \delta(x) \tag{1.49}$$

to rewrite the imaginary part as

$$\text{Im } A(s,t) = \pi \; G_{J00}^2 \; P_J(z_s) \; \delta(s - m_R{}^2) + \cdots \tag{1.50}$$

Now, at resonance, we have

$$\hat{a}_J(m_R{}^2) = \frac{i \; G_{J00}^2 \; k}{8\pi \; m_R{}^2 \; \Gamma_R(2J+1)} \tag{1.51}$$

so it is clear that if $\Gamma_R \rightarrow 0$ and $G_{J00}^2 \neq 0$ we go outside the EUC and unitarity is violated. Nevertheless, as we recall shortly, such a unitarity-violating narrow-resonance approximation leads to predictive power in superconvergence relations and finite energy sum rule bootstraps.

Now we turn to Regge poles. For equal mass scattering the t-channel centre-of-mass angle is given by

$$z_t = \cos\theta_t = \frac{2\nu}{t - 4m^2} \tag{1.52}$$

and we may write our scattering amplitude as $A(t,z_t)$ and let us assume an unsubtracted dispersion relation in ν (and z_t); subtractions can be handled in a trivial way if necessary. Then we write

$$a_\ell(t) = \frac{1}{2}\int_{-1}^{+1} dz_t\, P_\ell(z_t)\, A(t,z_t) \tag{1.53}$$

$$= \frac{1}{2\pi}\int_{-1}^{+1} dz_t \int_{-\infty}^{\infty} dz'_t\, \frac{\mathrm{Im}\, A(t,z'_t)}{z'_t - z_t} \tag{1.54}$$

and now use[21]

$$\frac{1}{2}\int_{-1}^{+1} \frac{dz\, P_\ell(z)}{x - z} = Q_\ell(x) \tag{1.55}$$

to write the form

$$a_\ell(t) = \frac{1}{\pi}\int_{-\infty}^{\infty} dz_t\, \mathrm{Im}\, A(t,z_t)\, Q_\ell(z_t) \tag{1.56}$$

and we wish to continue $a_\ell(t)$ to general ℓ in the form $a(\ell,s)$, such that $a(\ell,s) = a_\ell(s)$ at $\ell = 0, 1, 2, \ldots$. This can only be done uniquely if we continue separately even and odd partial waves because of the unfavorable asymptotic behaviour in ℓ. Let us recall[21]

(i) $$Q_\ell(z) = \frac{\sqrt{\pi}\;\Gamma(\ell+1)}{(2z)^{\ell+1}\;\Gamma(\ell+3/2)}\; F\left(1 + \frac{\ell}{2}, \frac{1}{2} + \frac{\ell}{2}; \frac{3}{2} + \ell; \frac{1}{z^2}\right)$$

$$\sim \frac{\sqrt{\pi}\,\Gamma(\ell+1)}{(2z)^{\ell+1}\,\Gamma(\ell+3/2)} \quad \text{as } |z| \to \infty \tag{1.57}$$

$$\sim \frac{C}{\sqrt{\ell}} \exp[(\ell + \tfrac{1}{2})\,\ell n(z - \sqrt{z^2 - 1})] \tag{1.58}$$

(ii) Carlson's theorem[22]: If $f(\ell,s)$ is regular and bounded by $|f(\ell,s)| < e^{a|\ell|}$ with $a < \pi$ for Re $\ell >$ some ℓ_0 and further if $f(\ell,s) = 0$ for $\ell = 0, 1, 2, \ldots$ then $f(\ell,s) = 0$ identically.

First note that the large z behaviour of $Q_\ell(z)$ ensures us that the z integral converges. However, the negative z part of the integral gives a term $\sim e^{\pi|\ell|}$ as $|\ell| \to \infty$, and this is eliminated by defining

$$a_\ell^{\pm}(t) = \frac{1}{\pi}\int_0^\infty dz_t\, Q_\ell(z_t)\,[\text{Im } A(t,z_t) \pm \text{Im } A(t,-z_t)] \tag{1.59}$$

which may be continued uniquely. Here the two Froissart-Gribov expressions $a_\ell^{\pm}(t)$ continue respectively the even and odd partial waves.

Leaving aside the signature problem for a moment let us make the Sommerfeld-Watson transformation

$$A(t,z_t) = \sum_{\ell=0}^{\infty} (2\ell + 1)\, a_\ell(t)\, P_\ell(z_t)$$

$$= -\frac{1}{2i}\int d\ell\, \frac{a(\ell,t)\, P_\ell(-z_t)\,(2\ell + 1)}{\sin\pi\ell} \tag{1.60}$$

with the contour taken clockwise around the positive real ℓ-axis. The use of $P_\ell(-z_s)$ would mean that we were unable to move the background integral below Re $\ell = -\frac{1}{2}$ because of the unfavorable symmetry

$$P_\ell(z) = P_{-\ell-1}(z) \tag{1.61}$$

Let us recall that[21]

$$P_\ell(z) = \frac{\tan \ell\pi}{\pi}\left(Q_\ell(z) - Q_{-\ell-1}(z)\right) \tag{1.62}$$

$$= \frac{(2z)^{-\ell-1}\,\Gamma(-\ell-\frac{1}{2})}{\Gamma(-\ell)\,\Gamma(1/2)} F(1+\frac{\ell}{2}, \frac{1}{2}+\frac{\ell}{2}; \frac{3}{2}+\ell; \frac{1}{z^2})$$

$$+ \frac{(2z)^{\ell}\,\Gamma(\ell+\frac{1}{2})}{\Gamma(1+\ell)\,\Gamma(1/2)} F(\frac{1}{2}-\frac{\ell}{2}, -\frac{\ell}{2}; \frac{1}{2}-\ell; \frac{1}{z^2}) \tag{1.63}$$

and which of the terms (by $\ell \to -\ell-1$ reflection) leads depends on Re $\ell \gtrless -\frac{1}{2}$. Therefore we proceed by

$$A(t,z_t) = \frac{1}{2i}\int d\ell\,(2\ell+1)a(\ell,t)\,\frac{Q_{-\ell-1}(-z_t)}{\pi\cos\ell\pi}$$

$$+ \frac{1}{2i}\int d\ell\,(2\ell+1)\,a(\ell,t)\,\frac{Q_\ell(-z)}{\pi\cos\pi\ell} \tag{1.64}$$

$$= \frac{1}{2i}\int d\ell\,(2\ell+1)\,a(\ell,t)\,\frac{Q_{-\ell-1}(-z)}{\pi\cos\pi\ell}$$

$$+ \frac{1}{\pi}\sum_{n=1}^{\infty}(2n)\,a(n-\frac{1}{2}, t)\,Q_{n-1/2}(-z)\,(-1)^n \tag{1.65}$$

Now open up the Sommerfeld-Watson contour and move it to Re$\ell = -L$ where $-N-\frac{1}{2} < -L < -N+\frac{1}{2}$ (N = integer), picking up any poles and cuts with Re$\ell > -L$.

Then

$$A(t,z_t) = \frac{1}{2i} \int_{-L-i\infty}^{-L+i\infty} d\ell \; (2\ell + 1) \; a(\ell,t) \; \frac{Q_{-\ell-1}(-z_t)}{\pi \cos\pi\ell}$$

$$+ \frac{1}{\pi} \sum_{n=1}^{\infty} (2n) \; a(n - \tfrac{1}{2}, t) \; Q_{n-1/2}(-z_t) \; (-1)^n$$

$$- \sum_{m=0}^{-N} (2m) \; \frac{Q_{-m-1/2}}{\pi} (-z) \; a(m - \tfrac{1}{2}, t) \; (-1)^m$$

$$+ \sum_{\text{poles}} + \sum_{\text{cuts}} \tag{1.66}$$

If we now input the physical assumption

$$a(\ell,t) = a(-\ell - 1, t) \tag{1.67}$$

there is a cancellation of the first N terms in the summations leaving a sum and a background integral behaving as z^{-2}. We now throw these away (for $|\nu| \to \infty$) and keep only pole terms where

$$a(\ell,t) = \frac{b_i(t)}{\ell - \alpha_i(t)} \tag{1.68}$$

giving the Regge pole representation (for large $|s|$)

$$A(s,t) = - \sum_i \frac{2}{\sqrt{\pi}} \frac{b_i(t) \; \Gamma(\alpha_i(t) + \frac{3}{2})(-q^2)^{-\alpha_i(t)} (s)^{\text{Re}\alpha_i(t)}}{\Gamma(\alpha_i(t) + 1) \; \sin\pi\alpha_i(t)} \tag{1.69}$$

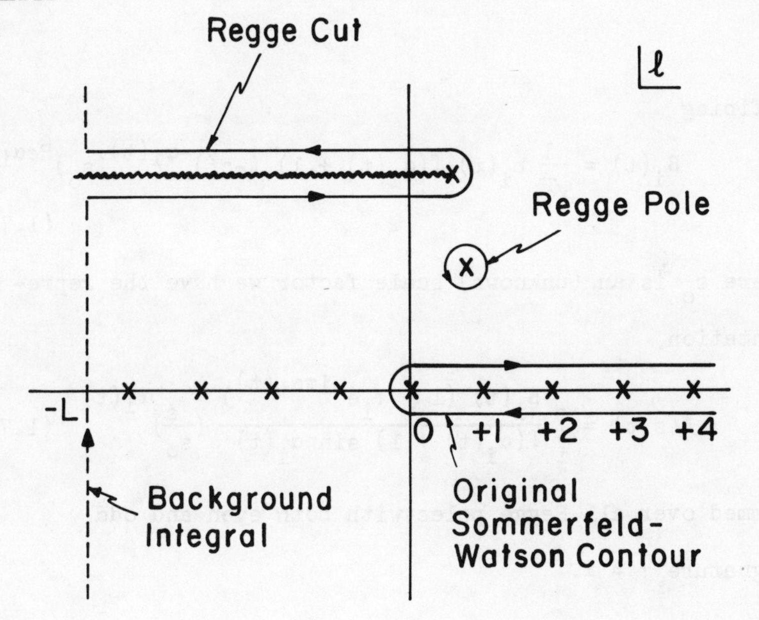

FIGURE 1.3

Complex ℓ-Plane

where we used $z_t \simeq s/2q^2$, $|s| \to \infty$. In Figure 3 we show the ℓ-plane contours, including one Regge pole and one Regge cut.

To include signature write

$$a(\ell,t) \to [a^+(\ell,t)\ \eta^+(\ell) + a^-(\ell,t)\ \eta^-(\ell)] \tag{1.70}$$

with

$$\eta^\pm(\ell) = \frac{1 \pm e^{-i\pi\ell}}{2} \tag{1.71}$$

whereupon our physical assumption becomes

$$a^\pm(\ell,t) = a^\mp(-\ell-1,t) \tag{1.72}$$

Defining

$$\beta_i(t) = \frac{1}{\sqrt{\pi}} b_i(t)\ \Gamma(\alpha_i(t) + 1)\ (-q^2)^{-\alpha_i(t)} (s_o)^{\mathrm{Re}\alpha_i(t)} \tag{1.73}$$

where s_o is an (unknown) scale factor we have the representation

$$A(s,t) = \sum_i \frac{\beta_i(t)\ (1 + \tau_i e^{-i\pi\alpha_i(t)})}{\Gamma(\alpha_i(t) + 1)\ \sin\pi\alpha_i(t)} \left(\frac{s}{s_o}\right)^{\alpha_i(t)} \tag{1.74}$$

summed over all Regge poles with both even and odd signature $\tau = \pm$.

Note that now

$$\mathrm{Im}\ A(s,t) = - \sum_i \frac{\beta_i(t)}{\Gamma(\alpha_i + 1)} \tau_i \left(\frac{s}{s_o}\right)^{\alpha_i(t)} \tag{1.75}$$

$$= - \sum_i \overline{\beta}_i(t)\ \tau_i \left(\frac{s}{s_o}\right)^{\alpha_i(t)} \tag{1.76}$$

with $\overline{\beta}_i(t) = \beta_i(t)/\Gamma(\alpha_i(t) + 1)$ as the reduced residue function.

Now we add some remarks about Regge poles:

(i) The constancy of total cross sections at high energy, together with the optical theorem, imply an ℓ-plane singularity with $\alpha(o) = 1$ (the Pomeron singularity).

(ii) The scale factor s_o is usually taken to be $s_o \approx 1\ \mathrm{GeV}^2$. Changing s_o means simply putting an exponential dependence into $\beta(t)$.

(iii) Trajectory functions can be determined for $t < 0$ from fits to data and for $t > 0$ from resonance spectroscopy. The striking observed feature is that trajectories (other than the Pomeron) are approximately straight lines $\alpha(t) = \alpha(o) + \alpha' t$ with $\alpha' \simeq 0.9\ \mathrm{GeV}^{-2}$, an approximately universal slope.

(iv) Since $s^{\alpha(t)} \propto e^{\alpha' t\ \ell n s}$ it leads to a forward peaking.

(v) An s-channel resonance of spin J contributes

$$A(s,t) = \frac{G_J{}^2\ P_J(z_t)}{m_R{}^2 - i m_R \Gamma_R - s} \simeq |t|^J \quad \text{as} |t| \to \infty \qquad (1.77)$$

Hence we cannot have $J > \alpha(m_J{}^2)$ where $\alpha(m_J{}^2)$ is the leading s-channel trajectory. This relates low energy in one channel to high energy in the crossed channel. Duality will relate low and high energies in the same channel.

(vi) Regge residues (at least for poles) should factorise. By the optical theorem, if the Pomeron factorises so should total cross-sections as $|s| \to \infty$. Hence $\sigma_{\pi\pi}^{total}\ \sigma_{pp}^{total} = (\sigma_{\pi p}^{total})$, $s \to \infty$, etc.

(vii) The phase of the Regge representation is given entirely by the signature factor $\eta_{\pm}$. This asymptotic phase is actually more general, and follows from s – u crossing symmetry; the virtue of the Regge theory is that it appears automatically when we insist on a unique continuation in ℓ. The signature phase will be crucial when we discuss local duality.

1.5 SUPERCONVERGENCE AND FINITE ENERGY SUM RULES

In the scattering of spinning particles, amplitudes have the Regge behaviour[23,24,25]

$$A_h(s,t) \sim (s)^{\alpha(t) - h} \tag{1.78}$$

where h is the t-channel helicity flip. For sufficiently high spins we may therefore find amplitudes such that $\alpha(t) - h < -1$ whereupon we may write unsubtracted dispersion relations for both $A(\nu,t)$ and $\nu A(\nu,t)$:

$$A(\nu,t) = \frac{1}{\pi} \int_{-\infty}^{\infty} \frac{d\nu'}{\nu' - \nu} \operatorname{Im} A(\nu',t) \tag{1.79}$$

$$\nu A(\nu,t) = \frac{1}{\pi} \int_{-\infty}^{\infty} \frac{d\nu'}{\nu' - \nu} \nu' \operatorname{Im} A(\nu',t) \tag{1.80}$$

Subtraction gives

$$\int_{-\infty}^{\infty} d\nu \operatorname{Im} A(\nu,t) = 0 \tag{1.81}$$

which is a superconvergence relation. It may be regarded as a dispersion relation with a negative number of subtractions. Making a narrow-resonance approximation to $\operatorname{Im} A(\nu,t)$ we arrive at a sum rule of the form

$$\sum_i g_i^2 \phi_i(m_i, p_i) = 0 \tag{1.82}$$

where ϕ is a kinematic factor. This relates hadron masses and strong interaction coupling constants.

Before going on to the more general finite energy

sum rules let us make two remarks about superconvergence relations.

(i) These sum rules were first derived indirectly from current algebra[23]. Sandwiching the commutator of axial charges between, say, rho meson states of fixed helicity leads to two independent Adler-Weisberger sum rules one of which is coincident with a superconvergence relation and is independent of the detailed form of the current commutator as long as it is local.

(ii) Related to (i) is the fact that taking only a set of narrow resonances corresponding to a single multiplet of, say, SU_6 leads to SU_6 relations between coupling constants and masses. Let us take for example $\pi(p_1) + \rho_\mu(p_2) \to \pi(-p_4) + \rho_\nu(-p_3)$ and define the vectors $P = (p_1 - p_4)$, $Q = (p_2 - p_3)$. Then the amplitudes defined in

$$\langle -p_4\ \pi;\ -p_3\ \lambda'\ \rho|T|\ p_1\ \pi;\ p_2\ \lambda\ \rho\rangle =$$

$$\varepsilon_\mu{}^{\lambda}(p_2)\ \varepsilon_\nu{}^{\lambda'}(-p_4)[\bar{A}(s,t)P_\mu P_\nu + \bar{B}(s,t)(P_\mu Q_\nu + Q_\mu P_\nu) + \bar{C}(s,t)Q_\mu Q_\nu + \bar{D}(s,t)g_{\mu\nu}] \qquad (1.83)$$

have the Regge behaviour for $|s| \to \infty$

$$\bar{A}(s,t) \sim s^{\alpha(t) - 2} \qquad \bar{B}(s,t) \sim s^{\alpha(t) - 1}$$

$$\bar{C}(s,t),\ \bar{D}(s,t) \sim s^{\alpha(t)} \qquad (1.84)$$

Assuming that for t-channel isospin $T_t = 0,1,2$ the leading trajectories satisfy $\alpha^{(o)}(o) = 1$, $\alpha^{(1)}(o) = \frac{1}{2}$, $\alpha^{(2)}(o) < 0$ gives rise to four superconvergence relations at t = 0 [26]

$$\int_{-\infty}^{\infty} d\nu \; \mathrm{Im}\, \overline{B}^{(2)}(\nu,0) = 0 \tag{1.85}$$

$$\int_{-\infty}^{\infty} d\nu \; \mathrm{Im}\, \overline{A}^{(2)}(\nu,0) = 0 \tag{1.86}$$

$$\int_{-\infty}^{\infty} d\nu \; \nu \; \mathrm{Im}\, \overline{A}^{(2)}(\nu,0) = 0 \tag{1.87}$$

$$\int_{-\infty}^{\infty} d\nu \; \mathrm{Im}\, \overline{A}^{(1)}(\nu,0) = 0 \tag{1.88}$$

The second one is trivially satisfied because $\mathrm{Im}\, A^{(2)}(\nu,t)$ is odd under s-u crossing. Inserting ρ, ω and ϕ intermediate states into the first and fourth, for example, gives rise to the famous successful SU_6 prediction

$$g^2_{\rho\omega\pi} = \frac{1}{m_\rho^{\;2}} \, g^2_{\rho\pi\pi} \tag{1.89}$$

where $g_{\rho\omega\pi}$ is defined through

$$\langle -p_1, \rho, \lambda'; -p_2 \; \pi \; |T| \; P, \omega, \lambda \rangle =$$

$$= i \; g_{\rho\omega\pi} \; \varepsilon_\mu^{(\lambda)} \; (P) \; \varepsilon_\nu^{(\lambda')} \; (-p_1) \; P_k \; p_{1\lambda} \; \varepsilon_{\mu\nu k\lambda} \tag{1.90}$$

Here this result has been obtained from a smaller and conceptually simpler set of assumptions.

Superconvergence relations obtain if $\nu \; A(\nu,t)$ vanishes for $|\nu| \to \infty$. For the majority of the most important S-matrix elements (for example those describing $\pi\pi$, πN scattering) this will not be the case. More generally we can subtract off the leading Regge pole contributions by writing at fixed t

$$A(\nu,t) - \sum_i \frac{\bar{\beta}_i(t)\, 2\eta_i(t)}{\sin\pi\alpha_i(t)} \left(\frac{\nu}{\nu_o}\right)^{\alpha_i(t)} \approx 0 \qquad (1.91)$$

$$\text{for } |\nu| > N$$

where N is a suitably chosen cut-off. Writing a superconvergence relation for this combination gives rise to the finite energy sum rule (FESR) [References 27 - 34]

$$\int_{-N}^{+N} d\nu \left(\frac{\nu}{\nu_o}\right)^n \mathrm{Im}\, A(\nu,t) = \sum_i \frac{\bar{\beta}_i(t)\, \tau_i}{\alpha_i(t) + n + 1} \left(\frac{N}{\nu_o}\right)^{\alpha_i(t)+n+1} \qquad (1.92)$$

The FESR relates the low energy contributions on the left-hand-side, which we may estimate by narrow-resonance approximation or even by detailed phase shifts in the case of πN scattering, to the high energy parameters on the right-hand-side where we take a small number of Regge poles. This leads to the programme of the FESR bootstrap[35,36] where one imposes the FESR as self-consistency conditions between direct-channel resonances and crossed-channel Regge poles. Concerning such a bootstrap philosophy it is important to remark that the FESR are valid for a continuous range of t values; yet the functional dependence on t is seen to be quite different on the two sides. Not surprisingly, it can be shown that to find an analytic solution of the FESR valid for all t an infinite number of resonances (in the narrow-resonance approximation) and correspondingly an infinite number of Regge poles is necessary. The same remark of course applies to the simpler superconvergence relations, where only resonances are involved.

1.6 FESR DUALITY; SCHMID LOOPS

Now we are ready to introduce the ideas of FESR duality (global duality) and of Schmid Loops (local duality).

As we have described, the low-energy region of $A(\nu,t)$ may be parametrised by direct-channel resonances and the high-energy region by Regge poles. Many years ago (1966-67) it seemed natural to write the full amplitude as the sum (for example, Reference 37)

$$A = A_{\text{Resonances}} + A_{\text{Regge Poles}} \tag{1.93}$$

Often in phenomenological fits the interference effects between the two terms (due to their different phases) were crucial in the intermediate energy regions for the success of the fit. Hence this was called the interference model.

We make now three remarks on why the interference model, as originally proposed, had to be abandoned. The third remark will involve the statement of FESR duality.

(i) There is no sharp borderline between low-energy and high-energy and there exist intermediate energy regions where phenomena characteristics both of resonances and of Regge poles are seen at the same energy value. To quote only one famous example, in K^-p elastic scattering with $s = 3.5 \sim 6\ \text{GeV}^2$ there are both strong resonances in the direct channel <u>and</u> forward peaking and fixed t structures.

(ii) Parametrisation of the resonances leads to a 1/s high energy tail (a fixed singularity in Regge language) and this would be detectable for $t \ll 0$ such that $\alpha(t) < -1$ for all Regge trajectories. No such phenomenon seems to occur at high energies. Actually the superconvergence relations already imply that it does not.

(iii) The Regge pole term is not negligible at low energies in the resonance region. Indeed there are strong indications that the Regge term extrapolated back to low energies provides an average of the resonance contributions. Let us quote the most well-known and clearest example of this[32-34], which is for the πN charge exchange forward amplitude $A(\nu,0)$ which by the optical theorem is related to the total $\pi^{\pm}p$ cross sections by

$$\mathrm{Im}\, A(\nu,0) = \sqrt{s_{12}^{+} s_{12}^{-}}\, [\sigma^{\text{total}}(\pi^{-}p) - \sigma^{\text{total}}(\pi^{+}p)] \tag{1.94}$$

and is hence measured directly in experiment. At high energy the ν-dependence is well-described by a single rho Regge pole. [NOTE: We refer here only to the ν dependence of Im $A(\nu,0)$ and not to details such as polarisation which imply further ℓ-plane structure.] We now extrapolate back the rho Regge term to the resonance region to obtain the striking picture of Figure 4.

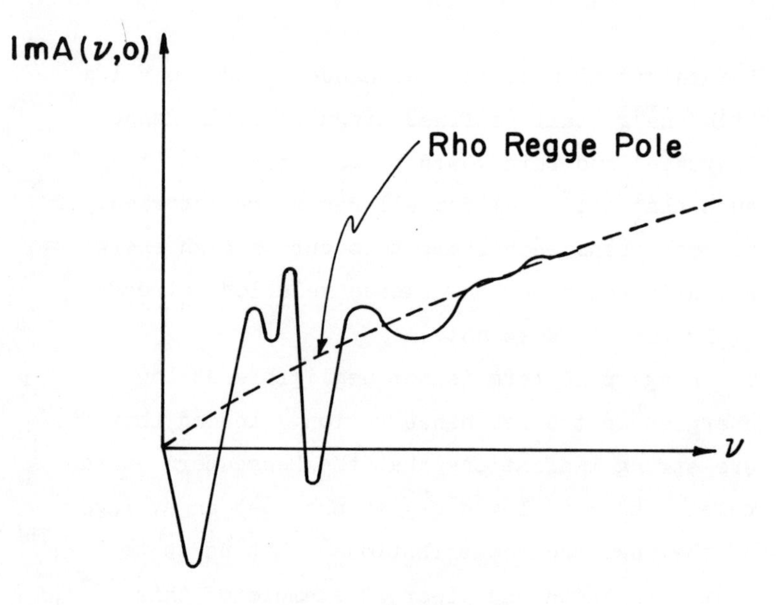

FIGURE 1.4
Duality of Rho Regge Pole and Resonances

This demonstrates that for this case the prescription

$$A(s,t) = A_{\text{Regge Pole}} + A_{\text{Resonance}} - \langle A_{\text{Resonance}} \rangle \tag{1.95}$$

is superior to the interference model. It implies that we can lower the cut-off N in the FESR through the intermediate energy region, at the same time replacing resonance contributions on the left-hand-side (at high N) by Regge pole contributions on the right-hand-side (at lower N).

We assume that such a global duality holds good in all processes with non-exotic direct-channel, and where pomeron exchange is disallowed in the cross-channel by quantum number considerations. This is the statement of FESR or global duality.

This idea can be pushed much further[38,39] by

looking in detail at the partial wave decomposition of the Regge pole representation extrapolated to low energies, to see whether it bears any resemblance in detail to the resonant behaviour of the observed partial wave amplitudes. Let us recall that we parametrised a resonance in spinless scattering by

$$A(s,t) = \frac{G_{J00}\ P_J(z_s)}{m_R^{\ 2} - i m_R \Gamma_R - s} = \sum_{\ell=0} (2\ell + 1)\ a_\ell(s)\ P_\ell(z_s) \tag{1.96}$$

$$\hat{a}_J(s) = \frac{k}{8\pi\sqrt{s}} \frac{G^2_{J00}}{(2J + 1)} \frac{1}{(m_R^{\ 2} - i m_R \Gamma_R - s)} \tag{1.97}$$

$$= \frac{\eta_J\ e^{2i\ J} - 1}{2i} \tag{1.98}$$

where we introduced the two real parameters δ_J, η_J the phase shift and elasticity ($0 \leq \eta_J \leq 1$) respectively. At resonance $\delta_J = \pi/2$ (modulo 2π). If the elasticity decreases monotonically with s (as we expect for more and more open channels) and the phase shift increases smoothly then for narrow non-overlapping resonances on linear trajectories we expect a behaviour of $\hat{a}_J(s)$ as

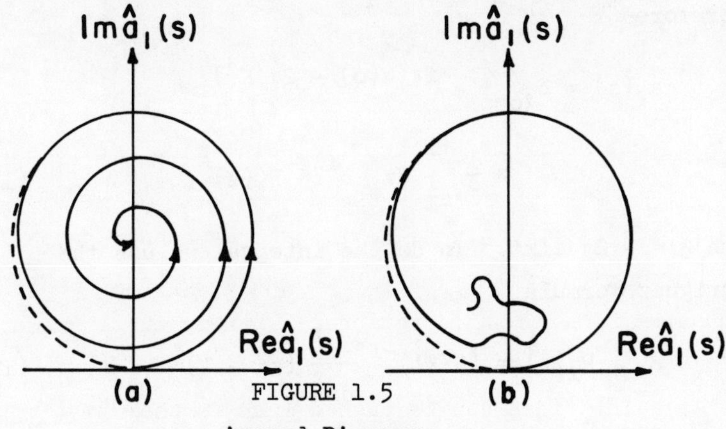

FIGURE 1.5

Argand Diagrams

in Figure 1.5(a). In reality, a typical behaviour is illustrated by Figure 1.5(b). Nevertheless, resonances are generally identified in πN scattering by the characteristic loop behaviour in the $\hat{a}_J(s)$.

Now if we partial wave analyse a Regge pole of the form

$$\frac{\beta(t)\ (1 \pm e^{-i\pi\alpha(t)})}{\Gamma(\alpha(t)+1)\ \sin\pi\alpha(t)} \left(\frac{s}{s_o}\right)^{(t)} \qquad (1.99)$$

the most significant factor is the rotating phase. To illustrate how resonance-like Argand loops are generated let us made the approximation[40,41]

$$\frac{\beta(t)}{\Gamma(\alpha(t)+1)\ \sin\pi\alpha(t)} \left(\frac{s}{s_o}\right)^{\alpha(t)} = \text{constant} \qquad (1.100)$$

for fixed s and $-1 \leq z_s \leq +1$. Then we have to partial wave analyse

$$1 \pm e^{-i\pi\alpha(t)} = \sum_{\ell=0}^{\infty} (2\ell+1)\ a_\ell(s)\ P_\ell(z) \qquad (1.101)$$

Assume linear trajectories and equal masses. Then $z_s = 1 + t/2K^2$ and $\alpha(t) = \alpha(o) + 2\alpha' K^2 (z_s - 1)$. Therefore

$$a_\ell(s) = \delta_{\ell o} \pm e^{-i\pi(\alpha(o) - 2\alpha' K^2)} \times$$

$$\times \frac{1}{2} \int_{-1}^{+1} dz_s\ e^{i\lambda z}\ P_\ell(z) \qquad (1.102)$$

with $\lambda = -2\pi\ \alpha' k^2$. To do the integral we use the Rodriguez formula

$$P_\ell(z) = (2^\ell \ell!)^{-1} \frac{d^\ell}{dz^\ell} (z^2 - 1)^\ell \qquad (1.103)$$

together with the integral representation of a Bessel function

$$J_\nu(z) = \frac{(\frac{1}{2} z)^\nu}{\sqrt{\pi}\,\Gamma(\nu + \frac{1}{2})} \int_{-1}^{+1} dt\,(1 - t^2)^{\nu - \frac{1}{2}} e^{izt} \quad (1.104)$$

whereupon one finds

$$a_\ell(s) = \delta_{\ell o} \pm e^{i\pi(\alpha(o) - 2\alpha' K^2)} (i)^{\ell+1} \times$$

$$\times \sqrt{\frac{\pi}{4\pi\alpha' k^2}}\, J_{\ell+\frac{1}{2}}(-2\pi\alpha' k^2) \quad (1.105)$$

In particular we see the rotating factor $e^{2i\pi\alpha' K^2}$ with phase monotonically increasing in K^2 or s.

By analysing the rho Regge pole description of the πN charge exchange scattering, Schmid[38] tried to identify specific resonances in the direct channel by this method.

We add now some general remarks about such an identification (local duality):

(i) One objection which can be made is that the Regge pole representation does not contain any second-sheet resonance poles. This is not a very convincing objection since the Regge description is only asserted for physical real s; there are then good analogies, for example, the function $\Gamma(-s)/\Gamma(-t-s)$ has the asymptotic expansion[21]

$$\frac{\Gamma(-s)}{\Gamma(-t-s)} = (-s)^t \left(1 + O(\frac{1}{s})\right) \quad \text{for } |s| \to \infty \quad (1.106)$$

provided we stay outside a wedge $|\arg s| < \varepsilon$ around the positive real axis, despite the fact that as we penetrate the wedge the expansion clearly does not contain the poles on the real axis. This is analogous to the physical situation where the Regge asymptotic expansion is valid along the real axis but not as we penetrate into the second sheet.

(ii) Accepting the resonance interpretation, the resonance poles in the different partial waves must conspire to give a smooth behaviour (when one partial wave has a maximum the others have minima). This cancellation is ensured by the smooth Regge starting point. Conversely when there are bumps in the total amplitude the local duality is untenable. We may say that smooth high energy behaviour represents the onset of local duality.

(iii) Partial waves with $\ell > \alpha(s)$ show (small) loops when one makes the analysis of a Regge term. These ancestors must be absorbed in error bars of the Regge fit to avoid an inconsistency discussed earlier between low energies in the direct channel and high energies in the crossed channel.

(iv) Taking the local duality very seriously we see that a factorisable t-channel Regge pole should lead to a factorisable s-channel resonance. For a single Regge pole and a particular resonance this is manifestly impossible for anything with nonzero spin so that considering different reactions must lead to inconsistencies. This implies once again that we need large (infinite) families of resonances and Regge poles in order to obtain consistency.

1.7 HARARI-FREUND ANSATZ

The observed approximate constancy of total cross-sections led us to introduce an ℓ-plane singularity with $\alpha(o) \approx 1$. The singularity must then contribute to all forward elastic amplitudes. Some amplitudes, such as $\pi^{\pm}N$, $K^{-}N$, $\bar{N}N$ show resonance behaviour at low energy while others such as $K^{+}N$, NN do not. Further no particle is known which fits on the Pomeron trajectory for time-like masses and its slope measured in very high energy proton-proton scattering is certainly smaller by at least a factor two than the approximately universal value of the other Regge trajectories.

All of these properties point to the fact that the Pomeron must be treated on a different footing to the other Regge trajectories. Clearly because of the examples already quoted it is not dual to resonances, and the most attractive first approximation is the two component duality of Harari and Freund who make the ansatz[42,43]

s-channel		t-channel
Resonances	⟷	Regge Poles
Non-resonant background	⟷	Pomeron

Thus our duality equation now reads in general

$$A = A_{\text{Resonances}} + A_{\text{Regge Poles}} - $$

$$- \langle A_{\text{Resonances}} \rangle + A_{\text{Pomeron}} \qquad (1.107)$$

This ansatz is consistent with the gross features of the data (although there are even theoretical arguments showing that it cannot be exact for $\pi\pi$ scattering), namely

(i) The great difference between, for example, K^+p and K^-p total cross-sections. The former has a very flat total cross-section, no resonances and hence only the energy-independent Pomeron contributing. The latter has a marked energy dependence of the total cross-section, has resonances, and hence has both Pomeron and Regge contributions.

(ii) Apart from examples like (i) (which generalises to $\pi^\pm N$, NN and $\bar{N}N$) it has been pointed out[44,45] that if we try out local duality (Schmid loops) in πN scattering we find that while there are rather clear resonance circles in the $T_t = 1$ case, for $T_t = 0$ (which includes Pomeron exchange) the resonances are seen above an appreciable background contribution.

1.8 EXCHANGE DEGENERACY

As already mentioned the striking feature of hadron spectroscopy is the absence of exotic states. Exotics are those mesons which cannot be made from $q\bar{q}$, or baryons (antibaryons) from qqq ($\overline{qqq}$). The quarks have of course the quantum numbers for q = p, n, λ

	Q	T	T_3	S	B
p	+ 2/3	1/2	+ 1/2	0	1/3
n	- 1/3	1/2	- 1/2	0	1/3
λ	- 1/3	0	0	- 1	1/3

First class exotics are those states which cannot be so made simply because of the isospin, strangeness and baryon number alone. Second-class exotics are mesons which have natural spin parity $J^P = 0^+, 1^-, 2^+, \ldots$ but odd CP = -1. Third-class exotics are mesons which are pseudoscalar, $J^P = 0^-$ but have negative charge conjugation C = -1. These last two classes cannot be made from $q\bar{q}$ with simple orbital excitations; no such states are well established in Nature.

Therefore let us consider the total cross-sections K N and the fact that

$$\sigma^{total}(K^+p) \simeq \sigma^{total}(K^+n) \simeq \text{constant} \simeq 18 \text{ millibarns} \quad (1.108)$$

over a wide range of energy. These are exotic baryonic channels ($\bar{\lambda}$ppnn, $\bar{\lambda}$ppnn respectively). In the crossed channel (t-channel) there are contributions from ρ, f, ω, A_2 in addition to the Pomeron, P. Let us denote by a single symbol the Regge contribution to the imaginary part, that is for example

$$\rho \leftrightarrow \beta_\rho(t)\ \tau_\rho\ (\frac{s}{s_o})^{\alpha_\rho(t)} \quad (1.109)$$

Then we may write for the total cross-sections

$$\sigma^{total}(K^-p) = P + f + \omega + A_2 + \rho \tag{1.110}$$

$$\sigma^{total}(K^-n) = P + f + \omega - A_2 - \rho \tag{1.111}$$

$$\sigma^{total}(K^+p) = P + f - \omega + A_2 - \rho \tag{1.112}$$

$$\sigma^{total}(K^+n) = P + f - \omega - A_2 + \rho \tag{1.113}$$

so that

$$\sigma(K^+p) - \sigma(K^+n) = 0 = 2(A_2 - \rho) \tag{1.114}$$

$$\sigma(K^+p) + \sigma(K^+n) = 2(P + f - \omega) = \text{constant} \tag{1.115}$$

For this to be true at all s and t we must have

$$\alpha_{A_2}(t) = \alpha_\rho(t) \tag{1.116}$$

$$\beta_{A_2}(t) = -\beta_\rho(t) \tag{1.117}$$

$$\alpha_f(t) = \alpha_\omega(t) \tag{1.118}$$

$$\beta_f(t) = -\beta_\omega(t) \tag{1.119}$$

which are exchange degeneracy relations[46,47,48] between the trajectory functions and for the residue functions. Note that when we have established exchange degeneracy for the trajectory functions it must hold good everywhere, whereas the residua must be considered for each reaction separately.

The name arises because one may say that the absence of s-channel resonances implies the absence of s-channel

Majorana exchange forces, which would contribute with alternating sign to the even and odd t-channel partial waves. Absence of such exchange forces means that there is no need to distinguish even and odd signatures in the t-channel, and they become a single degenerate Regge pole.

Exchange degeneracy was originally suggested by Arnold[46] who considered that mesons might be made out of nucleon-antinucleon pairs. Exchange forces between the constituents (being baryon number B = 2) would then be strongly damped relative to direct B = 0 forces; this was his explanation of the observed exchange degeneracy.

We can go on to apply the arguments to $\pi\pi$ scattering (here $\pi^+\pi^+$ is exotic, $pp\bar{n}\bar{n}$) and to $K\bar{K}$ scattering, using the same very simple methods. One finds that absence of exotics plus duality implies the exchange degenerate quartet of trajectories $\rho - f - \omega - A_2$, plus an additional trajectory coupled only to kaons (see Lipkin[48], for a clear analysis). This is in excellent agreement with experiment since $m_\rho = m_\omega$, $m_f = m_{A_2}$ and all lie on a trajectory $\alpha(s) \simeq 1/2 + 0.9\ s$; this together with the kind of result depicted in Figure 14 is perhaps the most compelling argument in favour of duality.

1.9 DUALITY DIAGRAMS

The rules for drawing a legal duality diagram are[49,50]

(1) There are three types of lines, corresponding to p, n, λ quarks and they retain their identity throughout the diagram.

(2) Every external baryon is represented by $\begin{smallmatrix}\rightarrow\\\rightarrow\\\rightarrow\end{smallmatrix}$

(3) Every external meson is represented by $\begin{smallmatrix}\rightarrow\\\leftarrow\end{smallmatrix}$

(4) In any B = 1 channel it is possible to cut the diagram into two by cutting only qqq (not $qqqq\bar{q}$, etc.).

(5) In any B = 0 channel we need cut only $q\bar{q}$ (not $qq\bar{q}\bar{q}$, etc.).

(6) No quark lines cross (planar duality diagrams).

(7) The two ends of a single line cannot belong to the same particle.

These rules correspond to the assumptions

(1) All baryons are in $\underline{1}$, $\underline{8}$, or $\underline{10}$ SU_3 representations.

(2) All mesons are in $\underline{1}$ or $\underline{8}$ SU_3 representations.

(3) The entire S-matrix element is given in any channel by a sum of single particle states (excepting the pomeron contribution.)

The prediction is then that when no legal diagram exists the imaginary part of the corresponding S-matrix element vanishes (except for the pomeron contributions).

There then follow many predictions (which can alternatively be derived from duality and exchange degeneracy). The diagrams automatically incorporate the essential features of the SU_3 crossing matrix. Their use is best illustrated by a few examples (see Figure 1.6).

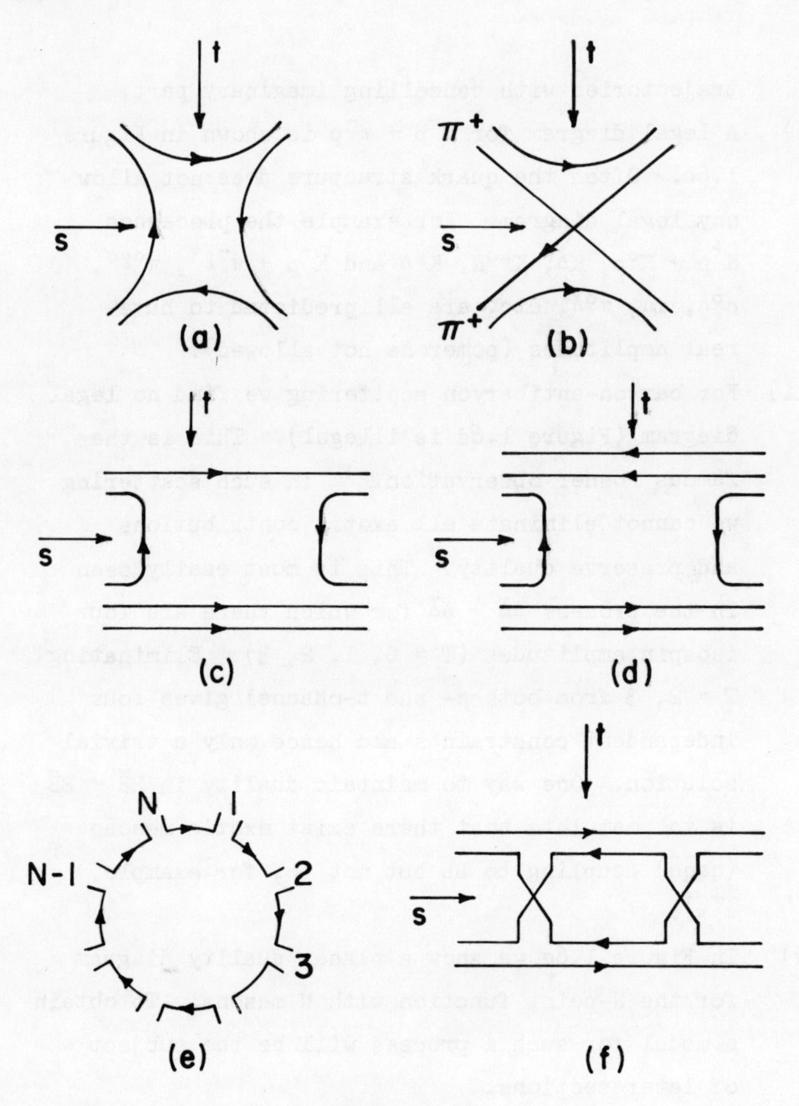

FIGURE 1.6

Quark Diagrams

(i) For meson-meson scattering the Figures 1.6a, 1.6b show a legal, illegal diagram ($\pi^+\pi^+$ scattering). Typically in the illegal diagram the crossed quark quark lines correspond to exchange degenerate trajectories with cancelling imaginary part.

(ii) A legal diagram for $\pi^- p \to \pi^- p$ is shown in Figure 1.6c. Often the quark structure does not allow any legal diagram. For example the processes $K^+ n \to K^o p$, $K\Delta$, $K^{*o}N$, $K^*\Delta$ and $K^- p \to \pi^-\Sigma^+$, $\pi^o\Sigma^o$, $\rho^o\Lambda$, $\omega\Lambda$, $\pi^o\Lambda$, etc. are all predicted to have real amplitudes (pomerons not allowed).

(iii) For baryon-antibaryon scattering we find no legal diagram (Figure 1.6d is illegal). This is the famous Rosner observation.[51] In such scattering we cannot eliminate all exotic contributions and preserve duality. This is most easily seen in the process $\Delta\bar{\Delta} \to \Delta\bar{\Delta}$ for which there are four isospin amplitudes (T = 0, 1, 2, 3). Eliminating T = 2, 3 from both s- and t-channel gives four independent constraints and hence only a trivial solution. One way to maintain duality in $B\bar{B} \to B\bar{B}$ is to postulate that there exist exotic mesons $(qq\bar{q}\bar{q})$ coupling to $B\bar{B}$ but not to, for example, $\pi^+\pi^+$.

(iv) In Figure 1.6e we show a planar duality diagram for the N-point function with N mesons. To obtain a model for such a process will be the subject of later sections.

(v) It is tempting to iterate these duality diagrams, as if they were Feynman diagrams. When we iterate the illegal $\pi^+\pi^+$ diagram we arrive at the much-discussed diagram, Figure 1.6f. It has been conjectured[52,53] that such a diagram may be associated with pomeron exchange in the t-channel; there are two reasons for this conjecture:

(i) the quantum numbers exchanged in the t-channel are those of the vacuum (except possibly for the subtle question of charge conjugation); (ii) this exchange is dual to non-resonant background in the s-channel.

Surprisingly enough, when this diagram is calculated in the Veneziano model a new ℓ-plane singularity is found in this channel and it is naturally tempting to identify it with the singularity responsible for high energy diffractive scattering.

The duality diagrams discussed here are often alternatively called quark diagrams or Harari-Rosner diagrams.

1.10 THE SITUATION OF MID-1968

Of course we pick the time mid-1968 because this was when Veneziano first proposed his beta function model. Since the course of the work has a fairly abrupt discontinuity at this point, it is instructive to summarise what was already established.

(i) We have mentioned the usefulness of the narrow resonance approximation in superconvergence relations and FESR. We wrote

$$A(s,t) = \frac{G_J^2\, P_J(z_s)}{m_R^2 - i m_R \Gamma_R - s}$$

$$\underset{\Gamma_R \to 0}{\to} PP\left[\frac{G_J^2\, P_J(z_s)}{m_R^2 - s}\right] + i G_J^2\, P_J(z_s)\, \delta(s - m_R^2) \tag{1.120}$$

This zero-width approximation was shown to violate unitarity, since it implies (i) that the partial wave amplitude $\hat{a}_J(s)$ goes outside of its elastic unitarity circle for $s \approx m_R{}^2$ and (ii) that the elastic width exceeds the total width.

(ii) It was remarked by Van Hove[54] and by Durand[55 †] that the zero width approximation can be combined consistently with Regge asymptotic behaviour if and only if the trajectories rise indefinitely. The observed trajectories appear to be approximately linear (except the pomeron). Assuming that this behaviour persists we may write for all s that

$$\alpha(s) = \alpha(o) + \alpha' s \qquad \text{(real)} \qquad (1.121)$$

Using purely real trajectory functions is essentially the same approximation as the zero-width approximation, and hence it is unitarity violating. Near a resonance of spin J we write

$$\alpha(s) = J + \alpha'(s - m_R{}^2) + \cdots\cdots + i \operatorname{Im} \alpha(s) \quad (1.122)$$

Hence near the resonance

$$\frac{1}{\sin\pi\alpha(s)} \approx \frac{(-1)^J}{\pi\alpha'(m_R{}^2 - i\alpha_{Im}(m_R{}^2)/\alpha' - s)} \qquad (1.123)$$

and comparison of the Regge pole and resonance representations gives

$$\operatorname{Im} \alpha(m_R{}^2) = \alpha' \, m_R \, \Gamma_R \qquad (1.124)$$

† See also Reference 36.

$$b(m_R^2) = \frac{\pi\alpha' G_{J00}^2}{(2J+1)} \qquad (1.125)$$

Unitarity therefore dictates that

$$\mathrm{Im}\ \alpha(m_R^2) \geq \frac{k}{8\pi^2 m_J} b(m_J^2) \qquad (1.126)$$

so that putting $\mathrm{Im}\ \alpha = 0$ while $b \neq 0$ violates unitarity.

(iii) Daughter trajectories

(iiia) Group theoretical analyses[56-59] at $t = 0$ had led to the suggestion of sequences of daughter trajectories spaced by two units at $t = 0$, although such analyses could say little about what happened to these trajectories for $t \neq 0$.

(iiib) The FESR bootstrappers[35,60-64] at the Weizmann Institute and elsewhere were finding that it is impossible for a single linear trajectory to maintain self-consistency, but that quite good bootstrap consistency could be obtained when parallel daughters spaced by two units of angular momentum were inserted.

(iv) On a technical point, which nevertheless provides considerable simplification, it had been emphasised that the amplitude for a process such as $\pi^a(p_1) + \pi^b(p_2) \to \pi^c(-p_3) + \omega(-p_4)$ was especially suitable for the FESR bootstrap since we may write[35,61,62,64]

$$\langle -p_3, \pi, c; -p_4, \omega, \lambda |T| p_1, \pi, a; p_2, \pi, b\rangle =$$

$$= \varepsilon^{abc}\ \varepsilon_{\mu\nu\rho\sigma}\ p_1^{\ \mu}\ p_2^{\ \nu}\ p_3^{\ \rho}\ \varepsilon^{\sigma}(-p_4,\lambda)\ A(s,t,u) \qquad (1.127)$$

whereupon $A(s,t,u)$ is fully symmetric in s,t,u. The complications of spin and isospin are removed. In particular there are no pomeron contributions; only the ρ trajectory, of the well-established trajectories, contributes.

(v) On a more philosophical level, we have so far not defined duality but rather discussed duality. The nearest to a definition was the equation

$$A = A_{\text{Resonances}} - \langle A_{\text{Resonances}} \rangle + A_{\text{Regge poles}} + \\ + A_{\text{Pomeron}} \tag{1.128}$$

Note, however, that none of the terms on the right-hand side is well-defined. In the Regge contributions there is an arbitrary residue function $b(t)$ for each Regge pole; in the resonance contribution the detailed shape of the resonance formula is not much restricted by unitarity (for a palliative to our simple Breit-Wigner form see Eq (11) of Reference 65).

Now that we have seen Figure 1.4 we may concoct a Regge term which vanishes at low energy and a resonance term which vanishes at high energy, and then re-instate a generalised interference model

$$A = A^{\text{mutilated}}_{\text{Resonances}} + A^{\text{mutilated}}_{\text{Regge}} + A_{\text{Pomeron}} \left\{\begin{array}{l}\text{generalised}\\ \text{interference}\\ \text{model}\end{array}\right\} \tag{1.129}$$

Here the resonance and Regge sets of parameters are independent. The principal advantage of duality over a generalised interference model is that the two sets are clearly interdependent, so that the number of free parameters is smaller.

In a model world (Veneziano model world) we will be able to give a precise mathematical meaning to duality. Although we should avoid any confusion between the model world and the real world, it will be clear that the precise definition (in a narrow resonance approximation) is motivated by the phenomenological facts.

1.11 MULTIPARTICLE PRODUCTION

Direct phenomenological evidence for validity of duality (local or global) in reactions $2 \to (N-2)$ particles ($N \geq 5$) is very scarce. Nevertheless, duality will soon be tested by inclusive reactions, and has already been invoked to justify usage of the multiperipheral bootstrap in regions where not all energies are large.

(i) In the multiregge picture multiparticle production

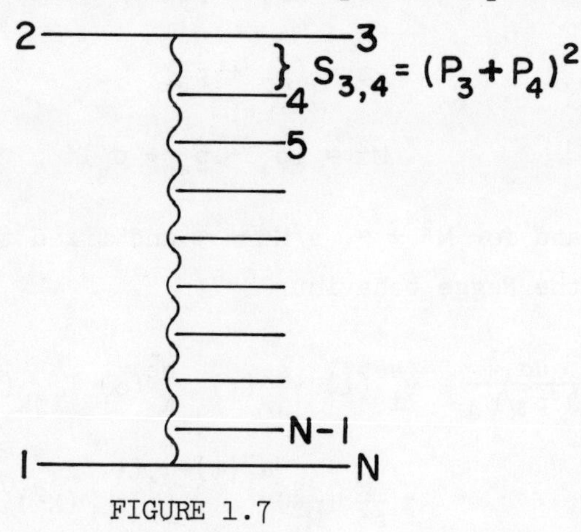

FIGURE 1.7

Multiregge Diagram

is assumed to be dominated by diagrams like Figure 1.7, characterised by peripherality (low momentum transfers) and generalised Regge asymptotics. When the sub-energies s_{34}, s_{45}, ... etc. are small, one might expect that resonances become important. Here one can invoke duality to argue that the resonance contributions are already counted, in an average sense, in the Regge exchanges (wiggly lines of Figure 1.7).

(ii) Phenomenological evidence about duality will be forthcoming in inclusive reactions where we detect and measure one, or at most two, final particle(s). For the reaction

$$1(p_1) + 2(p_2) \rightarrow 3(-p_3) + \text{anything} \tag{1.130}$$

we define variables

$$s = (p_1 + p_2)^2 \tag{1.131}$$

$$t = (p_2 + p_3) \tag{1.132}$$

$$M^2 = (p_1 + p_2 + p_3)^2 \tag{1.133}$$

and for $M^2 \rightarrow \infty$, $s/M^2 \rightarrow \infty$ and fixed t we expect the Regge behaviour[66,67]

$$\frac{d\sigma_{12}}{d^3p_3/E_3} \approx \gamma_i^{ac}(t)\ \gamma_j^{ac}(t)\ \gamma_k^{b\bar{b}}(o)\ \Gamma_{ijk}(t,t,o)\ \cdot$$

$$\cdot\ \frac{M}{s^2}\left(\frac{s}{M^2}\right)^{\alpha_i(t)+\alpha_j(t)} (M^2)^{\alpha_k(o)} \tag{1.134}$$

corresponding to the diagram of Figure 1.8d. We can

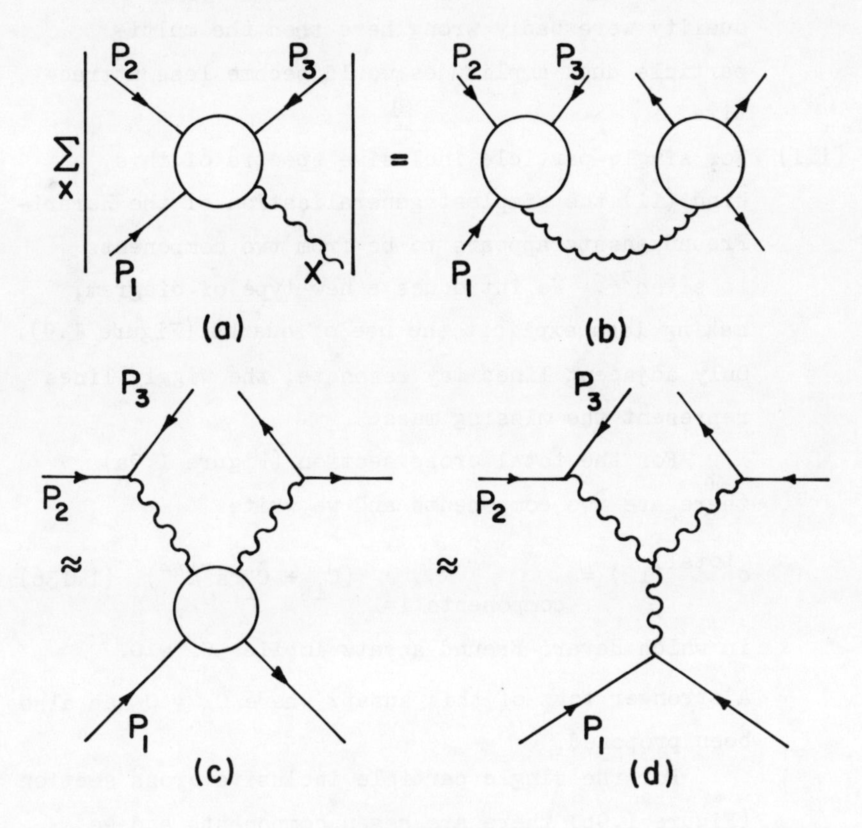

FIGURE 1.8

Regge-Limit for Inclusive Reaction

assume that the reggeon-particle forward amplitude

$$\alpha_i(p_2 + p_3) + 1(p_1) \to \alpha_j(-p_2 - p_3) + 1(-p_1) \quad (1.135)$$

has the usual two-body analyticity and write[68-71] a generalised FESR in the variable M^2. Hence we can check whether the triple Regge limit, extrapolated back to low M^2 gives an average of the reggeon-particle resonances.

This is an important programme, since if duality were badly wrong here then the multiparticle dual amplitudes would become less attractive.

(iii) For single-particle inclusive spectra of this kind (ii) the simplest generalisation of the Harari-Freund ansatz appears to be from two components to seven[72]. We introduce a new type of diagram, making less explicit the use of quarks (Figure 1.9). Only adjacent lines may resonate; the wiggly lines represent the missing mass.

For the total cross section (Figure 1.9a) there are two components and we write

$$\sigma^{total}(12) = \sum_{\text{components } i=1}^{2} (C_i + \tilde{C}_i\, s^{-1/2}) \quad (1.136)$$

in which Harari-Freund ansatz implies $C_1 = 0$. A stronger form of this ansatz where $\tilde{C}_2 = 0$ has also been proposed.

For the single particle inclusive cross section (Figure 1.9b) there are seven components and we write (analogously to the previous equation)

$$\frac{d\sigma_{12}}{d^3p_3/E_3} = \sum_{\text{components, } i=1}^{7} (d_i + \tilde{d}_i s^{-1/2}) \quad (1.137)$$

and then the problem is to understand the role of the pomeron in this case. For further details of this question we refer the reader to the recent literature. [Reference 73 and references cited therein, and Reference 74].

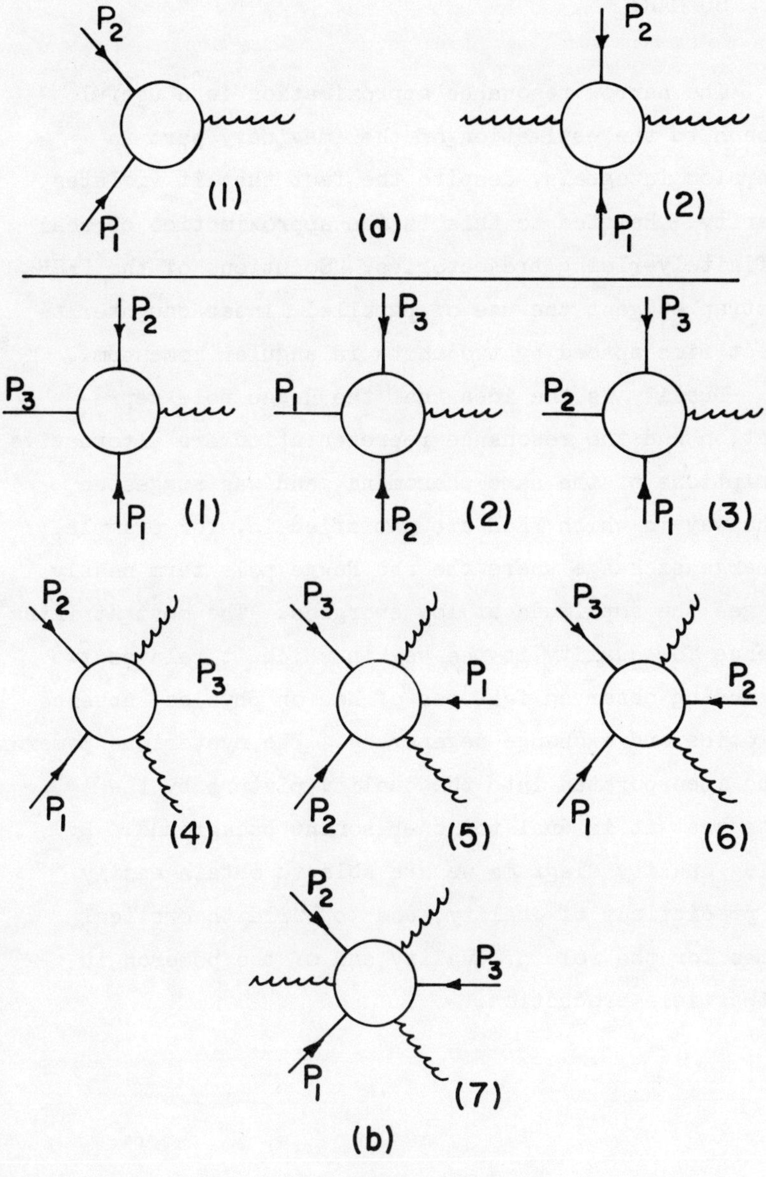

FIGURE 1.9

Seven-Component Duality

1.12 SUMMARY

The narrow resonance approximation is a useful approach to the estimation of the imaginary part in dispersion integrals, despite the fact that it violates unitarity. Related to this is the approximation of real indefinitely-rising trajectories. Solutions of the FESR bootstrap suggest the use of parallel linear daughter trajectories spaced by two units in angular momentum.

Duality is the idea that the Regge pole representation and the resonance representation are alternative descriptions of the same phenomena, and was suggested by the way in which FESR are satisfied in, for example, πN charge exchange where the rho Regge pole term neatly averages the amplitude at low energies. The most striking evidence for duality is the way in which it relates two outstanding observed features of hadron physics: absence of exotics and exchange degeneracy. The mysterious pomeron can be incorporated into the duality picture by the ansatz that it is dual to non-resonant background. By drawing duality diagrams we are able to obtain easily many predictions of duality, and to build theoretical schemes for the role of duality and of the pomeron in multiparticle production.

REFERENCES

1. J. M. Blatt and V. F. Weisskopf, Theoretical Nuclear Physics, Wiley (1962), Pages 379-441.
2. T. Regge, Nuovo Cimento 14, 951 (1959).
3. T. Regge, Nuovo Cimento 18, 947 (1960).
4. A. Bottini, A. M. Longini and T. Regge, Nuovo Cimento 23, 954 (1962).
5. S. Mandelstam, Ann. Phys. 19, 254 (1962).
6. G. F. Chew and S. C. Frautschi, Phys. Rev. Letters 7, 394 (1961).
7. G. F. Chew and S. C. Frautschi, Phys. Rev. Letters 8, 41 (1962).
8. S. C. Frautschi, M. Gell-Mann and F. Zachariasen, Phys. Rev. 126, 2204 (1962).
9. V. N. Gribov, J. Exptl. Theoret. Phys. (USSR) 41, 667, 1962 (1961).
10. G. F. Chew, S. C. Frautschi and S. Mandelstam, Phys. Rev. 126, 1202 (1962).
11. B. M. Udgaonkar, Phys. Rev. Letters 8, 142 (1962).
12. R. Blankenbecler and M. L. Goldberger, Phys. Rev. 126, 766 (1962).
13. V. N. Gribov and I. Ya. Pomeranchuk, Phys. Rev. Letters 8, 343, 412 (1962).
14. C. Lovelace, Nuovo Cimento 25, 730 (1962).
15. P. T. Matthews, Proc. Phys. Soc. 80, 1 (1962).
16. S. C. Frautschi, Regge Poles and S-Matrix Theory, Benjamin (1963).
17. G. F. Chew, Revs. Mod. Phys. 34, 394 (1962).
18. E. J. Squires, Complex Angular Momenta and Particle Physics, Benjamin (1963).

19. P. D. B. Collins and E. J. Squires, Regge Poles in Particle Physics, Springer Tracts in Modern Physics, Vol. 45, Springer (1968).
20. P. D. B. Collins, Physics Reports 1C, 105 (1970).
21. A. Erdélyi et al., Higher Transcendental Functions, Bateman Manuscript Project, McGraw Hill (1953); E. T. Whittaker and G. N. Watson, Modern Analysis, Cambridge University Press, Fourth Edition (1969).
22. E. C. Titchmarsh, The Theory of Functions, Oxford University Press, Second Edition (1939).
23. V. de Alfaro, S. Fubini, G. Rossetti and G. Furlan, Physics Letters 21, 576 (1966).
24. G. Mahoux and A. Martin, Phys. Rev. 174, 2140 (1968).
25. J. S. Bell, Nuovo Cimento 61A, 541 (1969).
26. P. H. Frampton and J. C. Taylor, Nuovo Cimento 49A, 152 (1967).
27. K. Igi, Phys. Rev. Letters 9, 76 (1962).
28. K. Igi, Phys. Rev. 130, 820 (1963).
29. K. Igi and S. Matsuda, Phys. Rev. Letters 18, 625 (1967).
30. K. Igi and S. Matsuda, Phys. Rev. 163, 1622 (1967).
31. A. A. Logunov, L. D. Soloviev and A. N. Tavkhelidze, Phys. Letters 24B, 181 (1967).
32. D. Horn and C. Schmid, CALT 16-127 (1967) (unpublished).
33. R. Dolen, D. Horn and C. Schmid, Phys. Rev. Letters 19, 402 (1967).
34. R. Dolen, D. Horn and C. Schmid, Phys. Rev. 166, 1768 (1968).
35. M. Ademollo, H. R. Rubinstein, G. Veneziano and M. A. Virasoro, Phys. Rev. Letters 19, 1402 (1967)
36. S. Mandelstam, Phys. Rev. 166, 1539 (1968).

37. V. Barger and M. Olsson, Phys. Rev. 151, 1123 (1966).
38. C. Schmid, Phys. Rev. Letters 20, 689 (1968).
39. C. Schmid, Nuovo Cimento 61A, 289 (1969).
40. C. B. Chiu and A. Kotanski, Nucl. Phys. B7, 615 (1968).
41. C. B. Chiu and A. Kotanski, Nucl. Phys. B8, 553 (1968).
42. H. Harari, Phys. Rev. Letters 20, 1395 (1968).
43. P. G. O. Freund, Phys. Rev. Letters 20, 235 (1968).
44. F. J. Gilman, H. Harari and Y. Zarmi, Phys. Rev. Letters 21, 323 (1968).
45. H. Harari and Y. Zarmi, Phys. Rev. 187, 2230 (1969).
46. R. C. Arnold, Phys. Rev. Letters 14, 657 (1965).
47. C. Schmid, Lettere al Nuovo Cimento 1, 165 (1969).
48. H. J. Lipkin, Nucl. Phys. B9, 349 (1969).
49. H. Harari, Phys. Rev. Letters 22, 562 (1969).
50. J. L. Rosner, Phys. Rev. Letters 22, 689 (1969).
51. J. L. Rosner, Phys. Rev. Letters 21, 951 (1968).
52. P. G. O. Freund and R. J. Rivers, Phys. Letters 29B 510 (1969).
53. D. J. Gross, A. Neveu, J. Scherk and J. H. Schwarz, Phys. Rev. D2, 697 (1970).
54. L. Van Hove, Phys. Letters 24B, 183 (1967).
55. L. Durand, Phys. Rev. 161, 1610 (1967).
56. G. Domokos and P. Suranyi, Nuclear Physics 54, 529 (1964).
57. M. Toller, Nuovo Cimento 37A, 631 (1965).
58. D. Z. Freedman and J. M. Wang, Phys. Rev. Letters 17, 569 (1966).
59. D. A. Freedman and J. M. Wang, Phys. Rev. 153, 1596 (1967).

60. M. Ademollo, H. R. Rubinstein, G. Veneziano and M. A. Virasoro, Nuovo Cimento 51A, 227 (1967).

61. M. Ademollo, H. R. Rubinstein, G. Veneziano and M. A. Virasoro, Phys. Letters 27B, 99 (1968).

62. H. R. Rubinstein, A. Schwimmer, G. Veneziano and M. A. Virasoro, Phys. Rev. Letters 21, 491 (1968).

63. M. Bishari, H. R. Rubinstein, A. Schwimmer and G. Veneziano, Phys. Rev. 176, 1926 (1968).

64. M. Ademollo, H. R. Rubinstein, G. Veneziano and M. A. Virasoro, Phys. Rev. 176, 1904 (1968).

65. G. J. Gounaris and J. J. Sakurai, Phys. Rev. Letters 21, 244 (1968).

66. G. F. Chew and A. Pignotti, Phys. Rev. Letters 22, 1219 (1969).

67. C. E. Detar, C. E. Jones, F. E. Low, J. H. Weis, J. E. Young and C. I. Tan, Phys. Rev. Letters 26, 675 (1971).

68. M. B. Einhorn, AIP Conf. Proc. No. 6 (1971) page 99.

69. A. I. Sanda, Phys. Rev. D6, 280 (1972).

70. J. Kwiecinsky, Nuov. Cim. Lett. 3, 619 (1972).

71. M. B. Einhorn, J. Ellis and J. Finkelstein, Phys. Rev. D5, 2063 (1972).

72. D. Gordon and G. Veneziano, Phys. Rev. D3, 2116 (1971).

73. P. H. Frampton, CERN preprint TH 1497 (unpublished).

74. S. H. H. Tye and G. Veneziano, Nuovo Cimento 14A, 711 (1973).

2

MULTIPARTICLE DUAL MODEL

2.1 INTRODUCTION

We introduce here the beta function model for four external particles as a closed-form solution of the FESR bootstrap in a narrow resonance approximation. The model possesses a precise form of duality: it can be written as a sum of poles (resonances) in either the direct or crossed channels and at the same time has complete Regge behaviour if we avoid the real axes. The model is extended to the 5 and then to the N point functions. A reformulation in terms of anharmonic ratios reveals an invariance group - the projective group ($SL(2R) \approx O(2,1) \approx SU(1,1)$). Using harmonic oscillators the N-point function is fully factorised, to reveal a level density increasing exponentially in the mass.

Alternative dual models are then discussed. Firstly we treat the addition of satellites which leads to infinite ambiguities in the four-point function, but factorisation of the N-point generalisation leads in general to a qualitatively different form of degeneracy. A model for pion-pion scattering shows some striking coincidences with predictions of chiral symmetry. A non-planar N-point function is then considered - the Shapiro-Virasoro model. Although of equal mathematical beauty, this model is rejected in favour of the planar model on the basis of the observed exchange degeneracy and absence of hadrons with exotic quantum numbers.

2.2 VENEZIANO'S BETA FUNCTION AND ITS PROPERTIES

In an already classic paper[1], Veneziano has proposed a beautiful closed-form solution of the FESR bootstrap, by writing down a crossing symmetric Regge behaved amplitude based on linearly-rising trajectories and a narrow resonance approximation.

Let us consider the fully crossing symmetric amplitude for elastic scattering of identical spinless particles (with no internal quantum numbers). Then the proposal is

$$\langle -p_3\ -p_4|T|p_1\ p_2\rangle = A_4(s,t)$$

$$= \bar{\beta}[B(-\alpha(s),\ -\alpha(t)) + B(-\alpha(t),\ -\alpha(u)) + $$

$$+ B(-\alpha(u),\ -\alpha(s))] \tag{2.1}$$

where

$$B(x,y) = \frac{\Gamma(x)\ \Gamma(y)}{\Gamma(x+y)} \tag{2.2}$$

is the Euler Beta function and

$$\alpha(s) = \alpha(o) + \alpha's, \text{ etc.} \tag{2.3}$$

are real linear trajectory functions.

Let us look in turn at the properties of $A_4(s,t)$. The most significant properties are that (1) each of the three terms can be completely represented by a sum of narrow-resonance poles in either of two channels with residua that are polynomial of the appropriate order in the other channel energy and (2) there is Regge behaviour in all channels. In more detail:

(i) In order to expand A_4 as a sum of poles note the following expressions for the gamma function[2]

$$\Gamma(z) = \int_0^\infty e^{-t}\, t^{z-1}\, dt \tag{2.4}$$

$$= \frac{1}{z} \prod_{n=1}^{\infty} \left(1 + \frac{1}{n}\right)^z \left(1 + \frac{z}{n}\right)^{-1} \tag{2.5}$$

$$= \sum_{n=0}^{\infty} \frac{(-1)^n}{n!} \frac{1}{z+n} + \int_1^\infty e^{-t}\, t^{z-1}\, dt \tag{2.6}$$

$$= \sum_{n=0}^{\infty} \frac{(-1)^n}{n!} \frac{1}{z+n} + \Gamma(1,z) \tag{2.7}$$

where $\Gamma(1,z)$ is an incomplete gamma function, which is analytic.

For the beta function we may write[2]

$$B(x,y) = \int_0^1 t^{x-1}\, (1-t)^{y-1}\, dt \tag{2.8}$$

$$= \sum_{n=0}^{\infty} \frac{(-1)^n}{n!} \frac{1}{x+n} \frac{\Gamma(y)}{\Gamma(-n+y)} \tag{2.9}$$

$$= \sum_{n=0}^{\infty} \frac{1}{n!} \frac{1}{x+n} (1-y)(2-y)\cdots(n-y) \tag{2.10}$$

$$= \frac{\Gamma(x)\,\Gamma(y)}{\Gamma(x+y)} \tag{2.11}$$

Notice an important distinction between these two meromorphic functions. The B function is completely determined by its poles and their residua while Γ has an added entire part. This is intimately related to their asymptotic behaviours being quite different as we shall see later in discussing the Regge limits.

Using the pole expansion of the beta function this leads to the beautiful identity (precise duality)

$$B(-\alpha(s), -\alpha(t)) = \sum_{n=0}^{\infty} \frac{R_n(t)}{n-\alpha(s)} = \sum_{n=0}^{\infty} \frac{R_n(s)}{n-\alpha(t)} \tag{2.12}$$

with

$$R_n(x) = \frac{1}{n!} (\alpha(x)+1)(\alpha(x)+2)\cdots\cdots(\alpha(x)+n) \tag{2.13}$$

which is an nth degree polynomial in $\alpha(x)$, and hence in x. The residue at $\alpha(s) = n$ is therefore a combination of spins n, n-1, n-2, ... down to zero (parallel daughter trajectories). For the full A_4 the residues are given by

$$A_4 = \overline{\beta} \sum_{n=0}^{\infty} \frac{1}{n - \alpha(s)} (R_n(t) + R_n(u)) \tag{2.14}$$

Bearing in mind that the centre of mass scattering angle obeys

$$z_s = \cos\theta_s = \frac{t - u}{s - 4\mu^2} \tag{2.15}$$

we see that only <u>even</u> spins are present. This is consistent with Bose statistics since in the absence of internal quantum numbers, an odd spin cannot couple to two identical scalars.

Now we introduce the important quantity

$$\sum = \alpha(s) + \alpha(t) + \alpha(u) = 3\alpha(o) + 4\alpha'\mu^2 \tag{2.16}$$

If we require that the external particles lie on the internal trajectory then (bootstrap condition)

$$0 = \alpha(o) + \alpha'\mu^2 \tag{2.17}$$

so that

$$\sum = - \alpha(o) \tag{2.18}$$

It is easy to show that $R_n(x)$ has the reflection property

$$R_n(t) = (-1)^n R_n(u) \tag{2.19}$$

if and only if $\sum = - 1$. Thus if we choose $\alpha(o) = 1$ the residua at $\alpha(s)$ = odd disappear, leaving only even poles and daughters space by two units. We may say therefore that the condition $\sum = - 1$ removes the odd daughters. (See Figure 2.1 . In Figure 2.1(a) is shown the Chew-Frautschi

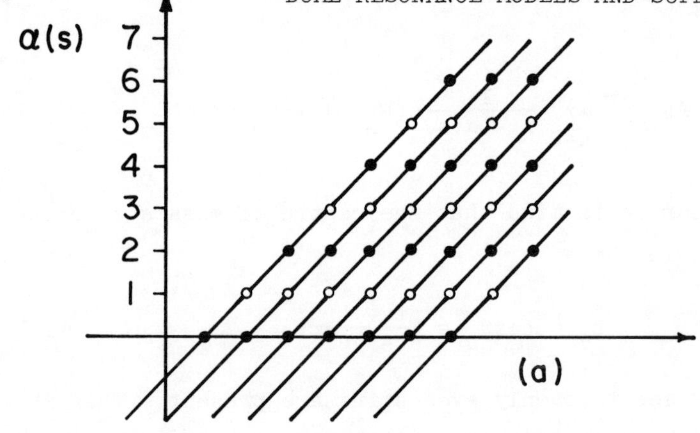

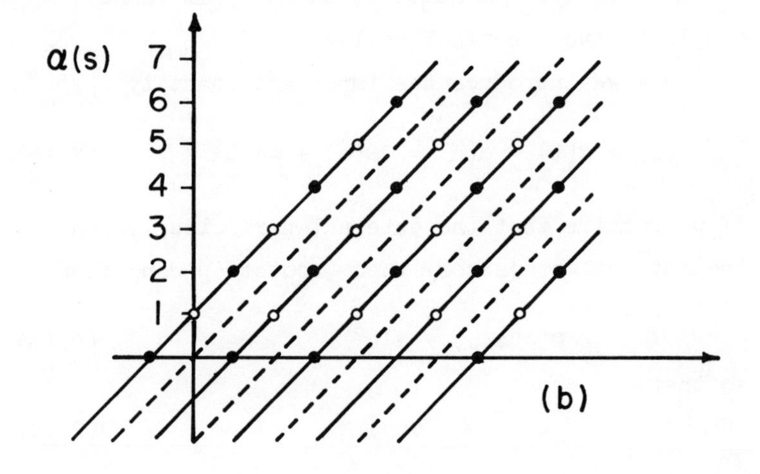

FIGURE 2.1

Chew-Frautschi Plots

plot for $\sum \neq -1$ and in Figure 2.1(b) for $\sum = -1$. The solid points are poles of A_4 while the open points are odd-spin poles of $B_4(-\alpha(s), -\alpha(t))$ and $B_4(-\alpha(s), -\alpha(u))$ separately which do not appear in A_4.)

Note that $\sum = -1$ and the bootstrap requirement

imply $\alpha(o) = 1$, whereupon the ground state on the leading trajectory becomes unphysical with negative squared mass (this is called a tachyon).

(ii) Consider now the behaviour of A_4 as $|s| \to \infty$ at fixed t. Now we need the asymptotic expansion[2]

$$\frac{\Gamma(z + a)}{\Gamma(z + b)} = z^{a-b} \left[1 + O\left(\frac{1}{z}\right)\right] \tag{2.20}$$

This expansion, as mentioned already, is quite different from that of the numerator and denominator separately since Stirling's formula gives[2]

$$\Gamma(z) = e^{-z} z^{z} \sqrt{2\pi z} \left[1 + O\left(\frac{1}{z}\right)\right] \tag{2.21}$$

The fact that the very strong blow up is avoided in the ratio can be related to the absence of an entire part, in its expansion as a sum of poles.

We shall need further the identity[2]

$$\Gamma(z)\, \Gamma(1 - z) = \frac{\pi}{\sin\pi z} \tag{2.22}$$

Use of this enables us to re-write

$$A_4 = -\beta\pi\left[\frac{1}{\Gamma(1 + \alpha(t))\sin\pi\alpha(t)} \frac{\sin\pi(\alpha(s) + \alpha(t))}{\sin\pi\alpha(s)} \times\right.$$

$$\times \frac{\Gamma(\alpha(s) + \alpha(t) + 1)}{\Gamma(\alpha(s) + 1)} + \frac{1}{\Gamma(1 + \alpha(t))\sin\pi\alpha(t)} \times$$

$$\times \frac{\Gamma(-\alpha(u))}{\Gamma(-\alpha(t) - \alpha(u))} + \frac{1}{\Gamma(1 + \alpha(s))\sin\pi\alpha(s)} \times$$

$$\left.\times \frac{\Gamma(\alpha(s) + \alpha(t) - \Sigma)}{\Gamma(\alpha(t) - \Sigma)}\right] \tag{2.23}$$

The first and second terms immediately combine rather neatly. Let us take Im $\alpha(s) \to \infty$ (we cannot go exactly along the real axis because of the poles). Then $\cot\pi\alpha(s) \to -i$, and the third term is damped exponentially. Using the asymptotic expansion for a ratio of two gamma functions gives then

$$A_4 \approx \frac{-\beta\pi}{\Gamma(1+\alpha(t))\sin(\pi\alpha(t))}(1+e^{-i\pi\alpha(t)})(\alpha(s))^{\alpha(t)} \tag{2.24}$$

exactly as expected for an even-signatured Regge pole. Bearing in mind that $\alpha(s) \simeq \alpha's$ we see that the Regge scale factor is uniquely determined as $s_o = (\alpha')^{-1}$.

Now an inspection of our expression for A_4 reveals that for the special case $\sum = -1$, the third term combines nicely with the rest of the formula, and this leads to a succession of different re-writings. The following are all for $\sum = -1$ only.

$$A_4 = -\frac{\beta\pi}{\Gamma(1+\alpha(t))}\frac{\Gamma(1+\alpha(s)+\alpha(t))}{\Gamma(1+\alpha(s))} \times$$

$$\times \left[\frac{1+\cos\pi\alpha(s)}{\sin\pi\alpha(s)} + \frac{1+\cos\pi\alpha(t)}{\sin\pi\alpha(t)}\right] \tag{2.25}$$

$$= -\frac{\beta}{\pi}\Gamma(-\alpha(s))\,\Gamma(-\alpha(t))\,\Gamma(-\alpha(u)) \times$$

$$\times [\sin\pi\alpha(s) + \sin\pi\alpha(t) + \sin\pi\alpha(u)] \tag{2.26}$$

$$= \frac{\sqrt{\pi}\beta\ \Gamma(-\frac{1}{2}\alpha(s))\ \Gamma(-\frac{1}{2}\alpha(t))\ \Gamma(-\frac{1}{2}\alpha(u))}{\Gamma(-\frac{1}{2}\alpha(s) - \frac{1}{2}\alpha(t))\ \Gamma(-\frac{1}{2}\alpha(t) - \frac{1}{2}\alpha(u))\ \Gamma(-\frac{1}{2}\alpha(u) - \frac{1}{2}\alpha(s))} \tag{2.27}$$

These expressions show the absence of poles at odd values of $\alpha(s)$ very clearly. To arrive at the second formula from the first uses only our relation between $\Gamma(z)$ and $\Gamma(1 - z)$. This is then a nicely symmetric form. To arrive at the rather surprising third form a couple of intermediate steps may be of use. We write

$$\sin\pi\alpha(s) + \sin\pi\alpha(t) + \sin\pi\alpha(u) =$$

$$-4 \cos\frac{\pi\alpha(s)}{2} \cos\frac{\pi\alpha(t)}{2} \cos\frac{\pi\alpha(u)}{2} \tag{2.28}$$

and then use the formula

$$\Gamma(2z) = 2^{2z-1}\ \pi^{-1/2}\ \Gamma(z)\ \Gamma(z + \tfrac{1}{2}) \tag{2.29}$$

to re-write

$$\cos\tfrac{1}{2}\pi\alpha(s)\ \Gamma(-\alpha(s)) = \sqrt{\pi}\ 2^{-\alpha(s)-1}\ \frac{\Gamma(-\frac{1}{2}\alpha(s))}{\Gamma(-\frac{1}{2}\alpha(t) - \frac{1}{2}\alpha(u))} \tag{2.30}$$

The third form will be discussed in detail later in the subsection on the Shapiro-Virasoro formula.

Before leaving the beta function expression, we examine further important properties.

(iii) Fixed angle behaviour. By using Stirling's asymptotic formula for fixed

$$z_s = 1 + \frac{2t}{s - 4\mu^2} \tag{2.31}$$

and $\alpha(s) \to +\infty$, $\alpha(t)$, $\alpha(u) \to -\infty$ we find

$$\alpha(t) \approx -\frac{\alpha(s)}{2}(1 - z_s) \qquad (2.32)$$

$$\alpha(u) \approx -\frac{\alpha(s)}{2}(1 - z_t) \qquad (2.33)$$

and

$$B(-\alpha(s), -\alpha(t)) \approx \exp(-\alpha(s)\, f(z_s)) \qquad (2.34)$$

where

$$f(x) = \left(\frac{1-x}{2}\right) \ell n\left(\frac{2}{1-x}\right) + \left(\frac{1+x}{2}\right) \ell n\left(\frac{2}{1+x}\right) \qquad (2.35)$$

For example, at $z_s = 0$ this becomes $B \approx \exp(-4\ \ell n\ 2 \cdot \alpha' k^2)$ where k is the centre-of-mass momentum.

Note that the famous Cerulus-Martin bound[3,4] which is a lower bound on the fixed angle behaviour ($\sim e^{-a\sqrt{s}}$), does not apply here since the derivation of that bound uses an assumption of polynomial boundedness of A(s,t) which is inapplicable in the presence of linearly-rising trajectories. For further details, see Chiu and Tan [5].

(iv) We know that the narrow resonance approximation violates unitarity, as we showed in detail earlier. Therefore the first temptation is to correct this by simply making the trajectory function imaginary above threshold. For example, account of elastic unitarity might be attempted by writing

$$\alpha(s) = \alpha(o) + \alpha' s + ic\sqrt{s - 4\mu^2}\ \theta(s - 4\mu^2) \qquad (2.36)$$

with c a constant. This sort of smearing is

essential in order to do phenomenology because the infinities at the poles are then avoided. At a deeper level, however, adding such an imaginary part is inconsistent because clearly $R_n(t)$ is now no longer an nth degree polynomial in t, but contains all powers. Hence there are high-spin low-mass ancestor components present. Thus the beta function is so tightly constrained that it is difficult to modify (except for the satellite ambiguity discussed later).

(v) The Regge behaviour is valid only along a ray at a finite angle to the real axis. An inconvenience of the narrow resonance approximation is that we cannot take $s \to \infty$ at real physical s. On the other hand, if we are able to displace the poles on to the second sheet by a more sophisticated approach than mentioned in (iv) (i.e. by iterating the tree approximation) we should hope to regain Regge behaviour along the real axis.

(vi) On the question of ghosts, when we make the partial wave analysis

$$R_n(t) = \sum_{\ell=0}^{n} G^2_{\ell oo} \; P_\ell(z_s) \tag{2.37}$$

we need to ensure positivity of the corresponding $g^2_{\ell oo}$ defined by

$$g^2_{\ell oo} = G^2_{\ell oo} \; C_\ell (k^2)^{-\ell} \tag{2.38}$$

These requirements are more appropriately discussed after we have dealt with full factorisation of the N-point function, and the full degeneracy has been exposed. Nevertheless, we should make one remark to which we can conveniently refer back later: the answer of the ghost question is the first point at which the dimensionality of space-time (d) is relevant, since the partial wave analysis is made in terms of irreducible representations of O(d - 1). For example, putting $\alpha(o) = 1$ we find

$$R_2(t) = \frac{1}{2}(\alpha(t) + 1)\ (\alpha(t) + 2) = \frac{25}{8}(z_s{}^2 - \frac{1}{25}) \quad (2.39)$$

where we recognise that $\alpha(t) = 1 + \alpha' t$ and $z_s = 1 + 2\alpha' t/5$. Now we write, in irreducible representations of O(d - 1)

$$R_2(t) = G^2_{200}\ (z_s{}^2 - \frac{1}{d-1}) + G^2_{000} \quad (2.40)$$

so that for $d > 26$ we find $g^2_{000} < 0$ and a ghost; for $d \leq 26$ there is no ghost. Later we shall see that these particular values $d \leq 26$ play a much more general role.

(vii) For the case $\alpha(o) = 1$, Fairlie and Jones[6] discovered that one can re-write A_4 in yet another way, namely

$$A_4 = \beta \int_{-\infty}^{\infty} dx\ |x|^{-\alpha(s)-1}\ |1 - x|^{-\alpha(t)-1} \quad (2.41)$$

since by considering separately the regions $-\infty < x < 0$, $0 < x < 1$, $1 < x < +\infty$ and making obvious changes of variables we regain the sum of three beta functions. We shall mention later the generalisation to the N-point function.

(viii) Finally, concerning the beta function, we should remark that the slopes of all input Regge trajectories must be equal. If this is not the case, then the beta function explodes exponentially for some fixed angles in the physical region. Indeed we find for fixed z_s

$$B(-\alpha(s), -\alpha(t)) \sim \exp(-\alpha(s)\, f(z_s)) \tag{2.42}$$

where now

$$f(x) = (\frac{R - Rx}{2})\, \ell n(\frac{2}{R - Rx}) + (\frac{2 - R + Rx}{2}) \times$$

$$\times\, \ell n(\frac{2}{2 - R + Rx}) \tag{2.43}$$

with $R = \alpha'_t/\alpha'_s$ the slope ratio. For $R > 1$, $f(-1) < 0$ which is disastrous. If $R < 1$ of course the same catastrophe happens at fixed z_t by crossing symmetry. Hence $R = 1$ is essential.

Therefore we will henceforth assume that the slope of the input Regge trajectories is universal. This requirement of universal slope may be regarded as a successful prediction of the Veneziano model, since apart from the pomeron (which, in any case, we will treat differently) the observed trajectories have a common slope within experimental errors.

2.3 FIVE-POINT FUNCTION

As in the case of the 4-point function, we shall assume that the N-point function can be written as a sum over $\frac{1}{2}(N - 1)!$ amplitudes for inequivalent permutations of the external lines (planar duality). For the five-point function we therefore fix the ordering p_1, p_2, ... p_5. Here we can have two simultaneous poles in a Feynman diagram with trilinear couplings. Adjacent channels such as (p_1p_2) and (p_2p_3) cannot have simultaneous poles (are dual or incompatible channels).

By analogy with the N = 4 case, where we may write

$$B_4(-\alpha(s), -\alpha(t)) = \int_0^1 dx_1\, x_1^{-\alpha(s)-1}\, x_2^{-\alpha(t)-1} \tag{2.44}$$

with $x_2 = (1 - x_1)$ we now write[7,8]

$$B_5 = \int_0^1 dx_1\, dx_4 \prod_{i=1}^{5} x_i^{-\alpha(s_i)-1}\, \rho(x_1,x_4) \tag{2.45}$$

where $0 \leq x_i \leq 1$ and $\rho(x_1, x_4)$ ensures cyclic symmetry. Further we make the constraint

$$x_i = 1 - x_{i-1}\, x_{i+1} \tag{2.46}$$

to avoid incompatible poles. A more general way to see which channels are incompatible is to draw a polygon to represent momentum conservation (Figure 2.2). The diagonals of this polygon represent the momenta of possible intermediate states; when two diagonals intersect we cannot draw a Feynman graph with poles in both channels simultaneously and they are therefore incompatible channels.

It is easy to find a solution of the duality constraints, namely

$$x_2 = \frac{1 - x_1}{1 - x_1x_4} \tag{2.47}$$

$$x_3 = \frac{1 - x_4}{1 - x_1x_4} \tag{2.48}$$

$$x_5 = 1 - x_4x_1 \tag{2.49}$$

Upon changing variables cyclically one finds a Jacobian

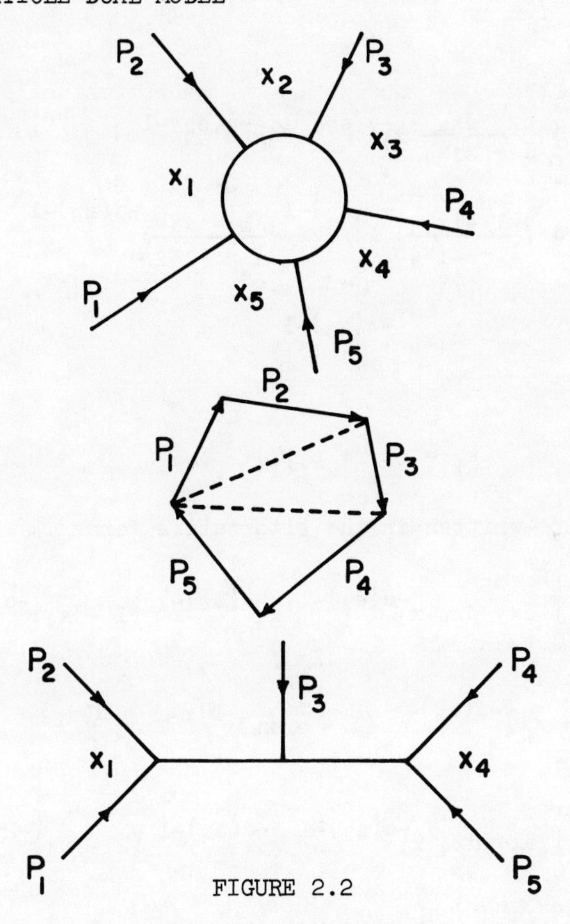

FIGURE 2.2

Five-Point Functior

$$\frac{\partial(x_2 x_5)}{\partial(x_1 x_4)} = \frac{x_1}{x_5} \tag{2.50}$$

and, therefore, cyclic symmetry is ensured if we use

$$\rho(x_1\ x_4) = \frac{1}{x_5} \tag{2.51}$$

whereupon

$$B_5 = \int_0^1 \frac{dx_1\, dx_4}{1 - x_1x_4}\, x_1^{-\alpha(s_1)-1}\, x_4^{-\alpha(s_4)-1} \cdot$$

$$\cdot \left(\frac{1 - x_1}{1 - x_1x_4}\right)^{-\alpha(s_2)-1} \left(\frac{1 - x_4}{1 - x_1x_2}\right)^{-\alpha(s_3)-1} \cdot$$

$$\cdot\ (1 - x_1x_4)^{-\alpha(s_5)-1} \tag{2.52}$$

with

$$s_i = (p_i + p_{i+1})^2 \tag{2.53}$$

This can be re-written in the alternative forms

$$B_5 = \int_0^1 dx_1\, dx_4\, x_1^{-\alpha(s_1)-1}\, x_4^{-\alpha(s_4)-1}\, (1 - x_1)^{-\alpha(s_3)-1} \cdot$$

$$\cdot\ (1 - x_2)^{-\alpha(s_4)-1}\, (1 - x_1x_4)^{-\alpha(s_5) + \alpha(s_2) + \alpha(s_3)} \tag{2.54}$$

$$= \int_0^1 dx_1\, dx_4\, x_1^{-\alpha(s_1)-1}\, x_4^{-\alpha(s_4)-1}\, (1 - x_1)^{-2p_2p_3-c_2} \cdot$$

$$\cdot\ (1 - x_4)^{-2p_3p_4-c_2}\, (1 - x_1x_4)^{-2p_2p_4-c_3} \tag{2.55}$$

where we have defined

$$c_n = \alpha_n(o) - 2\alpha_{n-1}(o) + \alpha_{n-2}(o) \tag{2.56}$$

with the conventions $\alpha_o(o) = 1$, $\alpha_1(o) = -\alpha'\mu^2$ with $p_i{}^2 = \mu^2$. Note that if all intercepts are equal then

$$c_n = \delta_{n2}\, (1 - \alpha(o)) \tag{2.57}$$

which vanishes for all n if $\alpha(o) = 1$.

Concerning the five-point function:

(i) The function was known in the mathematical literature[9] more than sixty years before its rediscovery in the present context.

(ii) Once we know how to deal with N = 5, it is a straight-forward matter to go on to general N.

2.4 N-POINT FUNCTION

We ascribe an energy s_{ij} and a channel variable u_{ij} (to be integrated between o and one) to each of the $\frac{1}{2}N(N-3)$ planar channels. We require that when any particular u_{ij} is zero all of the dual channels have $u_{k\ell} = 1$, to avoid double poles by analogy with the 4- and 5-point functions. This gives rise to the duality constraint equations[10-15]

$$u_{ij} = 1 - \prod_{\substack{1 \leq p < i \\ i \leq q < j}} u_{pq} \prod_{\substack{i < r \leq j \\ j < s \leq (N-1)}} u_{rs} \tag{2.58}$$

Let us choose as a set of (N - 3) independent variables u_{ij} (j = 2, 3, ..., N-2) which are associated with the poles in the multiperipheral configuration of Fig. 2.3.

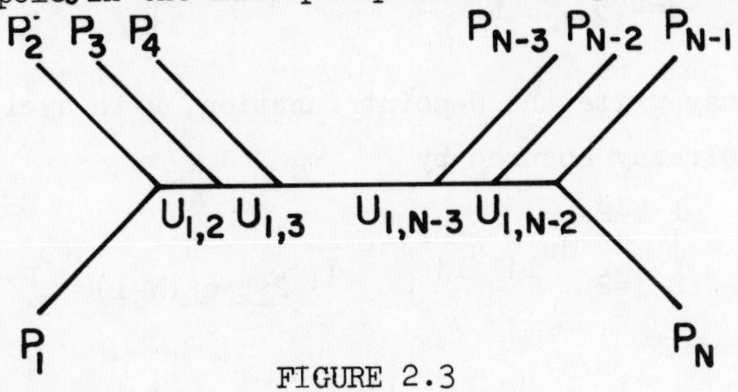

FIGURE 2.3

Multiperipheral Configuration

Then it is possible to find the solution to the duality constraints in the form

$$u_{pq} = \frac{(1 - u_{1p}u_{1p+1}\cdots u_{1q-1})\,(1 - u_{1p-1}u_{1p}\cdots u_{1q})}{(1 - u_{1p-1}u_{1p}\cdots u_{1q-1})\,(1 - u_{1p}u_{1p+1}\cdots u_{1q})} \tag{2.59}$$

for $p = 2, 3, \ldots, (N-2)$; $q = 3, 4, \ldots, (N-1)$ and $p < q$ (we define $u_{ii} = 0$). A geometrical interpretation of this solution will be given in the next subsection.

Now the Jacobian for the change of variables $u_{1j} \to u_{2j+1}$ must be calculated. It turns out to be given by

$$\frac{\partial(u_{23}, u_{34}, \ldots, u_{2,N-1})}{\partial(u_{12}, u_{13}, \ldots, u_{1,N-2})} = (-1)^{N-1}\,\frac{J_2}{J_1} \tag{2.60}$$

in which

$$J_1 = \prod_{\substack{i<j \\ i=2,3,\ldots,N-2 \\ j=3,4,\ldots,N-1}} (u_{ij})^{j-i-1} = \prod_{i=2}^{N-3} (1 - u_{1i}u_{1i+1}) \tag{2.61}$$

and

$$J_2 = \prod_{\substack{i<j \\ i=3,4,\ldots,N-1 \\ j=4,5,\ldots,N}} (u_{ij})^{j-i-1} = \prod_{i=3}^{N-2} (1 - u_{1i}\,u_{1i+1}) \tag{2.62}$$

Hence we may write the N-point function, with cyclic symmetry already ensured by

$$B_N = \int_0^1 \prod_{j=2}^{N-2} du_{1j}\, u_{1j}^{-\alpha_{1j}-1}\,\frac{1}{J_1} \prod_{2\leq p<q\leq(N-1)} (u_{p,q})^{-\alpha_{pq}-1} \tag{2.63}$$

Before going on to the properties of B_N, let us re-write it in a slightly different form due to Bardakci and Ruegg[13].

It is derived by recombining the terms and using the relation

$$-\alpha_{ij} + \alpha_{i+1,j} + \alpha_{i,j-1} - \alpha_{i+1,j-1} =$$

$$= -2p_i p_j - c_{j-i+1} \tag{2.64}$$

which we have already mentioned. It follows that

$$B_N = \int_0^1 \prod_{j=2}^{N-2} du_{1j}\, u_{1j}^{-\alpha_{1j}-1} \prod_{2\leq i<j\leq(N-1)} \times$$

$$\times\ (1 - u_{1i}u_{1i+1}\dots u_{1j-1})^{-2p_ip_j - c_{j-i+1}} \tag{2.65}$$

When all intercepts are equal (as we shall assume everywhere from here on, unless very explicitly stated otherwise) this becomes simply

$$B_N = \int_0^1 \prod_{j=2}^{N-2} du_{1j}\, u_{1j}^{-\alpha_{1j}-1} (1 - u_{1j})^{\alpha(o)-1} \times$$

$$\times \prod_{2\leq i<j\leq(N-1)} (1 - u_{1i}u_{1i+1}\dots u_{1j-1})^{-2p_ip_j} \tag{2.66}$$

Comparison of this formula with Figure 2.3 shows a simple pattern for the factors in this formula. Let us now enumerate three properties that follow easily for B_N

(i) Factorisation at $\alpha(s_{1j}) = 0$ (bootstrap property). This pole occurs when $u_{1j} = 0$ and hence when all the dual channel variables are one. This then gives

$$J_1 \underset{u_{1j}\to 0}{\to} \prod_{i=2}^{j-2} (1 - u_{1i}u_{1i+1}) \prod_{k=j}^{N-3} (1 - u_{1k}u_{1k+1}) \tag{2.67}$$

and

$$\prod_{2\le i<k\le(N-1)} u_{ik}^{-\alpha_{ik}-1} \underset{u_{1j}\to 0}{\to} \prod_{2\le p<q\le j-1} u_{pq}^{-\alpha_{pq}-1} \times$$

$$\times \prod_{(j+1)\le r<s\le(N-1)} u_{rs}^{-\alpha_{rs}-1} \tag{2.68}$$

$$\prod_{k=2}^{N-2} du_{1k}\, u_{1k}^{-\alpha_{1k}-1} \underset{u_{1j}\to 0}{\to} \prod_{\ell=2}^{j-1} du_{1\ell} u_{1\ell}^{-\alpha_{1\ell}-1} \times$$

$$\times \prod_{m=j+1}^{N-2} du_{1m}\, u_{1m}^{-\alpha_{1m}-1} \tag{2.69}$$

From these relations it follows that

$$B_N \underset{\alpha_{1j}\to 0}{\to} \frac{1}{\alpha_{1j}} B_{j+1}\, B_{N-j+1} \tag{2.70}$$

Thus satisfying the bootstrap constraint.

Full factorisation is most easily seen in the operator formalism, and we therefore defer it. It is, however, very easy to show at this stage that the higher poles are of the correct polynomial degree in momentum transfer (no ancestors). Consider the pole at $\alpha_{1j} = \ell$ = integer > 0. Write

$$B_N = \int_0^1 du_{ij}\, u_{1j}^{-\alpha_{1j}-1}\, \tilde{\tilde{B}}_N \tag{2.71}$$

where B_N is an (N-4)-fold integration. Then

$$B_N \underset{\alpha_{1j}\to\ell}{\to} \frac{1}{\alpha_{1j} - \ell} \frac{1}{\ell!} \left[\frac{d^\ell}{du_{ij}^\ell} \tilde{\tilde{B}}_N\right]_{u_{1j}=0} \tag{2.72}$$

and the residue can be seen to be a polynomial of degree ℓ in the dual $\alpha_{k\ell}$, when we remember that all dual $u_{k\ell} = 1$ at $u_{1j} = 0$. Hence there are no ancestors, and the pole at $\alpha_{1j} = \ell$ corresponds to a superposition of spins ℓ, $\ell-1$, $\ell-2$, ..., 1,0. By cyclic symmetry this is then true for all poles of B_N.

(ii) Recurrence relations[12] for B_N. We can write a relation for B_N in terms of a sum of products B_{N-1} B_4. By iteration we can write many equivalent forms of the general type

$$B_N \sim \sum \cdot\cdot B_{i_1} B_{i_2} \dots B_{i_k} \tag{2.73}$$

with

$$\sum_{j=1}^{k} i_j = N + 3(k-1) \tag{2.74}$$

For example, we can write B_N in terms of products of (N-3) B_4 functions. To illustrate this we write the Bardakci-Ruegg form of B_6 (according to Figure 2.4(a)) and expand factors of the integrand

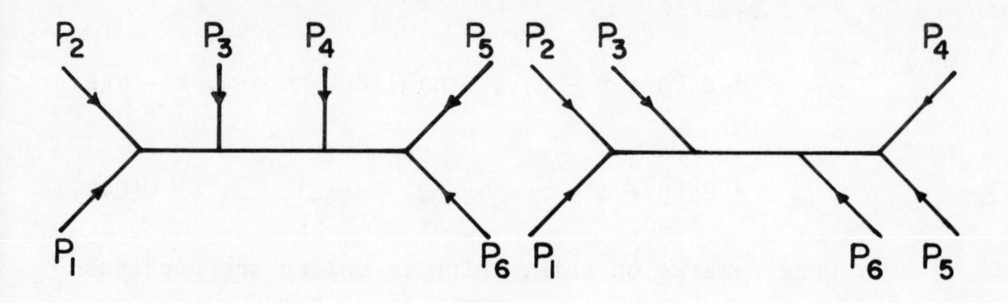

FIGURE 2.4
Rearrangement of B_6 Poles

as binomial expansions to obtain (for $\alpha(o) = 1$)

$$B_6 = \sum_{p=0}^{\infty} \sum_{r=o}^{\infty} \binom{-2p_2 \cdot p_4}{p} \binom{-2p_2 \cdot p_5}{r} B_4(p + r - \alpha_{12}, - \alpha_{23}) \cdot$$

$$\cdot \; B_5(p + r - \alpha_{123}, - \alpha_{34}, - \alpha_{45}, r - \alpha_{56}, - \alpha_{345}) \quad (2.75)$$

$$= \sum_{pq,r=o}^{\infty} \binom{-2p_2 \cdot p_4}{p} \binom{-2p \cdot p_5}{q} \binom{-2p_2 \cdot p_5}{r}$$

$$B_4(p + r - \alpha_{12}, - \alpha_{23})$$

$$B_4(p + q + r - \alpha_{123}, - \alpha_{34})$$

$$B_4(-\alpha_{45}, q + r - \alpha_{56}) \quad (2.76)$$

By a re-arrangement of the B_5 arguments we can display the poles of Figure 2.4(b).

$$B_6 = \sum_{p,q,r=o}^{\infty} \binom{-2p_2 \cdot p_4}{r} \binom{-2p_5 \cdot (p_1 + p_2)}{q} \binom{-2p_2 \cdot p_5}{r} \cdot$$

$$\cdot \; B_4(p + r - \alpha_{12}, - \alpha_{23}) \, B_4(q - \alpha_{45}, r - \alpha_{56}) \cdot$$

$$\cdot \; B_4(p + q + r - \alpha_{123}, - \alpha_{234}) \quad (2.77)$$

Some remarks on this, which is called the Hopkinson-Plahte representation,

(a) This recurrence relation is useful for numerical work on B_N using an electronic computer.

(b) It enables "by hand" inspection of some elementary properties without resort to the operator formalism. The operator formalism is nevertheless more powerful and is essential to make some properties demonstrable in practice.

(iii) Regge behaviour of B_N. We studied the asymptotics of B_4 already in detail. We illustrate here how the arguments for multiregge behaviour of B_N are obtained by considering N = 5. Take again Figure 2.4(a) and suppose $|\alpha_{51}|$, $|\alpha_{23}|$, $|\alpha_{34}| \to \infty$ (avoiding the real axis) with α_{12}, α_{45} fixed. Then from a Hopkinson-Plahte form we deduce that

$$B_5 \sim \Gamma(-\alpha_{12})\,\Gamma(-\alpha_{45}) \sum_k (-\alpha_{23})^{\alpha_{12}-k}\,(-\alpha_{34})^{\alpha_{45}-k} \cdot (-\alpha_{51})^k \qquad (2.78)$$

as expected for the double Regge limit. For further details of multiregge limits we refer to the literature[11,13].

2.5 KOBA NIELSEN FORM AND PROJECTIVE INVARIANCE

In our discussion of the N-point function, we have been able to satisfy a large number $\left(\frac{N}{24}(N-1)(N-2)(N-3)\right)$ of duality constraints on $\frac{N}{2}(N-3)$ channel variables, which seems to have been a miracle since one would expect no non-trivial solution. Another question is: can we express B_N in a form in which its cyclic symmetry is more manifest?

The answers to these questions are provided by the second paper on the N-point function by Koba and Nielsen[16,17]. In addition, this work reveals an invariance group (projective group) which will play a central role in our discussion of the operator formalism.

We take an arbitrary circle in a complex z-plane and place N points z_i, in cyclic order, around its circumference. Then the following identification of the channel variables satisfies all duality constraint equations

$$u_{ij} = (z_i, z_{i-1}; z_j, z_{j+1}) \tag{2.79}$$

where the notation defines an anharmonic ratio

$$(a, b; c, d) = \frac{(a - c)(b - d)}{(a - d)(b - c)} \tag{2.80}$$

In order to check this assertion it is worth writing down some of the most important properties of (a, b; c, d), namely its obvious symmetries

$$(a, b; c, d) = (b, a; d, c) = (c, d; a, b) = \text{etc.} \tag{2.81}$$

together with the fact that

$$(a, b; c, d) + (a, c; b, d) = 1 \tag{2.82}$$

and finally the multiplication rule

$$(a, b; c, d)(a, b; d, e) = (a, b; c, e) \tag{2.83}$$

or equivalently

$$(a, b; c, d)(b, e; c, d) = (a, e; c, d) \tag{2.84}$$

With these relations at hand it is agreeable to prove that

$$1 - \prod_{\substack{1 \le p < i \\ i \le q < j}} u_{pq} \prod_{\substack{i < r \le j \\ j < s \le (N-1)}} u_{rs} =$$

$$= 1 - \prod_{\substack{1 \le p < i \\ i \le q < j}} (z_p, z_{p-1}; z_q, z_{q+1}) \cdot$$

$$\cdot \prod_{\substack{i < r \le j \\ j < s \le N-1}} (z_r, z_{r-1}; z_s, z_{s+1}) \quad (2.85)$$

$$= 1 - \prod_{1 \le p < i} (z_p, z_{p-1}; z_i, z_j) \prod_{i < r \le j} (z_r, z_{r-1}; z_{j+1}, z_N) \quad (2.86)$$

$$= 1 - (z_{i-1}, z_N; z_i, z_j)(z_j, z_i; z_{j+1}, z_N) \quad (2.87)$$

$$= 1 - (z_i, z_j; z_{i-1}, z_{j+1}) \quad (2.88)$$

$$= (z_i, z_{i-1}; z_j, z_{j+1}) \quad (2.89)$$

$$= u_{ij} \quad (2.90)$$

as required.

Re-writing the B_N integrand in terms of anharmonic ratios we find for the momentum-dependent part (we return to the integration measure later)

$$\Pi (u_{ij})^{-\alpha_{ij}-1} = \prod_{1\leq i<j\leq N} |z_i - z_j|^{-2p_i \cdot p_j - c_{j-i+1}} \cdot$$

$$\cdot \prod_{i=1}^{N} |z_i - z_{i+2}| \qquad (2.91)$$

This expression, because it is expressable entirely in terms of cross ratios, is invariant under projective transformations of the kind

$$z \to z' = \frac{az + b}{cz + d} \qquad (2.92)$$

with ad - bc = 1 as normalisation. It is easy to see that any such projective transformation is a combination of dilation, translation and inversion operations; for example,writing SL(2,R) matrices for

$$z \xrightarrow{\Lambda_1} cz \xrightarrow{\Lambda_2} (cz + d) \xrightarrow{\Lambda_3} \frac{-1/c}{cz + d} \xrightarrow{\Lambda_4} \frac{-1/c}{cz + d} + \frac{a}{c} =$$

$$= \frac{az + b}{cz + d} = \Lambda z \qquad (2.93)$$

gives

$$\Lambda_4\Lambda_3\Lambda_2\Lambda_1 = \begin{bmatrix} 1 & a/c \\ 0 & 1 \end{bmatrix} \begin{bmatrix} 0 & -1/\sqrt{c} \\ \sqrt{c} & 0 \end{bmatrix} \begin{bmatrix} 1 & d \\ 0 & 1 \end{bmatrix} \begin{bmatrix} \sqrt{c} & 0 \\ 1 & 1/\sqrt{c} \end{bmatrix}$$

$$= \begin{bmatrix} a & b \\ c & d \end{bmatrix} \qquad (2.94)$$

For an anharmonic ratio the projective invariance is clear: dilation (zero dimensionality), translation (differences of z_i only), inversion (each z_i occurs only once in numerator and denominator).

The Lie algebra of this projective group is isomorphic to that of O(2,1), namely for the three generators

L_+, L_o, L_-

$$[L_+, L_-] = 2L_o \tag{2.95}$$

$$[L_\pm, L_0] = \pm L_\pm \tag{2.96}$$

which differs by one sign from the Lie algebra of O(3) [Note: for O(3) with $[J_i, J_j] = i\,\varepsilon_{ijk} J_k$ and $J_\pm = J_1 \pm J_2$ we have $[J_+, J_-] = 2J_o$, but $[J_\pm, J_o] = \mp J_\pm$].

The three-parameter projective group enables us[16] to keep three of the N points z_i fixed on the circumference of the Koba-Nielsen circle , and to integrate the remaining (N-3) such that the order is preserved. It remains to determine a projective invariant integration measure, and the correct choice is

$$\prod_{i=1}^{N} \frac{dz_i}{(z_i - z_{i+2})} \left[\frac{dz_a \, dz_b \, dz_c}{(z_a - z_b)(z_b - z_c)(z_c - z_a)}\right]^{-1} \tag{2.97}$$

where z_a, z_b, z_c are fixed values for three of the N points, chosen arbitrarily. The projective invariance of this measure follows from its invariance under dilation, translation and inversion. We now see that the essential role of the 3-parameter projective group is to reduce the number of integrations from N (the number of external particles) to (N-3) (the number of allowable simultaneous poles).

The N-point function can now be written

$$B_N = \int \prod_{i=1}^{N} dz_i \left[\frac{dz_a \, dz_b \, dz_c}{(z_a - z_b)(z_b - z_c)(z_c - z_a)}\right]^{-1} \cdot$$

$$\cdot \prod_{1 \leq i < j \leq N} |z_i - z_j|^{-2p_i \cdot p_j - c_{j-i+1}} \qquad (2.98)$$

which is manifestly cyclic symmetric. The domain of integration is such that the cyclic order of the z_i is maintained, and the points z_a, z_b, z_c fixed, but otherwise is unrestricted on the circumference of the circle.

This is a convenient point to remark the fact[6] that for intercept $\alpha(o) = 1$ we can sum over the inequivalent cyclic permutations in a trivial way since then $c_n = 0$. We have

$$A_N = \sum_{\substack{\text{inequivalent} \\ \text{permutations}}} B_N(p_1 p_2 \cdots p_N) \qquad (2.99)$$

and now A_N is obtained from B_N in the Koba-Nielsen form by the simple expedient of increasing the domain of integration, namely to allow the (N-3) moving z_i to be entirely unrestricted on the circumference of the circle. Note that if $\alpha(o) \neq 1$ there are nearest-neighbour factors of the form

$$|z_i - z_{i+1}|^{\alpha(o)-1} \qquad (2.100)$$

which depend on the cyclic ordering, and hence do not allow a simple expression for A_N (which can be written as a single integral only if these nearest-neighbour factors are each combined with an appropriate combination of step functions).

Finally we must demonstrate the equivalence to the other forms of the N-point function. To do this we choose the Koba-Nielsen circle to be the real axis and the fixed points to be $z_1 = 0$, $z_{N-1} = 1$, $z_N = \infty$. Making the identification

$$z_i = u_{1i} u_{1i+1} \cdots u_{1N-2} \quad (2 \leq i \leq N-2) \qquad (2.101)$$

it is straightforward to derive that

$$A_N = \int \prod_{i=2}^{N-2} du_{1i}\, u_{1i}^{-\alpha_{1i-1}} \prod_{2\leq j<k\leq(N-1)} (1 - u_{1j} u_{1j+1} \cdots u_{1k})^{-2p_j \cdot p_k} \qquad (2.102)$$

which is the Bardakci-Ruegg form, as required.

2.6 OPERATOR FACTORISATION AND LEVEL DENSITY

Historically the N-point function was first factorised without the use of operators[18,19], but the introduction of operators so much simplifies the discussion that we shall introduce them immediately here. We define harmonic-oscillator-like operators $a_\mu^{(n)}$, $a_\mu^{(n)+}$ ($n = 1, 2, 3, 4, \ldots$; μ = Lorentz index = 0, 1, 2, 3) which satisfy

$$[a_\mu^{(n)}, a_\nu^{(m)+}] = - g_{\mu\nu}\, \delta mn \qquad (2.103)$$

in which we recall that $g_{\mu\mu} = (+, -, -, -)$. Thus the space components act as a normal harmonic oscillator, while the time components have the property that any state with odd occupancy has negative norm, that is

$$\left\| \frac{(a_o^{(n)+})^{\ell}}{\sqrt{\ell!}} \, |0\rangle \right\|^2 = (-1)^{\ell} \qquad (2.104)$$

In the Fock space spanned by the operators we can set up a complete set of orthonormalised occupation number states of the general form

$$|\{\ell\}\rangle = \prod_{n=1}^{\infty} \prod_{\mu=0}^{3} \frac{(a_\mu^{(n)})^{\ell_{n,\mu}}}{\sqrt{\ell_{n,\mu}!}} \, |0\rangle \qquad (2.105)$$

But, for the purposes of factorisation, it will be technically very much more convenient to have in mind the coherent state basis of the Fock space.

Coherent states are defined to be eigenstates of the annihilation operator. [See, for example, Ref. 22.] To recall their properties let us simply consider a single oscillator satisfying

$$[a, a^+] = 1 \qquad (2.106)$$

Then we define, for α a complex number

$$|\alpha\rangle = e^{\alpha a^+} \, |0\rangle \qquad (2.107)$$

and then follow the properties

$$e^{\beta a^+} \, |\alpha\rangle = |\alpha + \beta\rangle \qquad (2.108)$$

$$a|\alpha\rangle = \alpha|\alpha\rangle \qquad (2.109)$$

$$\beta^{a^+ a} \, |\alpha\rangle = |\alpha\beta\rangle \qquad (2.110)$$

$$\langle\beta|\alpha\rangle = \exp(\beta^* \alpha) \qquad (2.111)$$

and for any function f and complex variable x

$$f(a)x^{a^+a} = x^{a^+a} f(ax) \tag{2.112}$$

$$x^{a^+a} f(a+) = f(a+x) x^{a^+a} \tag{2.113}$$

The resolution of the identity is given by

$$1 = \frac{1}{\pi} \int d \text{ Im } \alpha \, d \text{ Re } \alpha \, e^{-|\alpha|^2} |\alpha\rangle\langle\alpha| \tag{2.114}$$

Also we should note the identity

$$e^A e^B = e^B e^A e^{[A,B]} \tag{2.115}$$

valid when [A,B] commutes with A,B.

In the full Fock space spanned by $a_\mu^{(n)}$, $a_\mu^{(n)+}$ we correspondingly define

$$|\{\alpha\}\rangle = \prod_{n=1}^{\infty} \prod_{\mu=0}^{3} \exp(\alpha_\mu^{(n)} a_\mu^{(n)+}) |0\rangle \tag{2.116}$$

Now we are ready to proceed to the factorisation. We define a vertex operator by

$$V(p) = \exp(i\sqrt{2}p\cdot \sum_{n=1}^{\infty} \frac{a^{(n)+}}{\sqrt{n}}) \exp(i\sqrt{2}p\cdot \sum_{n=1}^{\infty} \frac{a^{(n)}}{\sqrt{n}}) \tag{2.117}$$

and a propagator by

$$D(s) = \int_0^1 dx \, x^{-\alpha(s)-1+R} (1 - x)^{\alpha(o)-1} \tag{2.118}$$

with

$$R = - \sum_{n=1}^{\infty} n \, a^{(n)+} a^{(n)} \tag{2.119}$$

and then make the following beautiful identification

$$B_N = \int \prod_{i=2}^{N-2} du_{1i}\, u_{1i}^{-\alpha_{1i}-1} (1 - u_{1i})^{\alpha(o)-1} \times$$

$$\times \prod_{2\leq j<k\leq(N-2)} (1 - u_{1j}\, u_{1j+1} \cdots u_{1k-1})^{-2p_j \cdot p_k} \tag{2.120}$$

$$= \langle 0|\, V(p_2)\, D(s_{12})\, V(p_3)\, D(s_{123}) \cdots\cdots V(p_{N-1})|0\rangle \tag{2.121}$$

which is a fully factorised form.

Let us check how this identity works for N = 4. There we have

$$\int_0^1 dx\, x^{-\alpha(s)-1} (1 - x)^{\alpha(o)-1}$$

$$\langle 0|\, \exp(i\sqrt{2}\, p_2 \cdot \sum_{n=1}^{\infty} \frac{a^{(n)}}{\sqrt{n}})\, x^R \times$$

$$\times \exp(i\sqrt{2}\, p_3 \cdot \sum_{n=1}^{\infty} \frac{a^{(n)+}}{\sqrt{n}})|0\rangle$$

$$= \int_0^1 dx\, x^{-\alpha(s)-1} (1 - x)^{\alpha(o)-1}$$

$$\langle 0|\, \exp(i\sqrt{2}\, p_2 \cdot \sum_{n=1}^{\infty} \frac{a^{(n)}x^n}{\sqrt{n}}) \times$$

$$\times \exp(i\sqrt{2}\, p_3 \cdot \sum_{n=1}^{\infty} \frac{a^{(n)+}}{\sqrt{n}})|0\rangle \tag{2.122}$$

$$= \int_0^1 dx\, x^{-\alpha(s)-1} (1 - x)^{\alpha(o)-1} \exp[2p_1 \cdot p_2 \sum_{n=1}^{\infty} \frac{x^n}{n}] \tag{2.123}$$

$$= \int_0^1 dx\, x^{-\alpha(s)-1} (1 - x)^{\alpha(o)-1-2p_1 \cdot p_2} \tag{2.124}$$

$$= \int_0^1 dx\, x^{-\alpha(s)-1} (1-x)^{-\alpha(t)-1} \tag{2.125}$$

as required.

To complete the check for B_N we need to confirm that

$$<0| V(p_2)\, u^R_{12}\, V(p_3)\, u^R_{13} \cdots u^R_{1,N-2}\, V(p_{N-1})\; 0 =$$

$$= \prod_{2 \le i < j \le (N-1)} (1 - u_{1i}\, u_{1i+1} \cdots u_{1j-1})^{-2p_i \cdot p_j} \tag{2.126}$$

which the reader can easily verify by first commuting all u^R_{ij} factors to one side and then evaluating by coherent state techniques.

Let us suppose for simplicity that $\alpha(o) = 1$, whereupon

$$D(s) = \int_0^1 dx\, x^{-\alpha(s)-1+R} = \frac{1}{R - \alpha(s)} = \frac{1}{L_o - 1} \tag{2.127}$$

where in the final form we wrote $L_o = R - p^2$ with P as the momentum operator.

Consider now the pole at $\alpha_{1j} = N$, an integer. Then we can span the Fock space with occupation number states satisfying

$$R\ |\lambda_N{}^i> \ = N|\lambda_N{}^i> \tag{2.128}$$

$$<\lambda_N{}^i|\ \lambda_N{}^j> \ = \delta_{ij} \tag{2.129}$$

$$\sum_{i=1}^{d(N)} |\lambda_N{}^i> <\lambda_N{}^i| = 1 \tag{2.130}$$

and we insert this completeness relation twice into B_N:

$$B_N = \sum_{n=0}^{\infty} \sum_{i=1}^{d(n)} <0|\ V(p_2)\ D(s_{12})\ \cdots\ V(p_j)\ |\lambda_n{}^i> \cdot$$

$$\cdot\ \frac{1}{n - \alpha(s)} <\lambda_n{}^i|\ V(p_{j+1})\ \cdots\cdot\ V(p_{N-1})\ |0> \tag{2.131}$$

How many states $d(N)$ are there at the level $\alpha(s) = N$, assuming for the moment no linear dependences? This is the number of ways of partitioning N into integers λ_r by

$$\sum_{r=1}^{N} r\lambda_r = N \tag{2.132}$$

and is solved by (for space-time dimensionality d)

$$\prod_{n=1}^{\infty} (1 - x^n)^{-d} = \sum_{n=0}^{\infty} d(N)\ x^N \tag{2.133}$$

The behaviour of $d(N)$ for large N has been solved by Hardy and Ramanujan[23] in their partitio numerorum papers. The precise leading term is

$$d(N) \underset{N\to\infty}{\sim} \frac{1}{\sqrt{2}} \left(\frac{d}{24}\right)^{\frac{d+1}{2}} n^{-(d+3)/4} \exp(2\pi \sqrt{\frac{dN}{6}}) \tag{2.134}$$

Thus the level density increases exponentially in $\sqrt{N}$;

that is, exponentially in the mass.

Let us make some comments on this result:

(i) We have taken all the states in the Fock space to be linearly independent. Projective invariance in fact leads to linear dependences, as we shall see later. Nevertheless, the main feature that $\ell n\, d(N) \sim \sqrt{N}$ will not be altered by the linear dependences.

(ii) All states with an odd number of time excitations are negative norm ghosts. Simple counting reveals that such states are in a majority, so if no linear dependences existed it would be equally good (or bad) to identify odd time occupancy with non-ghosts by changing the overall sign of B_N. Fortunately there are so many linear dependences (for $\alpha(o) = 1$) that all such ghost states are eliminated.

(iii) The exponentially growing degeneracy is physically reasonable since it has been invoked in the statistical model by Hagedorn[24-28], and later by Frautschi[29,30]. See also References 31, 32. In fact the exponent for $d = 4$ and the power of N outside are numerically near the values favored by these authors.

(iv) We have taken $\alpha(o) = 1$ to avoid the $(1 - x)^{\alpha(o)-1}$ factor. With $\alpha(o) \neq 1$ this factor must be expanded as a binomial series. This gives a slightly higher degeneracy, but the difference is negligible to leading order.

Before going more deeply into the study of the B_N spectrum, we first describe some alternative dual models, in order that the particular form we have chosen to describe so far is set in better perspective.

2.7 SATELLITE TERMS

We have seen how the representation

$$A_4 = B_4(-\alpha(s), -\alpha(t)) \tag{2.135}$$

is a closed form solution of the FESR bootstrap when $\alpha(s) = \alpha(o) + \alpha' s$ (let us agree to take an exotic u-channel, to simplify our discussions). Actually we can write a more general solution in the form[1]

$$A_4 = \sum_{\ell=0}^{\infty} \sum_{h=0}^{\ell} c_{\ell h} \frac{\Gamma(\ell - \alpha(s))\, \Gamma(\ell - \alpha(t))}{\Gamma(\ell + h - \alpha(s) - \alpha(t))} \tag{2.136}$$

In this way the residue for each spin at each mass can be made arbitrary; to prove this note that the coefficients c_{00}, c_{10}, c_{20}, ... each determine a new parent trajectory coupling, c_{11}, c_{21}, c_{31}, ... each determine a first daughter coupling, and so on. Thus all predictive power seems to be lost. In the present subsection we will indicate how the requirement of minimum degeneracy greatly resolves the satellite ambiguity.

Note that we do not need to write a triple infinite summation which includes <u>asymmetric</u> terms in A_4, since the asymmetric terms are linearly dependent on these already present. (That this is so, can actually be proved by sharpening the arguments of the previous paragraph, but in any case the dependences have been written explicitly in the literature[33].)

Now we wish to generalise the satellite terms to an N-point function. To obtain an indication, let us take only pure beta functions at the four particle level ($c_{\ell h} = c_\ell \delta_{h\ell}$). Then we may generalise to the following form[34,35]

$$B_N^{\text{Satellites}} = \int \tilde{B}_N \exp(G_N(u_{kj})) \tag{2.137}$$

where

$$\tilde{B}_N = \prod_{i=2}^{N-2} du_{1i}\, u_{1i}^{-\alpha_{1i}-1} \prod_{2\leq i<j\leq(N-1)} \times$$

$$\times\ (1 - u_{1j}u_{1j+1} \ldots u_{1k})^{-2p_j \cdot p_k} \tag{2.138}$$

is the non-satellite integrand and

$$G_N(u_{ij}) = \sum_{\text{all } u_{ij}} u_{ij}(1 - u_{ij})\, g(u_{ij}) \tag{2.139}$$

with g(x) an arbitrary analytic function, related to c_ℓ by

$$x(1 - x)\,[g(x) + g(1 - x)] = \ln[\sum_{\ell=0}^{\infty} c_\ell\, x^\ell (1 - x)^\ell] \tag{2.140}$$

To factorise[35] fully this form introduce operators $A_s^{(r)}$, $A_s^{(r)+}$ (r = 1, 2, 3, ...; s = 1, 2, 3, ..., r) satisfying

$$[A_s^{(r)}, A_{s'}^{(r')+}] = \delta_{rr'}\, G_{ss'}^r \tag{2.141}$$

and auxiliary operators $b_i^{(r)}$, $b_i^{(r)+}$ (r = 1, 2, 3, ... ; i = 1, 2) satisfying

$$[b_i^{(r)}, b_j^{(r')+}] = \delta_{rr'}\delta_{ij}(\delta_{i1} - \delta_{i2}) \tag{2.142}$$

Here $G_{ss'}^r = \pm\, \delta_{ss'}$ is a metric to be determined. To do this we write

$$G_N(u_{ij}) = \sum_{j=2}^{N-2} u_{1j}\ (1 - u_{1j})\ g(u_{1j}) +$$

$$\sum_{2\leq i<j\leq(N-2)} \sum_{r=1}^{\infty} \sum_{p,q=1}^{r} C_r^{pq}\ u_{1i}^p\ u_{1q}^q\ \cdot$$

$$\cdot\ (u_{1i+1},\ u_{1i+2},\ \ldots\ u_{1,j-1})^r \qquad (2.143)$$

in which the symmetric coefficient C_r^{pq} may be diagonalised by a similarity transformation to give

$$C_r^{pq} = \sum_{s=1}^{r} W_{rs}^p\ G_{ss}^r\ W_{rs}^q \qquad (2.144)$$

which defines our new metric in the r-dimensional Fock space of the rth mode. Now we define a new propagator (the vertex is unaltered) by

$$D^{\text{Satellites}}(s) = \int_0^1 dx\ x^{-\alpha(s)-1+R}\ {}_b\langle 0|\ \Gamma\ x^{H_1+H_1'}\Gamma^+|0\rangle_b\ \cdot$$

$$\cdot\ \exp[x(1 - x)g(x)] \qquad (2.145)$$

with

$$H_1 = \sum_{r=1}^{\infty} r\ A_s^{(r)+}\ A_s^{(r)}\ G_{ss}^r \qquad (2.146)$$

$$H_1' = \sum_{r=1}^{\infty} r(b_{1r}^+ b_{1r} - b_{2r}^+ b_{2r}) \qquad (2.147)$$

$$\Gamma = \exp\Big(\sum_{r=1}^{\infty} \sum_{t=1}^{r} \frac{1}{\sqrt{2}} (b_{1t} - b_{2t})\, A_s^{(r)+}\, W_{st}^{(r)} \cdot$$

$$\cdot \exp\Big(\sum_{r=1}^{\infty} A_s^{(r)}\, W_{so}^{(r)} + \frac{1}{\sqrt{2}} (b_{1r} + b_{2r})\Big) \quad (2.148)$$

whereupon

$$B_N^{\text{Satellites}} = \langle 0|\, V(p_2)\, D^{\text{Satellites}}(s_{12})\, V(p_3) \cdot$$

$$\cdot \;\cdots\cdots\; D^{\text{Satellites}}(s_{1,N-2})\, V(p_{N-1})|0\rangle \quad (2.149)$$

which follows from

$$\exp\Big[\sum_{j=2}^{N-2} u_{1j}(1 - u_{1j})\, g(u_{1j})\Big] \cdot \langle 0|\, \varepsilon(u_{11})\, \varepsilon(u_{12}) \cdot$$

$$\cdot \;\cdots\; \varepsilon(u_{1,N-2})\, |0\rangle = \exp(G_N(u_{ij})) \quad (2.150)$$

as the reader may easily check. (Here $\varepsilon(x) = {}_b\langle 0|\, \Gamma\, x^{H_1+H_1'}\, \Gamma^+ |0\rangle_b$).

The degeneracy is now given by the generating function (the full hamiltonian being $R + H_1 + H_1'$)

$$\prod_{r=1}^{\infty} (1 - x^r)^{-r-2-d} = \sum_{N=0}^{\infty} d^{\text{Satellites}}(N)\, x^N \quad (2.151)$$

and now one finds that

$$\ell n\, d(N) \underset{N\to\infty}{\sim} \Big[\frac{3\zeta(3)}{4}\Big]^{1/3} N^{2/3} \quad (2.152)$$

so that the form of the degeneracy is <u>qualitatively</u> different (c f. $\ell n\, d(N) \sim N^{1/2}$ for the non-satellite case).

More general terms may be included[34] in $G_N(U_{ij})$ whereupon in general the dependence of the degeneracy is (see Rivers, 36)

$$\ln d^{\text{Satellites}}(N) \sim N^{\gamma} \tag{2.153}$$

with $2/3 \leq \gamma < 2$.

These results illustrate that full factorisation is possible, but with a higher degeneracy in general. Since we have not considered linear dependences these estimates provide upper limits on the degeneracies; for very special choices of the $G_N(U_{ij})$ it may be possible to reduce to $\gamma = 1/2$ again as in the non-satellite case[37].

Factorisation of the N-point generalisation of the full double sum for A_4 (with $\ell \neq h$ terms included) has not been considered in the literature.

We close the subsection with two remarks:

(i) The Regge behaviour of $A_4(s,t)$ is ensured only if the sum over ℓ is finite. If it is infinite then the coefficients $c_{\ell h}$ must satisfy rather stringent convergence requirements[38].

(ii) Mandelstam[39] has focused attention on the following satellite form

$$A_4 = \int dx\, x^{-\alpha(s)-1} (1 - x)^{-\alpha(t)-1} (1 - x + x^2)^{\delta} \tag{2.154}$$

$$= \sum_{r=0}^{\infty} \binom{\delta}{r} (-1)^r B(r - \alpha(s), r - \alpha(t)) \tag{2.155}$$

$$= B(-\alpha(s), -\alpha(t))\ {}_3F_2(-\alpha(s), -\alpha(t), -\delta; -\frac{\alpha_s + \alpha_t}{2}, \frac{1}{2} - \frac{\alpha_s + \alpha_t}{2}; \frac{1}{4}) \tag{2.156}$$

which has the property that no odd daughters are present provided that

$$\sum = \alpha(s) + \alpha(t) + \alpha(u) = 2\delta - 1 \tag{2.157}$$

For the special case $\delta = 0$ we regain $\sum = -1$ for a single beta function, but in the more general case $\delta \neq 0$ we are enabled to remove odd daughter trajectories for any $\sum$ (and hence any intercept $\alpha(o)$).

2.8 AMPLITUDE FOR PION-PION SCATTERING

We introduce here a proposal for $\pi\pi$ scattering because (i) it seems to have some overlap with chiral symmetry predictions and perhaps with the real world, (ii) it is not a beta function, (iii) it gives us a first brush with an internal symmetry group (isospin in this case), and (iv) it can be obtained from a factorised N-point function, although this last point will be deferred until later.

For the reaction (superscripts are isospin labels)

$$\pi^a(p_1) + \pi^b(p_2) \to \pi^c(-p_3) + \pi^d(-p_4) \tag{2.158}$$

we write

$$\langle -p_3c;\ -p_4d|T|p_1a,\ p_2b\rangle =$$

$$= A_1(st)\,\delta_{ab}\delta_{cd} + A_2(st)\,\delta_{ac}\delta_{bd} + A_3(st)\,\delta_{ad}\delta_{bc} \tag{2.159}$$

$$= [3A_1(st) + A_2(st) + A_3(st)] \frac{1}{3} \delta_{ab}\delta_{cd}$$

$$+ [A_2(st) - A_3(st)] (\frac{1}{2} \delta_{ac}\delta_{bd} - \frac{1}{2} \delta_{ad}\delta_{bc})$$

$$+ [A_2(st) + A_3(st)] (\frac{1}{2} \delta_{ac}\delta_{bd} + \frac{1}{2} \delta_{ad}\delta_{bc} - \frac{1}{3} \delta_{ab}\delta_{cd}) \quad (2.160)$$

which defines $A^{(T_s=I)}(s,t)$ for I = 0, 1, 2 respectively as

$$A^{(T_s=0)} = 3A_1 + A_2 + A_3 \quad (2.161)$$

$$A^{(T_s=1)} = A_2 - A_3 \quad (2.162)$$

$$A^{(T_s=2)} = A_2 + A_3 \quad (2.163)$$

The proposed amplitude is now[40,41,42]

$$A_1 = C_{st} - C_{tu} + C_{us} \quad (2.164)$$

$$A_2 = C_{st} + C_{tu} - C_{us} \quad (2.165)$$

$$A_3 = -C_{st} + C_{tu} + C_{us} \quad (2.166)$$

where

$$C_{xy} = - g^2 \frac{\Gamma(1 - \alpha(x))\, \Gamma(1 - \alpha(y))}{\Gamma(1 - \alpha(x) - \alpha(y))} \quad (2.167)$$

$$= - g^2(1 - \alpha(x) - \alpha(y))\, B(1 - \alpha(x), 1 - \alpha(y)) \quad (2.168)$$

To normalise at the rho pole, we find

$$g^2 = g^2_{\rho\pi\pi} \quad (2.169)$$

where the $\rho\pi\pi$ coupling is defined by

$$L = g_{\rho\pi\pi}\, \rho_\mu^{\ a}\, \partial^\mu\, \pi^b\, \pi^c \varepsilon_{abc} \tag{2.170}$$

and hence $\Gamma(\rho \to 2\pi) = g^2_{\rho\pi\pi}\, k^3/(6\pi\, m_\rho^{\ 2})$.

We now turn to the properties

(i) According to Adler[43] the amplitude should vanish when any pion becomes soft ($p_2^{\ \mu} \to 0$). When this occurs, s, t, u $\to m_\pi^{\ 2}$ ($=p_i^{\ 2}$). The first pole in the denominator of C_{st}. etc. will coincide with this Adler point if we impose[41]

$$\alpha_\rho(m_\pi^{\ 2}) = \frac{1}{2} \tag{2.171}$$

or

$$\alpha_\rho(o) - \alpha_\pi(o) = \frac{1}{2} \tag{2.172}$$

which is consistent with the physical masses. Therefore we adopt

$$\alpha_\rho(s) = \frac{1}{2} + \alpha'(s - m_\pi^{\ 2}) \tag{2.173}$$

and $\alpha' = [2(m_\rho^{\ 2} - m_\pi^{\ 2})]^{-1}$.

(ii) Pion-pion s-wave scattering lengths $a_o^{(I)}$ are defined by

$$a_o^{(I)} = \lim_{k \to 0} \frac{1}{k\, \cot\delta_o^{(I)}} \tag{2.174}$$

where $\delta_o^{(I)}$ is the s-wave phase shift in

$$\hat{a}_o^{(I)}(s) = \frac{\eta_o\, e^{2i\pi\delta_o} - 1}{2i} \tag{2.175}$$

Now $\cot\delta_o^{(I)} \approx 1/\delta_o^{(I)}$ and bearing in mind that for identical particles we have derived the relation

$$\hat{a}_o^{(I)}(s) = \frac{k}{16\pi\sqrt{s}} a_o^{(I)}(s) \qquad (2.176)$$

we see that

$$a_o^{(I)} = \frac{1}{32\pi m_\pi} A^{(T_s=I)} \quad (s = 4m_\pi^2,\ t = u = 0) \qquad (2.177)$$

in our normalisations.

Now for small arguments

$$C_{st} \approx -(2m_\pi^2 - s - t)\ \pi g^2\alpha' \qquad (2.178)$$

whence

$$A^{(T_s=0)} \approx -\ \pi g^2\alpha'(10m_\pi^2 - 6s - 2t - 2u) \qquad (2.179)$$

$$A^{(T_s=2)} \approx -\ \pi g^2\alpha'\ (4m_\pi^2 - 2t - 2u) \qquad (2.180)$$

and therefore

$$a_o^{(0)} = \frac{7}{16} m_\pi g^2\alpha' \qquad (2.181)$$

$$a_0^{(2)} = \frac{1}{8} m_\pi g^2\alpha' \qquad (2.182)$$

In particular we note the ratio $a_o^{(0)}/a_o^{(2)} = -\frac{7}{2}$, which agrees with the prediction obtained from current algebra and PCAC assumptions[44]. We defer a more detailed discussion of chiral and dual models at low energy but make one remark here: it can be shown that this ratio follows only from the assumption that each planar component (C_{st}, C_{tu}, C_{us}) separately has the Adler zero[45,46].

(iii) The idea[41] that the Adler zero is due to a pole in the denominator gamma function has been generalised to other processes by Ademollo et al.[47]. These authors make simple ansatz for other reactions and arrive at the rule "Whenever particles on one trajectory can decay by pion emission into particles of opposite normality on another trajectory, the two trajectories have intercepts which differ by a half odd-integer."

This seems to be the case for several pairs of trajectories in Nature, for example ρ-π, Δ-N, K*-K, and Y_1*-Σ, Λ.

(iv) The $\pi\pi$ amplitude has been used in slightly modified forms to discuss[41] channels such as $\bar{p}n \to \pi^+\pi^-\pi^-$, since this annihilation at rest occurs predominantly in s-wave and hence the decaying state is rather like a very heavy pion. Some evidence for the line of fixed u zeros (at $1 - \alpha(s) - \alpha(t) = -N$, $N = 0, 1, 2, \ldots$) is seen in the Dalitz plot, although alternative explanations for the data can be equally successful. [References 48, 49, 50].

2.9 SHAPIRO-VIRASORO MODEL

We have already mentioned in (2.2) the formula

$$B(-\alpha(s), -\alpha(t)) + B(-\alpha(t), -\alpha(u)) + B(-\alpha(u), -\alpha(s))$$

$$= \frac{\sqrt{\pi}\,\Gamma(-\frac{1}{2}\alpha(s))\,\Gamma(-\frac{1}{2}\alpha(t))\,\Gamma(-\frac{1}{2}\alpha(u))}{\Gamma(-\frac{1}{2}\alpha(s) - \frac{1}{2}\alpha(t))\Gamma(-\frac{1}{2}\alpha(t) - \frac{1}{2}\alpha(u))\Gamma(-\frac{1}{2}\alpha(u) - \frac{1}{2}\alpha(s))} \tag{2.183}$$

valid for $\sum = -1$. Virasoro[51] considered the right-hand side for arbitrary $\sum$. For $\sum \neq -1$ the Virasoro amplitude exhibits non-planar duality, since we can no longer write it as a sum of three terms each of which separately displays one empty channel and exchange degeneracy (for more details, see Ref. 52).

Shapiro[53] discovered that there is an essentially non-planar extension of the Virasoro formula to N particles for the case $\sum = -2$ by using a generalisation of the Koba-Nielsen formula. His result is

$$B_N^{\text{Non-planar}} = \int \prod_{i=1}^{N} d^2 z_i \left[\frac{d^2 z_a \; d^2 z_b \; d^2 z_c}{(z_a - z_b)^2 (z_b - z_c)^2 (z_c - z_a)^2}\right]$$

$$\prod_{1 \leq i < j \leq N} |z_i - z_j|^{-2 p_i \cdot p_j} \tag{2.184}$$

where the three fixed points z_a, z_b, z_c are chosen arbitrarily, and now the domain of integration is unrestricted over the surface of a sphere (of arbitrary radius: for example, we can simply take the complex z-plane). The point here is that the integrand is invariant under (complex) projective transformations if and only if $p_i^2 = -2$ ($\sum = -2$ for the four particle case). Making a complex dilation

$$z \to z' = a\mathbf{z} \tag{2.185}$$

we pick up an additional factor

$$|a|^{2N - 2 \sum_{i<j} p_i \cdot p_j} = |a|^{N(2 + p_i^2)} = 1 \tag{2.186}$$

$$\text{for } p_i^2 = -2$$

as required. Once this condition is imposed invariance

under inversion (and, of course, translation) is assured.

Before factorising this non-planar model, let us show its equivalence to the Virasoro formula for four particles, since it is not obvious. Use the formula

$$a^{-\rho}\,\Gamma(\rho) = \int_0^\infty dt\ t^{\rho-1}\ e^{-at} \tag{2.187}$$

to re-write

$$B_4^{\text{Non-planar}} = \int d^2z\ |z|^{-2p_1\cdot p_2}\ |1 - z|^{-2p_2\cdot p_3} \tag{2.188}$$

$$= \frac{1}{\Gamma(p_1\cdot p_2)}\frac{1}{\Gamma(p_2\cdot p_3)}\int_0^\infty dt\ t^{p_1\cdot p_2-1}\int_0^\infty du\ u^{p_2\cdot p_3-1}$$

$$\int_{-\infty}^\infty dx \int dy\ \exp[-x^2t - y^2t - (1 - x)^2u - y^2u] \tag{2.189}$$

$$= \frac{\pi}{\Gamma(p_1\cdot p_2)\Gamma(p_2\cdot p_3)}\int_0^\infty dt\ t^{p_1\cdot p_2-1}\int_0^\infty du\ u^{p_2\cdot p_3-1}\ .$$

$$\cdot\ \frac{1}{t + u}\exp(-\ \frac{tu}{t + u}) \tag{2.190}$$

$$= \frac{\pi}{\Gamma(p_1\cdot p_2)\Gamma(p_2\cdot p_3)}\int_0^1 dq\ q^{-p_1\cdot p_2}\ (1 - q)^{-p_2\cdot p_3}\ .$$

$$\cdot \int_0^\infty dt\ t^{p_1\cdot p_2+p_2\cdot p_3-2}\ e^{-qt} \tag{2.191}$$

$$= \frac{\pi\ \Gamma(1 - p_1\cdot p_2)\ \Gamma(1 - p_2\cdot p_3)\ \Gamma(p_1\cdot p_2 + p_2\cdot p_3-1)}{\Gamma(2 - p_1\cdot p_2 - p_2\cdot p_3)\ \Gamma(p_1\cdot p_2)\ \Gamma(p_2\cdot p_3)} \tag{2.192}$$

$$= \frac{\pi\,\Gamma(-\frac{1}{2}\alpha_s)\,\Gamma(-\frac{1}{2}\alpha_t)\,\Gamma(-\frac{1}{2}\alpha_u)}{\Gamma(-\frac{1}{2}\alpha_s - \frac{1}{2}\alpha_t)\Gamma(-\frac{1}{2}\alpha_t - \frac{1}{2}\alpha_u)\,\Gamma(-\frac{1}{2}\alpha_u - \frac{1}{2}\alpha_s)} \quad (2.193)$$

if $\alpha_s + \alpha_t + \alpha_u = -2$. Here we have used the changes of variable according to $z = x + iy$, $u = tq/(1-q)$ respectively.

To factorise[54,55] $B_N^{\text{non-planar}}$ we choose as fixed points $z_1 = 0$, $z_c = 1$, $z_N = \infty$ and then order the z moduli according to

$$\infty > |z_{N-1}| > |z_{N-2}| > \cdots\cdot > |z_2| \quad (2.194)$$

We can write

$$B_N^{\text{Non-planar}} = \sum_{(N-2)!\text{permutations}} F_N(p_1 p_2 \cdots p_N) \quad (2.195)$$

(Note that each permutation separately will not be Regge behaved.)

Now we introduce variables resembling the planar Chan variables by writing

$$z_i = r_i r_{i+1} \cdots r_{i-1}\, e^{i\theta_i} \quad (2 \leq i \leq (c-1)) \quad (2.196)$$

$$z_i^{-1} = r_c r_{c+1} \cdots r_{i-1}\, e^{-i\theta_i} \quad (c+1 \leq i \leq n-1) \quad (2.197)$$

with $0 \leq r_i \leq 1$. This enables us to re-write

$$F_N(p_1 p_2 \cdots p_N) = \int_0^{2\pi} \prod_{i\neq 1,c,N} d\theta_i \prod_{j=2}^{N-2} dr_j\, r_j^{-\alpha(1j)-1}$$

$$\prod_{2\leq i<j\leq(N-1)} |1 - r_i r_{i+1} \cdots\cdot r_{j-1}\, e^{i(\theta_i-\theta_j)}|^{-2p_i\cdot} \quad (2.198)$$

This expression coincides exactly with the planar case, except for the angular integrations. Its factorisation is therefore done by analogy. We introduce two sets of operators $a^{(n)}$ and $\bar{a}^{(n)}$ satisfying

$$[a_\mu^{(n)}, a_\nu^{(m)+}] = - g_{\mu\nu} \delta_{mn} \quad (2.199)$$

$$[\bar{a}_\mu^{(n)}, \bar{a}_\nu^{(m)+}] = - g_{\mu\nu} \delta_{mn} \quad (2.200)$$

and define a propagator

$$D^{\text{Non-planar}}(s) = \int_0^1 dx\, x^{-\alpha(s)-1+R+\bar{R}}$$

$$= \frac{1}{R + \bar{R} - \alpha(s)} \quad (2.201)$$

with

$$R = - \sum_{n=1}^{\infty} n\, a^{(n)+} \cdot a^{(n)} \quad (2.202)$$

$$\bar{R} = - \sum_{n=1}^{\infty} n\, \bar{a}^{(n)+} \cdot \bar{a}^{(n)} \quad (2.203)$$

The appropriate vertex is

$$V^{\text{Non-planar}}(p) = \int_0^{2\pi} d\theta \, \cdot$$

$$\cdot \exp[ip\cdot \sum_{n=1}^{\infty} (\frac{a^{(n)+} e^{in\theta} + \bar{a}^{(n)+} e^{-in\theta}}{\sqrt{n}})] \cdot$$

$$\cdot \exp[ip\cdot \sum_{n=1}^{\infty} (\frac{a^{(n)} e^{-in\theta} + \bar{a}^{(n)} e^{in\theta}}{\sqrt{n}})] \quad (2.204)$$

Then we have

$$F_N(p_1\, p_2 \cdots p_N) = \frac{1}{2\pi} \langle 0|\, V^{\text{Non-planar}}(p_2) \cdot$$

$$\cdot\, D^{\text{Non-planar}}(s_{12}) \cdots\cdots V^{\text{Non-planar}}(p_{N-1})\, |0\rangle \tag{2.205}$$

as required.

The degeneracy of states grows very much as it does in the planar model. The fact that the number of oscillators is doubled means only that the leading term ($\sim \sqrt{N}$) of $\ell n\, d(N)$ is multiplied by $\sqrt{2}$. Of course, to exhibit the pole of $B_N^{\text{non-planar}}$ at $\alpha_{1j} = N$ we must combine all of the $[(j-1)!(N-j-1)!]$ permutations which contain it; this does not alter the degeneracy.

Finally let us make a general observation. The non-planar model discussed here and the planar model discussed earlier are on an equal footing mathematically; it is only the experimentally observed absence of exotics and exchange degeneracy that leads us to devote nearly all our attention to the planar model (Veneziano) rather than the non-planar one (Shapiro-Virasoro). To put the same statement differently: the dual resonance model formalism sheds no light on the mystery of why exotics are not observed.

2.10 SUMMARY

The generalisation of the beta function to N external particles involves an integrand invariant under a three-parameter projective group. The construction of the N-point function is best understood in terms of anharmonic ratios, invariant under the group, and such that the simultaneous poles in incompatible channels, and the cyclic invariance are easily guaranteed.

Factorisation of the N-point function requires a number of states increasing exponentially in the mass. It seems impossible to construct a consistent dual scheme with any lower degeneracy. In general, the addition of satellite terms involves an increase in the degeneracy, the logarithm of the degeneracy now increasing as a power of the $(\text{mass})^{\gamma}$ with $2/3 \leq \gamma < 2$. Thus requirement of minimum degeneracy resolves the satellite ambiguity, except possibly in very special cases.

A non-planar generalisation can be made for Regge intercept $\alpha(o) = 2$. This model is equally attractive from a mathematical viewpoint (indeed it possesses an even higher symmetry) but we are led to reject it because of the observed exchange degeneracy and absence of exotics. If exotics are subsequently discovered, the non-planar models will become of more physical interest.

REFERENCES

1. G. Veneziano, Nuovo Cimento 57A, 190 (1968).
2. A. Erdélyi et al., Higher Transcendental Functions, Bateman Manuscript Project, McGraw-Hill (1953); E. T. Whittaker and G. N. Watson, Modern Analysis, Cambridge University Press. Fourth Edition (1969).
3. F. Cerulus and A. Martin, Physics Letters 8, 80 (1964).
4. A. Martin, Nuovo Cimento 37, 671 (1965).
5. C. B. Chiu and C. I. Tan, Phys. Rev. 162, 1701 (1967).
6. D. B. Fairlie and K. Jones, Nucl. Phys. B15, 323 (1970).
7. K. Bardakci and H. Ruegg, Physics Letters 28B, 342 (1968).
8. M. A. Virasoro, Phys. Rev. Letters 22, 37 (1969).
9. A. C. Dixon, Proc. Lond. Math. Soc. 2, 8 (1905).
10. H. M. Chan, Physics Letters 28B, 425 (1969).
11. H. M. Chan and T. S. Tsun, Physics Letters 28B, 485 (1969).
12. J. F. L. Hopkinson and E. Plahte, Physics Letters 28B, 489 (1969).
13. K. Bardakci and H. Ruegg, Phys. Rev. 181, 1884 (1969).
14. C. J. Goebel and B. Sakita, Phys. Rev. Letters 22, 257 (1969).
15. Z. Koba and H. B. Nielsen, Nucl. Phys. B10, 633 (1969).
16. Z. Koba and H. B. Nielsen, Nucl. Phys. B12, 517 (1969)

17. Z. Koba and H. B. Nielsen, Z. Physik 229, 243 (1969).
18. S. Fubini and G. Veneziano, Nuovo Cimento 64A, 811 (1969).
19. K. Bardakci and S. Mandelstam, Phys. Rev. 184, 1640 (1969).
20. Y. Nambu, Symmetries and Quark Models (1969), edited by R. Chand. Gordon and Breach (1970), Page 269.
21. S. Fubini, D. Gordon and G. Veneziano, Phys. Letters 29B, 679 (1969).
22. J. R. Klauder and E. C. G. Sudarshan, Quantum Optics, W. A. Benjamin (1968).
23. G. H. Hardy and S. Ramanujan, Proc. Lond. Math. Soc. 17, 75 (1917).
24. R. Hagedorn, Nuovo Cimento Suppl. 3, 147 (1965).
25. R. Hagedorn, Nuovo Cimento 52A, 1336 (1967).
26. R. Hagedorn, Nuovo Cimento 56A, 1027 (1968).
27. R. Hagedorn, Nuovo Cimento Suppl. 3, 147 (1965).
28. R. Hagedorn and G. Ranft, Nuovo Cimento Suppl. 6, 169 (1968).
29. S. Frautschi, Phys. Rev. D3, 2821 (1971).
30. C. J. Hamer and S. C. Frautschi, Phys. Rev. D4, 2125 (1971).
31. K. Huang and S. Weinberg, Phys. Rev. Letters 25 895 (1970).
32. W. Nahm, Nucl. Phys. B45, 525 (1972).
33. R. E. Kreps and M. S. Milgram, Phys. Rev. D1, 2271 (1970).
34. D. J. Gross, Nucl. Phys. B13, 467 (1969)
35. P. H. Frampton, Physics Letters 32B, 195 (1970).
36. R. J. Rivers, Phys. Rev. D3, 363 (1971).

37. J. L. Gervais and A. Neveu, Orsay preprint LPTHE (1972).

38. P. H. Frampton and C. W. Gardiner, Phys. Rev. D2, 2378 (1970).

39. S. Mandelstam, Phys. Rev. Letters 21, 1724 (1968).

40. J. Shapiro and J. Yellin, Berkeley preprint (1968).

41. C. Lovelace, Physics Letters 28B, 264 (1968).

42. J. A. Shapiro, Phys. Rev. 179, 1345 (1969).

43. S. L. Adler, Phys. Rev. 137, B 1022 (1965).

44. S. Weinberg, Phys. Rev. Letters 17, 616 (1966).

45. L. Susskind and G. Frye, Phys. Rev. D1, 1682 (1970).

46. H. Osborn, Lettere al Nuovo Cimento 2, 717 (1969).

47. M. Ademollo, S. Weinberg and G. Veneziano, Phys. Rev. Letters 22, 83 (1969).

48. G. Altarelli and H. R. Rubinstein, Phys. Rev. 183, 1469 (1969).

49. R. Odorico, Phys. Letters 33B, 489 (1970).

50. S. Pokorski, R. O. Raitio and G. H. Thomas, Nuovo Cimento 7A, 828 (1972).

51. M. A. Virasoro, Phys. Rev. 177, 2309 (1969).

52. S. Mandelstam, Phys. Rev. 183, 1374 (1969).

53. J. A. Shapiro, Phys. Letters 33B, 361 (1970).

54. M. Yoshimura, Phys. Letters 34B, 79 (1971).

55. P. Di Vecchia and E. Del Guidice, Nuovo Cimento 5A, 90 (1971)

3

OPERATOR FORMALISM

3.1 INTRODUCTION

We probe more deeply into the properties of the multiparticle dual amplitude, by fully exploiting the operator formalism. In particular, by realising the projective group on the Fock space spanned by the harmonic-oscillator-like operators we find the gauge invariances of the theory; the latter are especially powerful if the intercept is exactly one which is the value for which we have already seen simplifications in the four-particle Euler B function. Combining the projective group and the operator formalism enables us to rewrite the N-point function in a manner such that both cyclic symmetry and factorisability are manifest.

For the case of unit intercept, it is shown that the large gauge invariance group enables elimination of all ghost states arising from the relativistic metric. This is essential if we are to preserve the usual quantum mechanical intrepretation of the resonances.

Finally we discuss other operator constructs. Specifically we introduce the twisting operator and the twisted propagator. These have no direct equivalent in local field theories. Then we discuss the multireggeon vertex, in which the projective group properties enable us to keep control over factorisation and cyclic symmetric properties.

3.2 PROJECTIVE GROUP

The cyclic symmetry and duality properties, including the absence of incompatible poles, of the N-point function are most evident in the Koba-Nielsen representation of the amplitude, where the integrand is invariant under a group of projective transformations. On the other hand, the factorisation properties and the density of levels are easily derived by using the harmonic oscillator operator formalism. Our objective now will be to study more deeply the properties of B_N by combining the two approaches such that eventually we can write B_N in a form where both cyclic symmetry and factorisation are evident; to do this we first need to construct unitary irreducible representations (UIR) of the projective group SU(1,1), or what is essentially the same thing up to a similarity transformation SL(2,R), realised on the Fock space spanned by the operators $a^{(n)}$, $a^{(n)+}$ satisfying

$$[a_\mu^{(n)}, a_\nu^{(m)+}] = - g_{\mu\nu} \delta_{mn} \tag{3.1}$$

Fortunately the UIR of SU(1,1) have been completely classified.[1,2] The first thing we learn from the literature is that because SU(1,1) is noncompact, all the UIR's are infinite dimensional; this is a happy event, since we have an infinite number of modes n = 1, 2, 3, ... on which to map these infinite-dimensional representations.

Pauli matrices are defined by

$$\sigma_1 = \begin{pmatrix} 0 & 1 \\ 1 & 0 \end{pmatrix}, \quad \sigma_2 = \begin{pmatrix} 0 & -i \\ i & 0 \end{pmatrix}, \quad \sigma_3 = \begin{pmatrix} 1 & 0 \\ 0 & -1 \end{pmatrix} \tag{3.2}$$

and putting angular momentum $\underline{J} = \frac{1}{2}\underline{\sigma}$, $J_\pm = (J_1 \pm iJ_2)$ we satisfy the SU(2) algebra

$$[J_+, J_-] = 2J_3 \tag{3.3}$$

$$[J_\pm, J_3] = \mp J_\pm \tag{3.4}$$

We can equally use the Pauli matrices to represent the SU(1,1) lie algebra by identifying

$$L_1 = -J_- = \begin{pmatrix} 0 & 0 \\ -1 & 0 \end{pmatrix} \tag{3.5}$$

$$L_{-1} = J_+ = \begin{pmatrix} 0 & 1 \\ 0 & 0 \end{pmatrix} \tag{3.6}$$

$$L_0 = J_3 = \frac{1}{2}\begin{pmatrix} 1 & 0 \\ 0 & -1 \end{pmatrix} \tag{3.7}$$

which satisfy

$$[L_1, L_{-1}] = 2L_0 \tag{3.8}$$

$$[L_{\pm 1}, L_0] = \pm L_\pm \tag{3.9}$$

as required. The reason for the identification of L_{-1} with J_+ rather than J_- is (i) that it conforms with the most common usage in the literature and (ii) that later we will want to use $L_{-1} = (L_1)^+$ as a raising operator (like $a^{(n)+}$) and L_1 as a lowering operator (like $a^{(n)}$) inside the Fock space.

We wish to deal with projective transformations on the Koba-Nielsen variables z according to

$$z' = \frac{az + b}{cz + d} \qquad ad - bc = 1 \tag{3.10}$$

We define $z = \xi_1/\xi_2$ whereupon

$$\xi_1' = a\xi_1 + b\xi_2 \tag{3.11}$$

$$\xi_2' = c\xi_1 + d\xi_2 \tag{3.12}$$

transform as the components of a two-component spinor. A suitable orthonormal basis for a given UIR can be written

$$|J\ k\ m\rangle = N(Jkm)(\xi_1\xi_2)^J\ (\xi_1/\xi_2)^{k+m} \tag{3.13}$$

where $J(J+1)$ is the eigenvalue of the Casimir operator $L_0(L_0+1) - L_1L_{-1}$ and k is a second label of the UIR. Finally m (related to the eigenvalue of L_0) labels states within a given UIR, and N(Jkm) is a normalization factor.

A detailed analysis of SU(1,1) reveals that there are three types of UIR: (i) $J \neq \pm k$, with eigenvalues of $L_0(=k+m)$ unbounded and integer spaced (ii) $J = +k$, with eigenvalues of L_0 bounded from above and integer spaced and (iii) $J = -k$, with eigenvalues of L_0 bounded from below and integer spaced. We will eventually identify L_0 with the squared mass operator of the dual model, so only the representations of the third type will be used, and the corresponding representation matrices (for $J=-k$) will be written $D^{(J,+)}_{mn}(X)$.

In general a representation matrix for the transformation

$$\Lambda = \begin{pmatrix} a & b \\ c & d \end{pmatrix} \tag{3.14}$$

is defined by

$$|J\ k\ m\rangle' = N(J\ k\ m)\ (\xi_1'\xi_2')^J\ (\xi_1'/\xi_2')^{k+m} \qquad (3.15)$$

$$= \sum_{n=0}^{\infty} D^{(J,k)}_{mn}(\Lambda)\ |J\ k\ n\rangle \qquad (3.16)$$

which becomes, for $J = -k$ and $\xi_1/\xi_2 = z$

$$N(J\ {-J}\ m)\ (az + b)^m\ (cz + d)^{2J-m} =$$

$$\sum_{n=0}^{\infty} D^{(J,+)}_{mn}(\Lambda)\ N(J\ {-J}\ n) z^n \qquad (3.17)$$

Now consider an infinitesimal transformation generated by L_0 according to

$$z \rightarrow z' = (1 + i\ \varepsilon\ L_0) z \qquad (3.18)$$

then, using our Pauli matrix representation for L_0,

$$(\xi_1'\xi_2') = z(1 + O(\varepsilon^2)) \qquad (3.19)$$

$$(\xi_1'/\xi_2') = (1 + i\varepsilon)z + O(\varepsilon^2) \qquad (3.20)$$

Therefore

$$N(J\ k\ m)\ z^{J+k+m}(1 + i\varepsilon)^{k+m} =$$

$$N(J\ k\ m) z^{J+k+m}\ (1 + i\varepsilon(k+m) + O(\varepsilon^2)) \qquad (3.21)$$

$$= \sum_{n=0}^{\infty} (\delta_{mn} + i\varepsilon\ D^{(Jk)}_{mn}(L_0))\ N(J\ k\ m)\ z^{J+k+n} +$$

$$+ O(\varepsilon^2) \qquad (3.22)$$

and it follows that (the normalisation factor N(Jkm) cancels for this case)

$$D^{(Jk)}_{mn}(L_0) = \delta_{mn}(m + k) \qquad (3.23)$$

and, in particular,

$$D_{mn}^{(J+)}(L_0) = \delta_{mn}(m - J) \tag{3.24}$$

Now consider the generator L_{+1}, to find

$$N(J\ k\ m)(z - i\varepsilon z^2)^J\ (z + i\varepsilon z^2)^{k+m}$$

$$= N(J\ k\ m)\ z^{J+k+m}\ (1 + (k + m - J)\ i\varepsilon z) + O(\varepsilon^2) \tag{3.25}$$

$$= \sum_{n=o}^{\infty} (\delta_{mn} + i\varepsilon\ D_{mn}^{(Jk)}(L_1))z^{J+k+n}\ N(J\ k\ n) \tag{3.26}$$

whence

$$D_{mn}^{(Jk)}(L_1) = \delta_{n,m+1}\ (k + m - J)\ \frac{N(J\ k\ m)}{N(J\ k\ n)} \tag{3.27}$$

The corresponding calculation for L_{-1} gives

$$D_{mn}^{(Jk)}(L_{-1}) = \frac{N(J\ k\ m)}{N(J\ k\ m-1)}\ \delta_{n,m-1}\ (k + m - J) \tag{3.28}$$

The normalization constants follow from the requirement that

$$(L_1)^+ = L_{-1} \tag{3.29}$$

or (for $J = -k$)

$$D_{m,m+1}^{(J+)}\ (L_1) = D_{m+1,m}^{(J+)}\ (L_{-1})^* \tag{3.30}$$

whereupon

$$(m - 2J)\ \frac{N(J\ -J\ m)}{N(J\ -J\ m+1)} = [\frac{N(J\ -J\ m+1)}{N(J\ -J\ m)}]^*\ (m+1) \tag{3.31}$$

and therefore

$$\left| \frac{N(J\ -J\ m)}{N(J\ -J\ 0)} \right|^2 = \frac{\Gamma\ (m - 2J)}{\Gamma(-2J)\ \Gamma(m + 1)} \tag{3.32}$$

Taking an arbitrary phase to be one, and defining

$$N(J\ -J\ 0) = \sqrt{\Gamma(-2J)} \tag{3.33}$$

one arrives at

$$N(J\ -J\ m) = \frac{\sqrt{\Gamma(m - 2J)}}{\sqrt{m!}} \tag{3.34}$$

Collecting together our results so far we have, for the UIR's bounded from below

$$D_{mn}^{(J+)}(L_0) = \delta_{mn}(-J + n) \tag{3.35}$$

$$D_{mn}^{(J+)}(L_1) = \sqrt{n(n - 1 - 2J)}\ \delta_{m,n-1} \tag{3.36}$$

$$D_{mn}^{(J+)}(L_{-1}) = \sqrt{(n + 1)\ (n - 2J)}\ \delta_{m,n+1} \tag{3.37}$$

Now we come to the identification of the realisation on the Fock space according to

$$L_i = - \sum_{m=0}^{\infty} \sum_{n=0}^{\infty} a_\mu^{(m-J)+} D_{mn}^{(J+)}(L_i)\ a_\mu^{(n-J)} \tag{3.38}$$

with the operators of the dual resonance model; in particular the identification of L_0 with the combination occurring in the propagator denominator, namely

$$R - p^2 = - \sum_{n=1}^{\infty} n\ a_\mu^{(n)+}\ a_\mu^{(n)} - p^2 \tag{3.39}$$

To make the identification $L_0 = R - p^2$, we take $J \to 0$ from below, that is $J = -\varepsilon$ ($\varepsilon > 0$) and $\varepsilon \to 0$. Note, however, that when $\varepsilon \to 0$ our normalisation constant blows up

$$N(-\varepsilon, +\varepsilon, 0) = \sqrt{\Gamma(2\varepsilon)} \underset{\varepsilon\to 0}{\to} \infty \tag{3.40}$$

This corresponds to the fact that no scalar $J = k = 0$ representation of SU(1,1) exists, strictly speaking. This technical difficulty was first discovered and overcome by Fubini and Veneziano in a well-known paper.[3] [Concerning the projective group properties see also References 4, 5]. A further difficulty is to identify the zeroth mode operators $a_\mu^{(\varepsilon)}$ and $a_\mu^{(\varepsilon)+}$ in the limit $\varepsilon \to 0$ with the momentum p_μ and its conjugate the position operator Q_μ.

For $J = -\varepsilon$ we have

$$L_0 = -\sum_{n=1}^{\infty} (n + \varepsilon)\, a^{(n)+}\cdot a^{(n)} - \varepsilon\, a^{(\varepsilon)+}\cdot a^{(\varepsilon)} \tag{3.41}$$

$$L_1 = -\sum_{n=1}^{\infty} \sqrt{(n+1)(n+2\varepsilon)}\, a^{(n)+}\cdot a^{(n+1)} - \sqrt{2\varepsilon}\,\cdot\, a^{(\varepsilon)+}\cdot a^{(1)} \tag{3.42}$$

$$L_{-1} = (L_1)^+ \tag{3.43}$$

The limit $\varepsilon \to 0+$ in the summations is harmless. We write

$$[a_\mu^{(\varepsilon)}, a_\nu^{(\varepsilon)+}] = -g_{\mu\nu} \tag{3.44}$$

and identify

$$a_\mu^{(\varepsilon)} = \frac{i}{\sqrt{2\varepsilon}} (P_\mu - i\varepsilon Q_\mu) \tag{3.45}$$

$$a_\mu^{(\varepsilon)} = \frac{-i}{\sqrt{2\varepsilon}} (P_\mu + i\varepsilon Q_\mu) \tag{3.46}$$

with

$$[Q_\mu, P_\nu] = -ig_{\mu\nu} \tag{3.47}$$

This then gives the projective gauge operators [References 6-9].

$$L_0 = -\sum_{n=1}^{\infty} na^{(n)+} a^{(n)} - p^2 \tag{3.48}$$

$$L_1 = -\sum_{n=1}^{\infty} \sqrt{n(n+1)}\, a^{(n)+} \cdot a^{(n+1)} + i\sqrt{2}\, p \cdot a^{(1)} \tag{3.49}$$

$$L_{-1} = (L_1)^+ \tag{3.50}$$

The fact that these satisfy the SU(1,1) algebra follows from our construction. We can of course easily check explicitly by

$$[L_1, L_{-1}] = \sum_{n,n'=1}^{\infty} \sqrt{n(n+1)\, n'(n'+1)} \cdot \\ \cdot [a^{(n)+} a^{(n+1)}, a^{(n'+1)+} a^{(n')}] + \\ + 2P_\mu P_\nu [a_\mu^{(1)}, a_\nu^{(1)+}] \tag{3.51}$$

$$= \sum_{n,n'=1}^{\infty} n(n+1)\, n'(n'+1) \cdot \\ \cdot (\delta_{nn'} a^{(n+1)+} a^{(n+1)} - \delta_{n+1,n'+1} a^{(n)+} \cdot a^{(n)}) \tag{3.52}$$

$$- 2p^2 = -2 \sum_{n=1}^{\infty} na^{(n)+} a^{(n)} - 2p^2 = 2L_0 \tag{3.53}$$

Similarly (we leave it as an exercise) one finds

$$[L_{\pm 1}, L_0] = \pm L_{\pm 1} \tag{3.54}$$

To proceed further recall that the expression for B_N, namely

$$B_N = \int \prod_{i=1}^{N} dz_i \left[\frac{dz_a\, dz_b\, dz_c}{(z_a - z_b)(z_b - z_c)(z_c - z_a)}\right]^{-1} \cdot$$

$$\cdot \prod_{i<j} (z_i - z_j)^{-2p_i \cdot p_j} \prod_{i=1}^{N} (z_i - z_{i+1})^{-1+\alpha(o)} \tag{3.55}$$

is projective invariant (SU(1,1) - invariant) for any intercept $\alpha(o)$. When $\alpha(o) = 1$, however, there is a higher symmetry - namely the integrand becomes totally symmetric under all N! permutations of the $\{z_i, p_i\}$ argument pairs. It turns out that for the unit intercept case $\alpha(o) = 1$ it is useful to introduce a generalised projective group with generators given by[10]

$$L_m = - \sum_{n=1}^{\infty} \sqrt{n(n+m)}\, a^{(n)+} \cdot a^{(n+m)} + i\sqrt{2m}\, P \cdot a^{(m)} +$$

$$+ \frac{1}{2} \sum_{n=1}^{m-1} \sqrt{n(m-n)}\, a^{(n)} \cdot a^{(m-n)} \tag{3.56}$$

$$L_{-m} = (L_m)^+ \tag{3.57}$$

These satisfy the algebra (of which SU(1,1) is a sub-algebra)

$$[L_m, L_n] = (m - n) L_{m+n} + \frac{d}{12} m(m^2 - 1)\, \delta_{m+n,0} \tag{3.58}$$

where in the anomaly term[11] for $m+n = 0$, d is the space-time dimension. We return to a detailed discussion of the anomaly term shortly.

It can be seen from our derivation of the representation matrices $D^{(J+)}_{mn}(L_n)$ for the SU(1,1) subgroup n = 0, ±1 that the generators $|n| \geq 2$ correspond to transformations of the type

$$z'^{n} = \frac{az^{n} + b}{cz^{n} + d} \tag{3.59}$$

It is important to emphasise, however, that the B_N integrand is not invariant under these transformations, which contain branch cut singularities. The only vestige of the higher symmetry is the total symmetry under permutations already mentioned.

Let us check the algebra for $m + n \neq 0$, where no anomaly term occurs. Then we have (m, n > 0, for example)

$$\begin{aligned}
[L_m, L_n] = & \sum_{p,q=1}^{\infty} \sqrt{p(p + m)q(q + n)} \cdot \\
& \cdot [a^{(p)+} \cdot a^{(p+m)}, a^{(q)+} \cdot a^{(q+n)}] - \\
& - i\sqrt{2m}\, P_\mu [a^{(m)}, \sum_{q=1}^{\infty} \sqrt{q(q + n)}\, a^{(q)+} \cdot a^{(q+n)}] \\
& - i\sqrt{2n}\, P_\mu [\sum_{p=1}^{\infty} \sqrt{p(p + m)}\, a^{(p)+} \cdot a^{(p+m)}, a_\mu^{(n)}] \\
& - \frac{1}{2} \sum_{p=1}^{m-1} \sum_{q=1}^{\infty} [a^{(m-p)} \cdot a^{(p)}, a^{(q)+} a^{(q+n)}] \cdot \\
& \cdot \sqrt{m(m - p)\, q(q+n)} \\
& + \frac{1}{2} \sum_{q=1}^{n-1} \sum_{p=1}^{\infty} [a^{(n-q)} a^{(q)}, a^{(p)+} a^{(p+m)}] \cdot \\
& \cdot \sqrt{q(n - q)\, p(p + m)}
\end{aligned} \tag{3.60}$$

$$= \sum_{p,q=1}^{\infty} \sqrt{p(p + m)q(q + n)} \cdot$$

$$\cdot [\delta_{p,q+n} \; a^{(p+m)} \cdot a^{(q)+} - \delta_{p+m,q} \; a^{(p)+} \cdot a^{(q+n)}]$$

$$+ i\sqrt{2m} \; a^{(m+n)} \cdot P\sqrt{m(m + n)}$$

$$- i\sqrt{2n} \; a^{(m+n)} \cdot P\sqrt{n(n + m)}$$

$$- \frac{1}{2} \sum_{p=1}^{m-1} \sum_{q=1}^{\infty} \sqrt{p(m - p) \; q(q + n)} \cdot$$

$$\cdot [-\delta_{m-p,q} \; a^{(p)} \cdot a^{(q+n)} - \delta_{p,q} \cdot a^{(m-p)} \cdot a^{(q+n)}] \tag{3.61}$$

$$+ \frac{1}{2} \sum_{q=1}^{n-1} \sum_{p=1}^{\infty} \sqrt{q(n - q) \; p(p + m)} \cdot$$

$$\cdot [-\delta_{n-q,p} a^{(q)} \cdot a^{(p+m)} - \delta_{q,p} a^{(n-q)} \cdot a^{(p+m)}] \tag{3.62}$$

$$= -(m - n) \sum_{p=1}^{\infty} \sqrt{p(p + m + n)} a^{(p)+} \cdot a^{(p+m+n)} +$$

$$+ i(m - n) \sqrt{2(m + n)} \; a^{(m+n)} \cdot P$$

$$+ \frac{1}{2}(m - n) \sum_{p=1}^{m+n-1} \sqrt{p(m + n - p)} \; a^{(p)} \cdot a^{(m+n-p)} =$$

$$= (m - n) \; L_{m+n} \tag{3.63}$$

When $m + n = 0$, an additional anomaly term arises from the commutation of the terms quadratic in a^+ and a; after commutation the resultant terms have to be normal ordered to obtain L_0 and the normal ordering gives rise to a trace $g^{\mu\mu} = d$ of the metric tensor. Consider the term contained in $[L_m, L_{-m}]$ of the form

$$\frac{1}{4}\sum_{p=1}^{m-1}\sum_{q=1}^{m-1}\sqrt{p(m-p)q(m-q)}\,\cdot$$

$$\cdot\,[a^{(p)}\cdot a^{(m-p)},\, a^{(q)+}\cdot a^{(m-q)+}]$$

$$= \frac{1}{4}\sum_{p=1}^{m-1}\sum_{q=1}^{m-1} p(m-p)\,[a^{(m-p)+}\cdot a^{(m-p)} +$$

$$+\, a^{(p)+}\cdot a^{(p)} + 2g_{\mu\mu}] \qquad (3.64)$$

The c-number anomaly term is then

$$\frac{d}{2}\sum_{p=1}^{m-1} p(m-p) = \frac{d}{12}\, m(m^2-1) \qquad (3.65)$$

as given before.

Since the presence of the anomaly term is so important let us show that its general structure follows from

i) the Jacobi identity

$$[[L_n, L_m], L_p] + [[L_m, L_p], L_n] + [[L_p, L_n], L_m] = 0 \qquad (3.66)$$

ii) absence of an anomaly in the projective subalgebra.

We can see this by writing

$$[L_m, L_n] = (m-n)\, L_{m+n} + C_{m,n} \qquad (3.67)$$

whereupon the anomaly $C_{m,n}$ satisfies

$$(n-m)\, C_{n+m,p} + (m-p)\, C_{m+p,n} + (p-n)\, C_{p+n,m} = 0 \qquad (3.68)$$

Further we assume that

$$C_{11} = C_{00} = C_{10} = C_{1,-1} = 0 \tag{3.69}$$

Then by putting $m = -1$, $p = 1-n$ for example we find

$$\frac{C_{n,-n}}{C_{n-1,-n+1}} = \frac{n+1}{n-2} \tag{3.70}$$

which implies that

$$C_{n,-n} = \frac{1}{6} n(n^2 - 1) C_{2,-2} \tag{3.71}$$

The fact that $C_{2,-2} = d/2$ follows from the explicit computation already given. Of course, these general considerations involving the Jacobi identity do not prove $C_{2,-2} \neq 0$; as we shall see later the nonvanishing of $C_{2,-2}$ and its dependence on d will lead to significant effects in the discussion of ghost elimination.

3.3 GAUGE INVARIANCE

To study the linear dependences implied by the projective gauge invariance of the theory, it is convenient to introduce a field operator $Q_\mu(z)$ defined by

$$Q_\mu(z) = \frac{1}{\sqrt{2}} Q_\mu + i\sqrt{2} P_\mu \ln z + \sum_{n=1}^{\infty} \left(\frac{a^{(n)} z^n}{\sqrt{n}} + \frac{a^{(n)+} z^{-n}}{\sqrt{n}}\right) \tag{3.72}$$

and then to define a vertex operator by the normal-ordered form

$$V(p,z) = :\exp(\sqrt{2}\, ip\cdot Q(z)): \tag{3.73}$$

Notice that this is related to the operator vertex defined

earlier by

$$V(p,1) = e^{ip\cdot Q}\, V(p) \tag{3.74}$$

$$V(p) = \exp(\sqrt{2}\, ip\cdot \sum_{n=1}^{\infty} \frac{a^{(n)+}}{\sqrt{n}}) \exp(\sqrt{2}\, ip\cdot \sum_{n=1}^{\infty} \frac{a^{(n)}}{\sqrt{n}}) \tag{3.75}$$

Further we define a conjugate to $Q_\mu(z)$ by

$$P_\mu(z) = -iz \frac{d}{dz} Q_\mu(z) \tag{3.76}$$

$$= \sqrt{2}\, P_\mu - i \sum_{n=1}^{\infty} \sqrt{n}\, (a_\mu^{(n)} z^n - a_\mu^{(n)+} z^{-n}) \tag{3.77}$$

Defining

$$Q_\mu^{(0)}(z) = \frac{1}{\sqrt{2}} Q_\mu + i\sqrt{2}\, P_\mu \ln z \tag{3.78}$$

$$Q_\mu^{(+)}(z) = \sum_{n=1}^{\infty} \frac{a_\mu^{(n)}}{\sqrt{n}} z^n \tag{3.79}$$

$$Q_\mu^{(-)}(z) = \sum_{n=1}^{\infty} \frac{a_\mu^{(n)+}}{\sqrt{n}} z^{-n} \tag{3.80}$$

we see that

$$[Q_\mu^{(0)}(z), Q_\nu^{(0)}(z')] = -g_{\mu\nu} \ln(z/z') \tag{3.81}$$

$$= -g_{\mu\nu} \lim_{\varepsilon\to 0} \frac{1}{2}\{\frac{1}{\varepsilon} (\frac{z}{z'})^{\varepsilon} + \frac{1}{(-\varepsilon)} (\frac{z}{z'})^{-\varepsilon}\} \tag{3.82}$$

$$[Q_\mu^{(+)}(z), Q_\nu^{(-)}(z')] = -g_{\mu\nu} \sum_{n=1}^{\infty} \frac{1}{n} (\frac{z}{z'})^n \tag{3.83}$$

and hence

$$[Q_\mu(z), Q_\nu(z')] = - g_{\mu\nu} \left[\sum_{n=-\infty}^{\infty} \frac{1}{n} \left(\frac{z}{z'}\right)^n + \right.$$

$$\left. + \lim_{\varepsilon\to 0} \frac{1}{2} \left\{ \frac{1}{\varepsilon} \left(\frac{z}{z'}\right)^{\varepsilon} + \frac{1}{-\varepsilon} \left(\frac{z}{z'}\right)^{-\varepsilon} \right\} \right] \tag{3.84}$$

$$= - 2\pi i \, g_{\mu\nu} \, \varepsilon(\theta - \theta') \tag{3.85}$$

where we wrote $z = e^{i\theta}$, $z' = e^{i\theta'}$. Hence $Q_\mu(z)$ does not commute with itself at different z-values. By differentiating, however, we find that the commutator of $Q_\mu(z)$ with $P_\mu(z')$ and of $P_\mu(z)$ with $P_\nu(z')$ are local namely (using $iz \frac{d}{dz} = \frac{d}{d\theta}$)

$$[Q_\mu(z), P_\nu(z')] = - 2\pi i \, g_{\mu\nu} \, \delta(\theta - \theta') \tag{3.86}$$

$$[P_\mu(z), P_\nu(z')] = 2\pi i \, g_{\mu\nu} \, \delta'(\theta - \theta') \tag{3.87}$$

In order to find the commutators of L_n with $P_\mu(z)$, $Q_\mu(z)$ it is convenient to rewrite L_n as an expectation value.[12] With the notation

$$\langle A(z) \rangle = \frac{1}{2\pi} \int_{-\pi}^{\pi} d\theta \, A(e^{i\theta}) \tag{3.88}$$

We can check that

$$L_n = - \frac{1}{2} \langle z^{-n} : P^2(z) : \rangle \tag{3.89}$$

by using

$$\frac{1}{2\pi} \int_{-\pi}^{\pi} d\theta \, e^{i(m-n)\theta} = \delta_{m-n,0} \tag{3.90}$$

as follows:

$$- \frac{1}{2} <:P^2(z):> = - \frac{1}{4\pi} \int_{-\pi}^{\pi} d\theta \cdot$$

$$\cdot :[\sqrt{2}\, P_\mu - i \sum_{n=1}^{\infty} \sqrt{n}(a^{(n)} e^{in\theta} - a^{(n)+} e^{-in\theta}]^2: \tag{3.91}$$

$$= - P^2 - \sum_{n=1}^{\infty} n\, a^{(n)+} \cdot a^{(n)} = L_0 \tag{3.92}$$

$$- \frac{1}{2} <z^{-n} :p^2(z):> = - \frac{1}{4\pi} \int_{\pi}^{\pi} d\theta\, e^{-in\theta} \cdot$$

$$\cdot :[\sqrt{2}\, P_\mu - i \sum_{n=1}^{\infty} \sqrt{n}(a^{(n)} e^{in\theta} - a^{(n)+} e^{-in\theta})]^2: \tag{3.93}$$

$$= i\sqrt{2n}\, P \cdot a^{(n)} - \sum_{m=1}^{\infty} \sqrt{m(m+n)}\, a^{(m)+} \cdot a^{(m+n)}$$

$$+ \frac{1}{2} \sum_{m=1}^{n-1} \sqrt{m(n-m)}\, a^{(m)} \cdot a^{(n-m)} \tag{3.94}$$

$$= L_n \tag{3.95}$$

as required. We can now easily compute

$$[L_n, Q_\mu(z)] = \frac{1}{2} <z'^{-n} [P^2(z), Q_\mu(z)]> \tag{3.96}$$

$$= - \frac{1}{2} <z'^{-n}\, 2P_\mu\, \delta(\theta - \theta')\, 2\pi i> \tag{3.97}$$

$$= - i\, e^{-in\theta}\, P_\mu(z) \tag{3.98}$$

$$= - z^{-n} \left(z \frac{d}{dz}\right) Q_\mu(z) \tag{3.99}$$

and, similarly

$$[L_n, P_\mu(z)] = - z \frac{d}{dz} (z^{-n} P)_\mu \tag{3.100}$$

$$= - z(-nz^{-n-1} P + z^{-n} \frac{d}{dz} P)_\mu \tag{3.101}$$

$$= - z^{-n}(z \frac{d}{dz} - n) P_\mu(z) \tag{3.102}$$

Let us define generalised projective spin, S, of an operator field $O^S(z)$ by

$$[L_n, O^S(z)] = - z^{-n}(z \frac{d}{dz} + nS) O^S(z) \tag{3.103}$$

Then we see that $Q_\mu(z)$ is a generalised projective scalar ($S = 0$) and $P_\mu(z)$ is a generalised projective vector ($S = -1$).

We can now study how the vertex operator $V(p,z)$ transforms under L_n. Here great care with normal ordering is essential (unless $p^2 = 0$ in which special case the normal ordering is trivial). It turns out to be convenient[13] to define $V(p,z)$ as the limit

$$V(p,z) = :\exp(\sqrt{2}\, ip\cdot Q(z)): \tag{3.104}$$

$$= \exp(\sqrt{2}\, ip\cdot Q^{(-)}(z)) \exp(\sqrt{2}\, ip\cdot Q^{(0)}(z)) \cdot$$

$$\cdot \exp(\sqrt{2}\, ip\cdot Q^{(+)}(z)) \tag{3.105}$$

$$= \lim_{z'\to z} (1 - \frac{z'}{z})^{p^2} \exp(\sqrt{2}\, ip\cdot Q(z)) \tag{3.106}$$

where in the last step we used the Baker-Hausdorff relation

$$e^A e^B = e^{A+B} e^{\frac{1}{2}[A, B]} \tag{3.107}$$

and observed that

$$[Q_\mu^{(-)}(z), Q_\nu^{(+)}(z')] = - g_{\mu\nu} \ \ell n(1 - \frac{z'}{z}) \tag{3.108}$$

To find $[L_n, V(p,z)]$ we make an infinitesimal transformation

$$V(p,z) + \varepsilon[L_n, V(p,z)] = \lim_{z'\to z} e^{\varepsilon L_n} (1 - \frac{z'}{z})^{p^2} e^{-\varepsilon L_n} \exp(\sqrt{2}\, ip_\mu \cdot \{e^{\varepsilon L_n} Q_\mu(z) e^{-\varepsilon L_n}\}) \tag{3.109}$$

Now

$$e^{\varepsilon L_n} z e^{-\varepsilon L_n} = z + \varepsilon[L_n, z] \tag{3.110}$$

$$= z + \delta z \tag{3.111}$$

where $\delta z = - \varepsilon z^{-n+1}$.

Hence

$$\lim_{z'\to z} e^{\varepsilon L_n} (1 - \frac{z'}{z})^{p^2} e^{-\varepsilon L_n} = \lim_{z'\to z} [(1 - \frac{z'}{z}) (1 - \frac{z' + \delta z'}{z + \delta z})]^{p^2} \tag{3.112}$$

$$= 1 - p^2 n\varepsilon \, z^{-n} + O(\varepsilon^2) \tag{3.113}$$

Using the fact that $Q_\mu(z)$ is a generalised projective scalar then gives for the full expression

$$[L_n, V(p,z)] = -z^{-n}(z \frac{d}{dz} - np^2) \, V(p,z) \tag{3.114}$$

Hence the result may be summarised by stating that V(p,z)

transforms as a generalised projective spin $S = p^2$. In particular it is a generalised projective vector, $s = -1$, if $p^2 = -1$ or equivalently $\alpha(o) = 1$. (Note: we always take slope $\alpha' = 1$).

All of the technical apparatus needed to find the unit intercept gauge identities are now available. Recall that we wrote

$$B_N = \langle 0|\ V(p_2,1)\ \frac{1}{L_0 - 1}\ V(p_3,1)\ \cdots\ V(p_{N-1},1)\ |0\rangle \tag{3.115}$$

By using the generalised projective algebra it is immediate to derive that, for any $f(x)$

$$L_n\ f(L_0) = f(L_o + n)\ L_n \tag{3.116}$$

and hence

$$(L_o - L_n - 1)\ \frac{1}{L_0 - 1} = \frac{1}{L_0 + n - 1}\ (L_0 + n\text{-}1 - L_n) \tag{3.117}$$

With $p^2 = -1$ we also know that

$$(L_o + n\text{-}1 - L_n)\ V(p,1) = V(p,1)\ (L_0 - L_n - 1) \tag{3.118}$$

By repeating this process and noting that for an on-shell ground state

$$(L_0 - L_n - 1)\ |0\rangle = 0 \tag{3.119}$$

we deduce that

$$(L_0 - L_n - 1)\ [\frac{1}{L_0 - 1}\ V(p_1,1)\ \frac{1}{L_0 - 1}\ V(p_2,1)\ \cdots\ \cdot$$

$$\cdot\ V(p_{N-1},1)\ |0\rangle] = 0 \tag{3.120}$$

for an arbitrary physical state, i.e., a state which can

be made from some number of ground state particles.

On mass shell we have the additional condition

$$(L_0 - 1) |\phi\rangle = 0 \qquad (3.121)$$

and hence

$$L_n |\phi\rangle = 0 \quad n = 1, 2, 3, \ldots\ldots (3.122)$$

are the on-shell gauge conditions.[14] The infinite number of gauge conditions opens up the possibility of removing the ghost negative norm states with odd time occupancy and we shall prove that this does in fact happen shortly. The crucial point in the rigorous proof[15,16] is the full recognition of the role played by the anomaly term in the generalized projective algebra. Once this point is understood, the proof becomes beautiful and straightforward. There were many earlier papers which discussed the properties of the generalised projective algebra. [References 13, 14, 17-22].

3.4 OPERATORIAL DUALITY

By using the formalism now developed, combining both the projective invariance and the operator formalism we can now rewrite the N-point function in a form[3] which exhibits explicitly both cyclic symmetry and factorisability as follows:

Recall that (for $\alpha(o) = 1$)

$$B_N = \langle 0| \, V(p_2,1) \frac{1}{L_0 - 1} V(p_3,1) \cdots \frac{1}{L_0 - 1} \, .$$

$$\cdot\ V(p_{N-1},1)\ |0\rangle \tag{3.123}$$

$$= \int \prod_{i=2}^{N-2} du_{1i}\ \langle 0|\ V(p_2,1)\ u_{12}^{L_0-2}\ V(p_3,1) \cdots \cdot$$

$$\cdots\cdot\ V(p_{N-1},1)\ |0\rangle \tag{3.124}$$

Now put

$$u_{1j} = \frac{z_j}{z_{j+1}} \qquad j = 2,\ 3,\ \ldots,\ N-2 \tag{3.125}$$

with the corresponding Jacobian

$$\partial\begin{pmatrix} u_{12} & u_{13} & \cdots & u_{1,N-2} \\ z_2 & z_3 & \cdots & z_{N-2} \end{pmatrix} =$$

$$= (z_3 z_4 \cdots z_{N-2})^{-1} \tag{3.126}$$

to obtain

$$B_N = \int \prod_{i=2}^{N-2} dz_i\ (z_3 z_4 \cdots z_{N-2})^{-1}$$

$$\langle 0|\ V(p_2,1)\left(\frac{z_2}{z_3}\right)^{L_0-2} V(p_3,1) \cdots \left(\frac{z_{N-2}}{z_{N-3}}\right)^{L_0-2} \cdot$$

$$\cdot\ V(P_{N-1},1)\ |0\rangle \tag{3.127}$$

The Jacobian is obtained by noticing that

$$\partial\begin{pmatrix} u_{12} & u_{13} & \cdots & u_{1,N-2} \\ z_2 & z_3 & \cdots & z_{N-2} \end{pmatrix} = \left|\frac{\partial u_{1i}}{\partial z_j}\right| \tag{3.128}$$

$$
= \left\| \begin{array}{cccccc}
\frac{1}{z_3} & -\frac{z_2}{z_3{}^2} & 0 & 0 & ---- & 0 \\
0 & \frac{1}{z_4} & -\frac{z_3}{z_4{}^2} & 0 & ---- & 0 \\
0 & 0 & \frac{1}{z_5} & -\frac{z_4}{z_5{}^2} & ---- & 0 \\
\cdot & & & & & \cdot \\
\cdot & & & & & \cdot \\
\cdot & & & & & \cdot \\
\cdot & & & & \frac{1}{z_{N-2}} & -\frac{z_{N-2}}{z_{N-3}^2} \\
0 & 0 & ------ & & 0 & 1
\end{array} \right\| \tag{3.129}
$$

$$
= (z_3 z_4 z_5 \cdots z_{N-2})^{-1} \tag{3.130}
$$

Now by noticing that

$$
z^{P^2} \exp(\sqrt{2}\ i\ p\cdot Q^{(0)}(1))\ z^{-P^2} =
$$

$$
= e^{ip\cdot Q} \exp[-\ ip_\mu\ \ell n\ z\ [Q_\mu,\ P^2]] \tag{3.131}
$$

$$
= \exp\ [i\sqrt{2}\ p.\ (\frac{Q}{\sqrt{2}} + i\sqrt{2}\ P\ \ell n\ z)]\ e^{-p^2 \ell n\ z} \tag{3.132}
$$

$$
= \exp\ (\sqrt{2}\ i\ p.\ Q^{(0)}\ (z))\ z^{-p^2} \tag{3.133}
$$

we deduce that

$$z^{-L_0}\, V(p,1)\, z^{L_0} =$$

$$= z^{-L_0} \exp(\sqrt{2}\, ip\cdot Q^{(-)}(1)) \exp(\sqrt{2}\, ip\cdot Q^{(0)}(1)) \cdot$$

$$\cdot \exp(\sqrt{2}\, ip\cdot Q^{(+)}(1))\, z^{L_0} \tag{3.134}$$

$$= \exp(\sqrt{2}\, ip\cdot Q^{(-)}(z)) \exp(\sqrt{2}\, ip\cdot Q^{(0)}(z)) \cdot$$

$$\cdot \exp(\sqrt{2}\, ip\cdot Q^{(+)}(z))\, z^{-p^2} \tag{3.135}$$

$$= V(p,z)z \tag{3.136}$$

for $\alpha(o) = 1$, i.e., $p^2 = -1$.

Also we need to use the following limits for the bra and ket in the vacuum expectation value

$$<0|\, e^{ip_1\cdot Q} = <0,p_1| = \lim_{z_1\to 0} <0|\, V(p_1,z_1) \tag{3.137}$$

$$e^{ip_N\cdot Q}\, |0> = |0,p_N> = \lim_{z_N\to\infty} V(p_N,\, z_N)\, |0> \tag{3.138}$$

Collecting together the factors one deduces that

$$B_N = \lim_{\substack{z_1\to 0\\ z_{N-1}\to 1\\ z_N\to\infty}} \int \prod_{i=1}^{N} dz_i \left[\frac{dz_a\, dz_b\, dz_c}{(z_1 - z_{N-1})(z_{N-1} - z_N)(z_N - z_1)}\right]^{-1} \cdot$$

$$\cdot <0|\, V(p_1 z_1)\, V(p_2 z_2) \cdots V(p_N z_N)\, |0> \tag{3.139}$$

If we take $\alpha(o) \neq 1$ the same procedure goes through, except that in the integrand there is one new factor, namely

$$B_N = \int_{\substack{z_1 \to 0 \\ z_{N-1} \to 1 \\ z_N \to \infty}} \prod_{i=1}^{N} dz_i \, (dV_3)^{-1} \prod_{i=1}^{N} |z_i - z_{i+1}|^{\alpha(o)-1} .$$

$$\cdot \ \langle 0| \, V(p_1 z_1) \, V(p_2 z_2) \cdots V(p_N z_N) \, |0\rangle \tag{3.140}$$

We can check that this is the appropriate Koba-Nielsen form by evaluating directly that

$$\langle 0| \, V(p_1 z_1) \, V(p_2 z_2) \cdots V(p_N z_N) \, |0\rangle =$$

$$= \prod_{i<j} |z_i - z_j|^{-2p_i \cdot p_j} \tag{3.141}$$

The form we have arrived at, for B_N, possesses operatorial duality in the following sense. Let us make a general projective transformation

$$z_i \to z_i' = \frac{az_i + b}{cz_i + d} \quad \text{or} \quad z_i = \frac{dz_i' - b}{a - cz_i'} \tag{3.142}$$

so that

$$(z_i - z_{i+1}) = (z_i' - z_{i+1}')(a - cz_i')^{-1}(a - cz_{i+1}')^{-1} \tag{3.143}$$

Hence $$\prod_{i=1}^{N} (z_i - z_{i+1})^{\alpha(o)} = \prod_{i=1}^{N} (z_i' - z_{i+1}')^{\alpha(o)} \cdot$$

$$\cdot \prod_{i=1}^{N} (a - cz_i')^{-2\alpha(o)} \tag{3.144}$$

The factor $\prod_{i=1}^{N} dz_i \, (dV_3)^{-1} \prod_{i=1}^{N} |z_i - z_{i+1}|$ is projective invariant by simple inspection of translation, dilation and inversion on z.

The vertex transformation property is most easily

seen by using the limiting procedure introduced earlier, namely

$$V(p, z) = \lim_{\tilde{z}\to z} [\exp(\sqrt{2}\, ip \{Q^{(-)}(z) + Q^{(0)}(z) + Q^{(+)}(z)\})(1 - \frac{\tilde{z}}{z})^{p^2}] \quad (3.145)$$

and hence

$$\Lambda(T)\, V(p, z)\, \Lambda^{-1}(T) =$$

$$= V(p, z') \lim_{\substack{\tilde{z}\to z \\ \tilde{z}'\to z'}} (\frac{z - \tilde{z}}{z' - \tilde{z}'})^{p^2} \quad (3.146)$$

$$= (a - cz')^{-2p^2}\, V(p, z') \quad (3.147)$$

$$= (a - cz')^{2\alpha(o)}\, V(p, z') \quad (3.148)$$

Hence projective invariance of the full integrand follows, because the factor

$$\prod_{i=1}^{N} (a - cz_i')^{-2\alpha(o)} \quad (3.149)$$

arising from the nearest-neighbor factor is precisely compensated by a similar factor in the vertex transformation.

To see the cylic symmetry we simply make a transformation which takes z_{N-1} to $z'_{N-1} = \infty$, z_{N-2} to $z'_{N-2} = 1$ and z_N to $z_N' = 0$. (These three constraints completely specify the transformation.) We then arrive at

$$B_N = \int \prod_{i=1}^{N} dz_i'\, (dV_3)^{-1} \prod_{i=1}^{N} |z_i' - z'_{i+1}|^{\alpha(o)} (-1)^{2\alpha(o)}$$

$$\langle 0|\, V(p_1, z_N')\, V(p_2, z_1') \cdots V(p_N, z_1')|0\rangle \quad (3.150)$$

where we exhibit a phase arising from the rearrangement of

two factors $(z_2 - z_1)$ and $(z_1 - z_N)$. The vertex $V(p_1, z_N')$ may now be commuted by using the Baker-Hausdorff relation in

$$V(p, z)\, V(p', z') = V(p', z')\, V(p, z)\, X \qquad (3.151)$$

with

$$X = \exp[-2p_\mu p_\nu'\{[Q_\mu^{(+)}(z), Q_\nu^{(-)}(z')] + [Q_\mu^{(-)}(z), Q_\nu^{(+)}(z')] + [Q_\mu^{(0)}(z), Q_\nu^{(0)}(z')]\}] \qquad (3.152)$$

$$= \exp[-2p_\mu p_\nu'\{g_{\mu\nu}\, \ell n(1 - \frac{z}{z'}) - g_{\mu\nu}\, \ell n(1 - \frac{z'}{z}) - \ell n(\frac{z'}{z})\}] \qquad (3.153)$$

$$= (-1)^{-2p\cdot p'} \qquad (3.154)$$

The commutation picks up an additional factor

$$(-1)^{-2p_1 \sum_{j=2}^{N} p_j} = (-1)^{2p_1^2} = (-1)^{-2\alpha(o)} \qquad (3.155)$$

giving

$$B_N = \int \prod_{i=1}^{N} dz_i' \,(dV_3)^{-1} \prod_{i=1}^{N} (z_i' - z_{i+1}')^{\alpha(o)}$$

$$\langle 0|\, V(p_2, z_1')\, V(p_3, z_2') \cdots V(p_1, z_N')\, |0\rangle \qquad (3.156)$$

which we may now refactorise (in a new multiperipheral configuration) by the use of

$$V(p, z) = z^{L_0}\, V(p, 1)\, z^{-L_0}\, z^{p^2} \qquad (3.157)$$

To make more clear what we have succeeded in doing,

compare Figures 3.1(a) and 3.1(b). We have obtained an

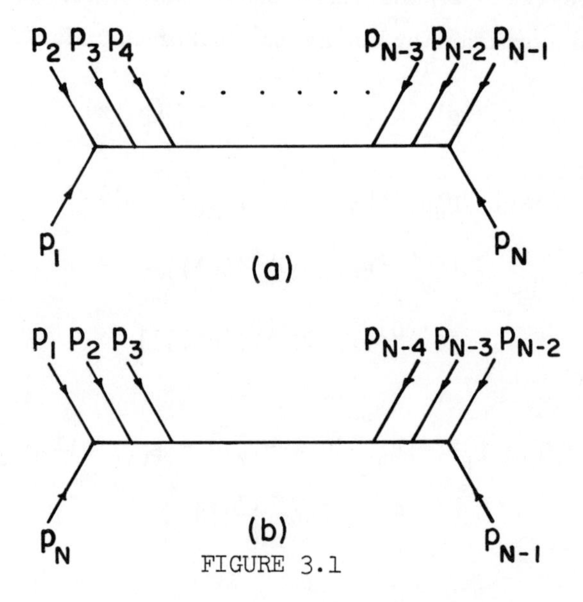

FIGURE 3.1

Operatorial Duality

expression for B_N which enables us to go from factorisation of the first configuration to factorisation in the second very directly by exploiting the cyclic symmetry.

This operatorially dual form for B_N thus combines the advantages of all earlier forms, where there was a choice of which one of the two principal properties (cyclic symmetry and factorisability) to exhibit explicitly.

3.5 UNIT INTERCEPT

We have derived the on-shell gauge conditions

$$L_n \; |\phi\rangle = 0 \qquad n = 1, 2, \ldots \tag{3.158}$$

only for the case $\alpha(o) = 1$. Actually the unit intercept condition (which is unphysical since it gives rise to a tachyon at $\alpha_s = 0$ on the leading trajectory) is essential in order to obtain the full set of generalised projective gauges, and only in this special case are there enough linear dependences to remove the ghosts which arise from the indefinite metric.

Because the unit intercept condition is so important, it is instructive to examine[23] why the rather unimportant-seeming factor

$$\prod_{i=1}^{N} |z_i - z_{i+1}|^{-1+\alpha(o)} \tag{3.159}$$

is enough to spoil the generalised projective invariance. Note that this factor is the only factor in the B_N integrand which does not have the property of total symmetry under all N! permutations made on the N momenta p_i and their corresponding Koba-Nielsen z_i variables.

The N-point function B_N may be written for general $\alpha(o)$ as

$$B_N = \int \prod_{i=2}^{N-2} du_{1i} \ (1 - u_{1i})^{-1+\alpha(o)} \ u_{1i}^{-1-\alpha(o)} \cdot$$

$$\langle 0| \ V(p_2) \ u_{12}^{L_0} \cdots V(p_{N-1}) \ |0\rangle \tag{3.160}$$

One simple way to factorise this expression was given earlier, namely tc define a propagator by

$$D(s) = \int_0^1 dx \ x^{L_0 - 1 - \alpha(o)} (1 - x)^{\alpha(o)-1} =$$

$$= B(L_0 - \alpha(o), \ \alpha(o)) \tag{3.161}$$

and then write

$$B_N = \langle 0| \ V(p_2) \ D(s_{12}) \ V(p_3) \cdots D(s_{1,N-2}) \cdot$$

$$\cdot \ V(p_{N-1}) \ |0\rangle \tag{3.162}$$

This method of factorisation is really too trivial in order to study the gauge problem. It is better to introduce a fifth dimension[24] in the oscillator operators $a_5^{(n)}$, $a_5^{(n)+}$ $(n = 1, 2, 3, \ldots)$ satisfying

$$[a_5^{(m)}, a_5^{(n)+}] = + \delta_{mn} \tag{3.163}$$

and then define

$$\hat{L}_0 = - \sum_{r=1}^{\infty} r\, \hat{a}^{(r)+} \cdot \hat{a}^{(r)} - \hat{p}^2 \tag{3.164}$$

$$\hat{L}_n = i\sqrt{2n}\, \hat{P} \cdot \hat{a}^{(n)} - \sum_{r=1}^{\infty} \sqrt{r(r + n)}\, \hat{a}^{(r)+} \cdot \hat{a}^{(r+n)} + \frac{1}{2} \sum_{r=1}^{n-1} \sqrt{r(n - r)}\, \hat{a}^{(r)} \cdot \hat{a}^{(n-r)} \tag{3.165}$$

where the scalar product is now $\hat{A} \cdot \hat{B} = A_o B_o - \underline{A} \cdot \underline{B} - A_5 B_5$ and we define $p_5 = \lambda/2$ with $\lambda^2 = (1 - \alpha(o))$.

We now define a propagator and vertex by

$$\hat{D} = (\hat{L}_0 + \frac{1}{2}\lambda^2 - 1)^{-1} \tag{3.166}$$

$$\hat{V}(p) = V(p)\, V_5 \tag{3.167}$$

$$V_5 = \exp(\lambda \sum_{n=1}^{\infty} \frac{a_5^{(n)+}}{\sqrt{n}})\, |0\rangle_5\ {}_5\langle 0| \exp(\lambda \sum_{n=1}^{\infty} \frac{a_5^{(n)}}{\sqrt{n}}) \tag{3.168}$$

to find

$$B_N = \langle 0|\ \hat{V}(p_2) \frac{1}{\hat{L}_o + \frac{1}{2}\lambda^2 - 1} \hat{V}(p_3) \cdots\cdots \hat{V}(p_{N-1})\ |0\rangle \tag{3.169}$$

as is easily proved by using the identity

$$ {}_5\langle 0| \exp(\lambda \sum_{n=1}^{\infty} \frac{a_5^{(n)}}{\sqrt{n}})\, x^{\sum_{n=1}^{\infty} n a_5^{(n)+} \cdot a_5^{(n)}} \cdot $$

$$ \cdot \exp(\lambda \sum_{n=1}^{\infty} \frac{a_5^{(n)+}}{\sqrt{n}})\, |0\rangle_5 = (1 - x)^{-1+\alpha(o)} \tag{3.170} $$

Notice that the vacuum projection of the fifth dimension in V_5 is essential to avoid factors of the form

$$ (1 - u_{1i} u_{1j})^{-1+\alpha(o)} \tag{3.171} $$

in B_N. This ugly vacuum projection is, therefore, linked to the absence of total symmetry in the integrand, and we will see that it is precisely what leads to loss of generalized projective invariance.

Consider now the gauge operator ($\hat{L}_0 = \frac{\lambda^2}{2} - 1 - \hat{L}_n$) acting on the new propagator. We have

$$ (\hat{L}_0 + \frac{\lambda^2}{2} - 1 - \hat{L}_n) \frac{1}{(\hat{L}_0 + \frac{\lambda^2}{2} - 1)} = $$

$$ = \frac{1}{(\hat{L}_0 + n + \frac{\lambda^2}{2} - 1)} (\hat{L}_0 + n + \frac{\lambda^2}{2} - 1 - \hat{L}_n) \tag{3.172} $$

To commute with the full vertex $\hat{V}(p)$ note that, by writing in an obvious notation

$$ \hat{L}_n = L_n + L_n^5 \tag{3.173} $$

$$ \hat{L}_0 = L_0 + L_0^5 \tag{3.174} $$

then

$$ [\hat{L}_0 - \hat{L}_n, \hat{V}(p)] = [L_0 - L_n, V(p)] V_5 + $$

$$ + [L_0^5 - L_n^5, V_5]\, V(p) \tag{3.175} $$

$$= np^2 \hat{V}(p) + [L_0{}^5 - L_n{}^5, V_5] V(p) \tag{3.176}$$

$$= - n\hat{V}(p) + V(p)[V_5 n\lambda^2 + [L_0{}^5 - L_n{}^5, V_5]] \tag{3.177}$$

where we used our previous result for $[L_0 - L_n, V(p)]$.

To calculate the fifth dimension commutator write

$$[L_0{}^5, V_5] = \sum_{n=1}^{\infty} n[a_5{}^{n+} a_5{}^n, V_5] \tag{3.178}$$

$$= \lambda \sum_{n=1}^{\infty} \sqrt{n} \, (a^{(n)+} V_5 - V_5 \, a^{(n)}) \tag{3.179}$$

and similarly

$$[L_1{}^5, V_5] = \sum_{n=1}^{\infty} \sqrt{n(n+1)} \, [a_5{}^{(n)+} a_5{}^{(n+1)}, V_5] +$$

$$+ [\lambda a_5{}^{(1)}, V_5] \tag{3.180}$$

$$= \lambda \sum_{n=1}^{\infty} \sqrt{n(n+1)} \, \Big(\frac{1}{\sqrt{n+1}} a_5{}^{(n)} V_5 -$$

$$- \frac{1}{\sqrt{n}} V_5 \, a^{(n+1)}\Big) + \lambda^2 V_5 + \lambda V_5 a_5{}^{(1)} \tag{3.181}$$

$$= \sum_{n=1}^{\infty} \sqrt{n} \, (a^{(n)+} V_5 - V_5 a^{(n)}) + \lambda^2 V_5 \tag{3.182}$$

Hence

$$[L_0{}^5 - L_1{}^5, V_5] = - \lambda^2 V_5 \tag{3.183}$$

In these equations we have used the general identity

$$[a_5, f(a_5{}^+)] = f'(a_5{}^+) \tag{3.184}$$

and the related forms. Now consider, by the same method

$$[L_2^{\;5}, V_5] = \sum_{n=1}^{\infty} \sqrt{n(n+2)}\,[a_5^{(n)+}\cdot a_5^{(n+2)}, V_5] + $$
$$+ [\sqrt{2}\,\lambda a_5^{(2)} + a_5^{(1)}\cdot a_5^{(1)}, V_5] \tag{3.185}$$

$$= \lambda \sum_{n=1}^{\infty} \sqrt{n(n+2)}\,\left(\frac{1}{\sqrt{n+2}}\,a_5^{(n)+} V_5 - \frac{1}{\sqrt{n}}\,V_5 a_5^{(n+2)}\right) + $$
$$+ 2\lambda^2 V_5 - \sqrt{2}\,V_5\,a_5^{(2)} - V_5\,a_5^{(1)}\cdot a_5^{(1)} \tag{3.186}$$

$$= \lambda \sum_{n=1}^{\infty} \sqrt{n}\,(a_5^{(n)+} V_5 - V_5 a_5^{(n)}) + 2\lambda^2 V_5 + \lambda V_5 a_5^{(1)} - $$
$$- V_5\,a_5^{(1)}\cdot a_5^{(1)} \tag{3.187}$$

and hence

$$[L_0^{\;5} - L_2^{\;5}, V_5] = -\,2\lambda^2 V_5 + (\text{anomaly terms}) \tag{3.188}$$

The anomaly terms arise precisely from the presence of the fifth dimension vacuum projection.

Now, collecting results we have

$$[\hat{L_0} - \hat{L_1}, \hat{V}(p)] = -\,\hat{V}(p) \tag{3.189}$$

which implies that

$$(\hat{L_0} + \frac{\lambda^2}{2} - 1 - \hat{L_1})\,\frac{1}{\hat{L_0} + \frac{\lambda^2}{2} - 1}\,\hat{V}(p) =$$

$$\frac{1}{(\hat{L_0} + \frac{\lambda^2}{2})}\,\hat{V}(p)\,(\hat{L_0} + \frac{\lambda^2}{2} - 1 - \hat{L_1}) \tag{3.190}$$

which together with the fact that for an on-shell ground state

$$(\hat{L_0} + \frac{\lambda^2}{2} - 1 - \hat{L}_1)\,|0\rangle = 0 \tag{3.191}$$

enables us to derive that physical states satisfy (on-shell)

$$(\hat{L}_0 + \frac{\lambda^2}{2} - 1)\ |\phi\rangle = 0 \tag{3.192}$$

$$\hat{L}_1\ |\phi\rangle = 0 \tag{3.193}$$

The higher gauges fail to work because of the anomaly terms; that is, in general

$$\hat{L}_n\ |\phi\rangle \neq 0 \quad n = 2, 3, 4, \ldots. \tag{3.194}$$

The L_1 gauges give linear dependences which reflect that underlying projective invariance. They are not sufficient in number to allow ghost elimination.

The only case in which the generalized projective gauge invariance is possible is therefore when $\alpha(o) = 1$; thus we are led to associate total symmetry of the integrand under all N! permutations of the $\{z_i, p_i\}$ variable pairs (i.e., absence of a preferred status for nearest neighbour z_i differences) with the full set of subsidiary conditions

$$L_n\ |\phi\rangle = 0 \quad n = 1, 2, 3, 4, \ldots. \tag{3.195}$$

3.6 NULL STATES

To discuss the way in which null states decouple[14] in the dual resonance model, it is convenient first to recall some features of quantum electrodynamics, in particular the ways in which the formalism ensures that only the two transverse components of the massless photon survive, with non-zero couplings, on mass shell.

To make calculations in quantum electrodynamics, we may take one of two quite different techniques. Either we may use a non-manifestly-covariant method where we define a three-vector field $\underline{A}$, in, for example, the Coulomb gauge

$\underline{\nabla}\cdot\underline{A} = 0$; or we may use a fully covariant method with a four vector $A\mu$ and the Lorentz gauge condition.

In the Coulomb gauge method, we may make a Fourier decomposition, assuming periodic boundary conditions inside a large cubic box of volume $V = L^3$

$$\underline{A} = \frac{1}{\sqrt{V}}\sum_{\underline{k}}\sum_{\alpha=1,2}\frac{1}{\sqrt{2E_{\underline{k}}}}[a_{\underline{k},\alpha}\,\underline{\varepsilon}^{\alpha}\,e^{i\underline{k}\cdot\underline{x}-iE_{\underline{k}}t} +$$

$$+ a^{+}_{\underline{k},\alpha}\,\underline{\varepsilon}^{\alpha}\,e^{-i\underline{k}\cdot\underline{x}+iE_{\underline{k}}t}] \tag{3.196}$$

where $\underline{\varepsilon}^{\alpha}$ ($\alpha = 1, 2$) are the transverse polarisation vectors with $\underline{\varepsilon}^{1}, \underline{\varepsilon}^{2}$ and $\underline{k}/|\underline{k}|$ forming an orthogonal triad of unit vectors. The Coulomb gauge $\underline{\nabla}\cdot\underline{A} = 0$ has been imposed and the presence of only transverse components is manifest. Going to an infinitely large box one obtains

$$A = \sum_{\alpha=1,2}\frac{1}{(2\pi)^3}\int\frac{d^3k}{2E_k}\{a_{\alpha}(\underline{k})\underline{\varepsilon}^{\alpha}\,e^{i\underline{k}\cdot\underline{x}-iE_k t} +$$

$$+ a_{\alpha}^{+}(\underline{k})\,\underline{\varepsilon}^{\alpha}\,e^{-i\underline{k}\cdot\underline{x}+iE_k t}\} \tag{3.197}$$

and then, to quantize, one imposes the canonical commutation relations

$$[a_{\alpha}(\underline{k}),\, a_{\beta}^{+}(\underline{k}')] = (2\pi)^3\,\delta_{\alpha\beta}\,\delta(\underline{k} - \underline{k}') \tag{3.198}$$

In such a theory of quantized transverse electromagnetic fields, the transversality or Coulomb gauge condition is noncovariant and, in general, after any Lorentz transformation we must adopt a new gauge. This is the method developed by Dirac and Fermi in the late 1920's and which is discussed in many text books [25, 26, 27].

Some twenty years later Feynman, Schwinger and Tomonaga used a fully covariant approach, in which the transversality property while not very explicitly displayed is not too difficult to extract. [See the original papers in Reference 27.] It is this last point which we wish to explain in detail here since exactly the same mechanism is operative in the (fully covariant) dual resonance model.

Consider the diagram for a single photon exchange indicated in Figure 3.2. In terms of photon creation and

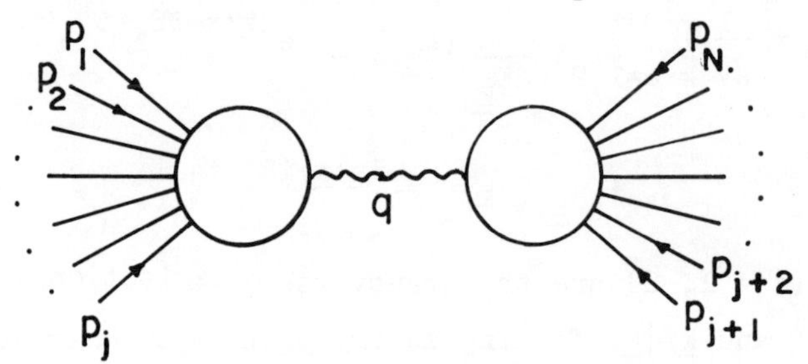

FIGURE 3.2

One-photon Exchange

annihilation operators this diagram corresponds to an amplitude

$$T_{fi} = - \langle f|\, a_\mu^+\, |0\rangle \frac{g_{\mu\nu}}{q^2} \langle 0|a_\nu|i\rangle \tag{3.199}$$

where $|i\rangle$, $|f\rangle$ are the initial, final multiparticle states respectively and

$$q_\mu = \sum_{i=1}^{j} p_{i\mu} = - \sum_{i=j+1}^{N} p_{i\mu} \tag{3.200}$$

In the photon propagator $- g_{\mu\nu}\, (q^2)^{-1}$ there appear four components of which the time component would give a negative (ghost) contribution. Gauge invariance of the first kind i.e., charge conservation is expressed by the Lorentz condition that

$$\langle f|\, q_\mu\, a_\mu^+\, |0\rangle = 0 \tag{3.201}$$

for any physical state $|f\rangle$. Thus the state

$$q_\mu a_\mu^+ |0\rangle = L^+ |0\rangle \tag{3.202}$$

is spurious and does not couple. Here we defined $L = q_\mu a_\mu$. Alternatively we may represent the Lorentz gauge condition by stating that any physical state must satisfy

$$L |\phi\rangle = 0 \tag{3.203}$$

Let us take $q_\mu = (1;0, 0, 1)$ along the 3-axis. Then the most general physical state becomes

$$|\phi\rangle = \alpha_1 |\lambda_1\rangle + \alpha_2 |\lambda_2\rangle + \alpha_3 |\mu_1\rangle \tag{3.204}$$

where we have chosen a linearly independent basis

$$|\lambda_1\rangle = a_1^+ |0\rangle \tag{3.205}$$

$$|\lambda_2\rangle = a_2^+ |0\rangle \tag{3.206}$$

$$|\mu_1\rangle = \frac{1}{\sqrt{2}} (a_0^+ - a_3^+) |0\rangle \tag{3.207}$$

$$|\mu_2\rangle = \frac{1}{\sqrt{2}} (a_0^+ + a_3^+) |0\rangle \tag{3.208}$$

and $|\mu_2\rangle$ is omitted because it is spurious ($L |\mu_2\rangle \neq 0$). If we take the norms of these four states we find that $\langle\lambda_1/\lambda_1\rangle = \langle\lambda_2/\lambda_2\rangle = 1$ but that $\langle\mu_1/\mu_1\rangle = \langle\mu_2/\mu_2\rangle = 0$. We may therefore say that $|\mu_1\rangle$ is a null physical state.

We seem to have three components remaining after applying the Lorentz gauge condition whereas we know there are only the two transverse ones. How does a second state decouple?

The answer is very simple when we recognize that $|\mu_1\rangle$ and $|\mu_2\rangle$ are linked together as conjugate null states and that the Hilbert space of one-photon states has the unit operator (using $<\mu_1/\mu_2> = 1$)

$$1 = |\lambda_1><\lambda_1| + |\lambda_2><\lambda_2| + |\mu_1><\mu_2| + |\mu_2><\mu_1| \quad (3.209)$$

This means that both the null spurious state $|\mu_2>$ and the null physical state $|\mu_1>$ decouple, leaving just the two transverse states $|\lambda_1>$, $|\lambda_2>$ as required.

Before proceeding let us make the following definitions:

i) A physical state is a state which satisfies the gauge constraints.

ii) A spurious state is a state which is orthogonal to all physical states, and which does not satisfy the gauge conditions.

iii) A null state is a state with zero norm. The conjugate of a null physical state will be a null spurious state.

With this terminology the four photon states comprise three physical states and one spurious state. The spurious state is null, as is one of the physical states; this leaves the two non-null (transverse) physical states which couple.

In the dual resonance model the on-shell physical states satisfy (for unit intercept)

$$(L_0 - 1)\ |\phi> = 0 \quad (3.210)$$

$$L_n\ |\phi> = 0 \quad (3.211)$$

At a given mass shell we have a completeness relation

$$1 = \sum_{\lambda_+} |\lambda_+><\lambda_+| - \sum_{\lambda_-} |\lambda_-><\lambda_-| + \sum_{\mu} [\,|\mu><\mu_c| + |\mu_c><\mu|\,] \quad (3.212)$$

where the $|\mu\rangle$ are null states and the subscript c means conjugate. The task is then to prove that all the $|\lambda_-\rangle$ are spurious.

Consider the state $L_1^+(p)\ |0\rangle$ for $p_\mu = p(1;0,0,1)$. This state is a null spurious state. Its conjugate is a null physical state. There remain two transverse positive-norm states. Explicitly we have

$$L_1^+(p)\ |0\rangle = (a_0^{(1)+} - a_3^{(1)+})\ |0\rangle \quad \text{null spurious state} \tag{3.213}$$

$$(a_0^{(1)} + a_3^{(1)+})\ |0\rangle \quad \text{null physical state} \tag{3.214}$$

$$\left.\begin{array}{l} a_1^{(1)+}\ |0\rangle \quad (3.215) \\ a_2^{(1)+}\ |0\rangle \end{array}\right\} \quad \text{physical states}$$

in precise analogy to the photon states of quantum electrodynamics.

More generally, in the dual resonance model note that any state of the form

$$L_n^+\ |f,\ N - n\rangle \tag{3.217}$$

is spurious, as a direct consequence of the physical gauge condition. Here N - n is the eigenvalue of $R = (L_0 + p^2)$.

In particular, any state made by L_1^+ acting on a physical state is a null spurious state since

$$||L_1^+\ |\phi,\ N - 1\rangle||^2 = \langle\phi,\ N - 1|\ [L_1,\ L_{-1}]\ |\phi,\ N - 1\rangle \tag{3.218}$$

$$= 0. \tag{3.219}$$

Thus the null states proliferate at least as fast as the number of physical states.

3.7 CRITICAL DIMENSION

We have already stated that the physical state conditions, on mass shell are

$$L_n \; |\phi, N> = 0 \tag{3.220}$$

$$(L_0 - 1) \; |\phi, N> = 0 \tag{3.221}$$

and that the gauge operators satisfy

$$[L_m, L_n] = (m - n) \, L_{m+n} + \frac{d}{12} m(m^2 - 1) \, \delta_{m+n,0} \tag{3.222}$$

with d = space-time dimension. Here N is the eigenvalue of $R = L_0 + p^2$. We will now see how the anomaly term, proportional to d, enables us to prove the required absence of ghosts, provided $d \leq 26$. In particular, for the realistic value d = 4 there is such a theoretical consistency.

We first set up a basis for the complete Fock space, F, spanned by $a_\mu^{(n)}$, $a_\mu^{(n)+}$ by defining

$$k_n = k \cdot a^{(n)}/\sqrt{n} \qquad n = 1, 2, \ldots \tag{3.223}$$

$$k_{-n} = k_n^{\;+} \tag{3.224}$$

where k_μ is some standard vector. For the moment let us take k_μ to be a light-like vector, $k_\mu = (1;1,0,0,\ldots..)$ so that $k_n = (a_0^{(n)} - a_1^{(n)})/\sqrt{n}$, and we have chosen the 1-axis to be longitudinal. There are a further (d-2) spatial transverse directions 2, 3, 4, ..., d-1.

Now we may set up a linearly-independent basis in the form

$$|\{\lambda, \mu\}, N\rangle = \prod_{m,n} (L_m^+)^{\lambda_m} (k_n^+)^{\mu_n} |\lambda = \mu = 0, N'\rangle \tag{3.225}$$

where $N' = (N - \sum_m m\lambda_m - \sum_n n\mu_n)$ is the eigenvalue of $R = L_o + p^2$. We set up such a basis iteratively as follows: denote a state with $\lambda = \mu = 0$ by $|\psi, N'\rangle$. Then we start from

$$|\psi, 0\rangle = |0\rangle \tag{3.226}$$

then set up

$$|\{\lambda_m = \delta_{m1}, \mu_n = 0\}, 1\rangle = L_1^+ |0\rangle \tag{3.227}$$

$$|\{\lambda_m = 0, \mu_n = \delta_{n1}\}, 1\rangle = K_1^+ |0\rangle \tag{3.228}$$

To complete the basis at $R = 1$ we find two states $|\psi, 1\rangle$ orthogonal to these two states. Then we proceed to $R = 2$, and so on. This iteratively defines the ψ-states which have the properties

$$L_n |\psi, N\rangle = 0 \quad n = 1, 2, \ldots \tag{3.229}$$

$$k_n |\psi, N\rangle = 0 \quad n = 1, 2, \ldots \tag{3.230}$$

and also are of positive norm

$$\langle\psi|\psi\rangle > 0 \tag{3.231}$$

This norm property follows because combinations of excitations of the type $(a^{(n)+} + a^{(n)})$ are absent (otherwise k_n would not annihilate $|\psi, N\rangle$), as are the combinations $(a^{(n)+} - a^{(n)+}) = nk_n^+$ since by definition $\mu_n = 0$ for a ψ-state.

Now all the states with $\lambda_m \neq 0$ for at least one m are spurious states and let us define by S_{ℓ_0} the spurious subspace with eigenvalue of L_o equal to ℓ_0. That is, S_1 is the on-shell spurious subspace. S_0 is the spurious subspace one unit below mass-shell, S_{-1} is two units below, and so on.

Incidentally, note that the spurious states with $\lambda_i = \delta_{i1}$ are null states, as proved at the end of the previous section.

Now take a general state $|f, N\rangle \in F$. It can be decomposed into (we take $|f, N\rangle$ on-shell, i.e., $\ell_0 = 1$)

$$|f, N\rangle = |s, N\rangle + |\phi, N\rangle \tag{3.232}$$

where $|s, N\rangle \in S_1$ and $|k, N\rangle$ is a state, in the complementary subspace to S_1, which we may call K_1.

Since $|f, N\rangle$ is on mass-shell it satisfies

$$(L_0 - 1)\ |f, N\rangle = 0 \tag{3.233}$$

Let us now consider the most general physical state, that is let $|f, N\rangle$ satisfy in addition the gauge conditions

$$L_n\ |f, N\rangle = L_n(\ |s, N\rangle + |k, N\rangle) = 0 \tag{3.234}$$

It is not true for arbitrary space-time dimension that we can deduce from this that L_n annihilates separately $|s, N\rangle$ and $|f, N\rangle$. We shall show, however, that such a deduction is possible if and only if $d = d_c = 26$ and this will provide a shortcut to the proof of absence of ghosts.

First we must prove that L_1 maps $S_1 \to S_0$ for arbitrary d. The most general state belonging to S_1 may be written

$$|s, N\rangle = L_1^+\ |f, N\text{-}1\rangle + L_2^+\ |f, N\text{-}2\rangle \tag{3.235}$$

To check this assertion, observe that the generalised projective algebra may be used to put all spurious states into this form, for example

$$L_3^+ \,|f, N-3\rangle = [L_1^+, L_2^+]\,|f, N-3\rangle \tag{3.236}$$

We act with L_1 on $|s, N\rangle$ to find

$$L_1|s, N\rangle = L_1^+ L_1 \,|f, N-1\rangle + (L_2^+ L_1 + 3L_1^+)\,|f, N-2\rangle \tag{3.237}$$

$$= |s, N-1\rangle \in S_0 \tag{3.238}$$

This is a spurious state belonging to S_0, as required. Thus we may deduce that the expressions

$$L_1 \,|f, N\rangle = L_1 \,|s, N\rangle = L_1 \,|k, N\rangle = 0 \tag{3.239}$$

separately vanish.

We now wish to find a special operator

$$\tilde{L}_2 = \alpha L_1 L_1 + \beta L_2 \tag{3.240}$$

that will map directly $S_1 \to S_{-1}$. Once this has been achieved the proof of the absence of ghosts will follow almost immediately. We need the commutators

$$[L_1^+ L_1^+, L_1 L_1] = L_1^+[L_1^+, L_1]L_1 + L_1[L_1^+, L_1]L_1^+ + [L_1^+, L_1]\,L_1^+ L_1 + L_1 L_1^+\,[L_1^+, L_1] \tag{3.241}$$

$$= -2[L_1^+ L_0 L_1 + L_0 L_1^+ L_1 + L_1 L_1^+ L_0 + L_1 L_0 L_1^+) \tag{3.242}$$

$$= 4(2L_0 - 1)\,L_0 - 8L_o L_1 L_1^+ \tag{3.243}$$

Further we can find that

$$[L_1^+ L_1^+, L_2] = [L_2^+, L_1 L_1] = -6L_1 L_1^+ + 6L_0 \tag{3.244}$$

$$[L_2^+, L_2] = -4L_0 - \frac{1}{2}d \tag{3.245}$$

and hence

$$[\tilde{L}_2, L_1^+] = 4\alpha L_1 L_0 + (3\beta - 2\alpha)L_1 \quad (3.246)$$

$$[\tilde{L}_2, \tilde{L}_2^+] = \alpha^2(8L_0L_1L_1^+ - 4(2L_0 - 1)L_0) + \alpha\beta(6L_1L_1^+ - 6L_0) + \beta^2(4L_0 + \frac{1}{2} d) \quad (3.247)$$

Let us rewrite the most general spurious state in S_1 as

$$|s, N\rangle = L_1^+ \, |f, N-1\rangle + \tilde{L}_2^+ \, |f, N-2\rangle \quad (3.248)$$

then we see that

$$\tilde{L}_2 \, |s, N\rangle = (3\beta - 2\alpha)L_1 \, |f, N-1\rangle + L_1^+ \, L_2 \, |f, N-1\rangle + [(-8\alpha^2 + 12\alpha\beta) \, L_1^+L_1 + (4\alpha^2 - 12\alpha\beta - 4\beta^2 + \frac{1}{2} d\beta^2)] \, |f, N-2\rangle + \tilde{L}_2^+\tilde{L}_2 \, |f, N-2\rangle \quad (3.249)$$

In order to achieve the desired result we put $3\beta - 2\alpha = 0$ e.g. by putting $\beta = 1$, $\alpha = 3/2$ such that

$$\tilde{L}_2 = L_2 + \frac{3}{2} L_1L_1 \quad (3.250)$$

whereupon

$$\tilde{L}_2 \, |s, N\rangle = \tilde{L}_1^+\tilde{L}_2 \, |f, N-1\rangle + \tilde{L}_2^+\tilde{L}_2 \, |f, N-2\rangle + \frac{1}{2} (d - 26) \, |f, N-2\rangle \quad (3.251)$$

If, now, we put $d = 26$ then

$$\tilde{L}_2 \, |s, N\rangle = |s, N-2\rangle \in S_{-1} \quad (3.252)$$

and we have proved that L_2 maps $S_1 \rightarrow S_{-1}$, as required. Note the peculiar way in which d enters through the anomaly term occurring in $[L_2, L_{-2}]$. If the anomaly term were absent, we could never obtain such a simple result. The result, for d = 26, is that the two components, spurious and non-spurious, in our general physical state satisfy

$$L_n |s, N\rangle = L_n |k, N\rangle = 0 \quad n = 1, 2, \ldots \tag{3.253}$$

The fact that $|s, N\rangle$ is annihilated by the L_n implies that we have a state which is simultaneously spurious and physical; it is therefore a null state, that is

$$\langle s, N|s, N\rangle = 0 \tag{3.254}$$

Concerning the state $|k, N\rangle$ we may write its most general form as

$$|k, N\rangle = \sum_{\{\mu_m\}} C_{\{\mu_m\}} \prod_m (k_m^+)^{\mu_m} |\mu, N - \sum_m m\mu_m\rangle \tag{3.255}$$

Now acting with L_n it is not difficult to show that $L_n|k, N\rangle = 0$ implies that all $C_{\{\mu_m\}} = 0$ ($\mu_m \neq 0$) and therefore

$$|k, N\rangle = |\psi, N\rangle \tag{3.256}$$

Thus we have the result that any physical state may be written (for $d = d_c = 26$)

$$|f, N\rangle = |\psi, N\rangle + |\text{null state}\rangle \tag{3.257}$$

The null state may be absorbed by redefinition since it does not alter the properties. Since $|\psi, N\rangle$ has been shown to have positive norm, this proves that all physical states have positive norm.

To discuss a space-time dimension $d < 26$, note that we may simply add $(26 - d)$ extra space-like dimensions, with zero momentum components and treat the smaller d as a subspace of the d = 26 case. Hence, an immediate corollary of our proof is that all physical states have positive norm for $d \leq 26$.

Let us write down the two necessary and sufficient conditions for absence of imaginary coupling constants in the generalised Veneziano model:

(i) The intercept must be $\alpha(o) = 1$.

(ii) The space-time dimension must be an integer satisfying $d \le d_c = 26$.

We should immediately add that the choice of k_μ as a light-like vector has no deep significance. We can equally choose k_μ to have only a timelike component ($k_\mu = (1, 0, 0, 0, \ldots)$). We can then see that the physical states, all of positive norm, may be taken as pure angular momentum eigenstates in the rest system, because of the explicit rotational invariance of the construction procedure.

This last point is important because to prove the absence of ghosts it is not sufficient simply to prove the decoupling of the time excitations. We must further prove (as we now have done) that the partial wave analysis into irreducible representations of $O(d - 1)$, in the rest frame, leads to positive partial-wave coefficients.

The above proof of absence of ghosts can be re-interpreted in terms of null states. We have already noted that any state

$$|s, N\rangle = L_1^+ |\phi, N\text{-}1\rangle \tag{3.258}$$

is null, where $|\phi, N\text{-}1\rangle$ is physical. For the case $d = 26$ it is straightforward to prove that any state of the form

$$|s, N\rangle = L_1^+ |\phi, N\text{-}1\rangle + \tilde{L}_2^+ |\phi, N\text{-}2\rangle \tag{3.259}$$

is a null state, where $|\phi, N\text{-}1\rangle$ and $|\phi, N\text{-}2\rangle$ are both physical as follows:

$$||L_1^+ |\phi, N\text{-}1\rangle + L_2^+ |\phi, N\text{-}2\rangle|^2 =$$

$$= \langle\phi, N\text{-}1| [L_1, L_1^+] |\phi, N\text{-}1\rangle + 2\text{Re}\langle\phi, N\text{-}1| \cdot$$

$$\cdot [L_1, \tilde{L}_2^+] |\phi, N\text{-}2\rangle +$$

$$+ \langle\phi, N\text{-}2| [\tilde{L}_2, \tilde{L}_2^+] |\phi, N\text{-}2\rangle \tag{3.260}$$

$$= \frac{1}{2}(d - 26) \langle\phi, N\text{-}2|\phi, N\text{-}2\rangle \tag{3.261}$$

$$= 0 \text{ for } d = d_c = 26, \tag{3.262}$$

where we have used the expressions for the commutators worked out earlier.

Note that for $d < d_c$ these new null states do not occur, and it can be shown that the null states generated by L_1^+ acting on a physical state are the only ones.

Let us finally indicate how to count the numbers of physical, spurious and null states both for $d = d_c$ and for $d < d_c$. Define $p(x)$ and $T^{(m)}(N)$ by

$$(p(x))^m = [\prod_{r=1}^{\infty} (1 - x^r)^{-1}]^m = \sum_{N=o}^{\infty} T^{(m)}(N)\, x^N \tag{3.263}$$

Then in the full Fock space the total number of states at $R = N$ is given by $T^{(d)}(N)$ where d is the space-time dimension.

The number of physical states at $d = d_c$ is equal to the number of our ψ-states and this is straightforwardly seen to be $T^{(d_c-2)}(N)$. On the other hand, the number of states satisfying the gauge conditions is $T^{(d_c-1)}(N)$. This then gives the breakdown at the level $R = N$:

$T^{(d_c-2)}(N)$ non-null (positive-norm) physical states (3.264)

$$T^{(d_c-1)}(N) - T^{(d_c-2)}(N) \quad \text{null physical states} \tag{3.265}$$

$$T^{(d_c)}(N) - 2T^{(d_c-1)}(N) + T^{(d_c-2)}(N) \quad \text{non-null spurious states} \tag{3.266}$$

$$T^{(d_c-1)}(N) - T^{(d_c-2)}(N) \quad \text{null spurious states} \tag{3.267}$$

$$T^{(d_c)}(N) \quad \text{Total number of states in the Fock space.} \tag{3.268}$$

For $d \neq d_c$ the breakdown of the total number $T^{(d)}(N)$ of states at $R = N$ (on mass shell) becomes

$$T^{(d-1)}(N) - T^{(d-1)}(N - 1) \quad \text{non-null physical states} \tag{3.269}$$

$$T^{(d-1)}(N - 1) \quad \text{null physical states} \tag{3.270}$$

$$T^{(d)}(N) - T^{(d-1)}(N) - T^{(d-1)}(N-1) \quad \text{non-null spurious states} \tag{3.271}$$

$$T^{(d-1)}(N - 1) \quad \text{null spurious states} \tag{3.272}$$

If $d < d_c$ the non-null physical states all have positive norm. If $d > d_c$ there are physical states of both positive and negative norms, as can be seen, most easily by considering the level $\alpha_s = 2$ in

$$B(-\alpha_s, -\alpha_t) = \sum_{N=0}^{\infty} R_N(\alpha_t)\,(N - \alpha_s)^{-1} \tag{3.273}$$

$$R_2(\alpha_t) = \frac{25}{8}\left(z^2 - \frac{1}{25}\right) \tag{3.274}$$

which implies a scalar ghost for all $d \geq 27$ as we found much earlier in discussing the four-particle function.

3.8 PHYSICAL STATE CONSTRUCTION

We have amply demonstrated the usefulness of the operators $a^{(n)}$, $a^{(n)+}$ in discussing the factorisation and duality properties. Further, even the absence of ghosts could be demonstrated. In studying the spectrum, however, we see that these covariant operators create an enormous number of redundant spurious states. We shall now find new operators[20] $A_i^{(n)}$, $A_i^{(n)+}$ $(i = 2, 3, 4, \ldots, (d-1))$ which create all the physical states in the critical dimension, and then briefly discuss the situation for a sub-critical case. The new operators are not manifestly covariant and their use is strictly analogous to the use of the Coulomb gauge in quantum electrodynamics.

Let us first construct the physical states coupled to two ground states. Recall that

$$V(p, z) = \exp(\sqrt{2}i\ p\cdot Q^{(-)}(z))\ \exp(\sqrt{2}\ i\ p\cdot Q^{(0)}(z))\ \cdot \\ \cdot\ \exp(\sqrt{2}\ ip\cdot Q^{(+)}\ (z)) \tag{3.275}$$

$$\exp(\sqrt{2}\ ip\cdot Q^{(0)}(z)) = e^{ip\cdot Q}\ z^{-2p\cdot P}\ z^{p^2} \tag{3.276}$$

$$\langle 0, p_1| = \lim_{z_1\to 0} \langle 0|\ V(p_1, z_1) \tag{3.277}$$

Take $p_{1\mu}$ and a second momentum $p_{2\mu}$ such that

$$1 + (p_1 + p_2)^2 = n, \text{ an integer} \tag{3.278}$$

$$2p_1\cdot p_2 = n + 1 \tag{3.279}$$

and consider

$$\lim_{z_1\to 0} \langle 0|\ V(p_1, z_1)\oint \frac{dz_2}{z_2} V(p_2, z_2) \tag{3.280}$$

where we have taken a contour which encircles the origin $z_2 = 0$.

This becomes

$$<0, (p_1 + p_2)| \oint \frac{dz_2}{z_1^{n+1}} \exp(\sqrt{2}\, i\, p_2 \cdot Q^{(+)}(z_2)) \quad (3.281)$$

where we used

$$z^{-2p_2 \cdot (p_1+p_2)} z^{-1} = z^{-2p_1 \cdot p_2 + 1} = z^{-n} \quad (3.282)$$

It is clear that the integral is single-valued (because we are precisely on mass shell) and that we are creating a physical state of momentum $(p_1 + p_2)_\mu$ and eigenvalue of R equal to n.

By using the commutator

$$[L_n, V(p, z)] = -z^{-n}(z \frac{d}{dz} + np^2) V(p, z) \quad (3.283)$$

which implies for $p^2 = -1$,

$$[L_n, \frac{V(p, z)}{z}] = - \frac{d}{dz} (z^{-n} V(p, z)) \quad (3.284)$$

we can check that the gauge conditions are satisfied. We write

$$<0, p_1| \oint \frac{dz_2}{z_2} V(p_2, z_2) L_n^+ \quad (3.285)$$

$$= <0, p_1| L_n^+ (\oint \frac{dz_2}{z_2} V(p_2, z_2) + \oint dz_2 \frac{d}{dz_2} \cdot$$

$$\cdot (z_2^n V(p_2, z_2)) \quad (3.286)$$

$$= 0 \text{ if the integral is single-valued.} \quad (3.287)$$

To make further physical states consider the more general N-particle expression.

$$\lim_{z_1 \to 0} <0| V(p_1 z_1) \oint \frac{dz_2}{z_2} V(p_2, z_2) \oint_{z_2} \frac{dz_3}{z_3} V(p_3 z_3) \cdot$$

$$\cdot \quad \cdots \oint_{z_{N-1}} \frac{dz_N}{z_N} V(p_N, z_N) \tag{3.288}$$

with

$$1 + (p_1 + p_2)^2 = n_{12}$$

$$1 + (p_1 + p_2 + p_3)^2 = n_{13}$$

$$\cdot$$

$$\cdot$$

$$1 + (p_1 + p_2 + \cdots + p_N)^2 = n_{1N} \tag{3.289}$$

The integration contour for dz_k encircles the point $z_k = z_{k-1}$, as indicated. That this multiparticle state satisfies the gauge conditions may also be checked explicitly.

It turns out to be very useful to consider the vertex for emission of the massless vector state of the unit intercept theory. The reason is that because $p^2 = 0$ the normal ordering difficulties completely disappear, as we have seen earlier in discussing the commutator $[L_n, V(p, z)]$.

It is straightforward to derive the vertex for massless vector emission. Putting

$$(p_1 + p_2)^2 = 0 \tag{3.290}$$

$$\therefore \; p_1 \cdot p_2 = 1 \tag{3.291}$$

we consider forming the massless vector from two tachyons by

$$\frac{V(p_1, z_1)}{z_1} \oint_{z_1} dz_2 \, \frac{V(p_2, z_2)}{z_2} \tag{3.292}$$

Use the commutators derived earlier to find

$$\exp(\sqrt{2}\, i\, p_1 \cdot Q^{(+)}(z_1)) \exp(\sqrt{2}\, i\, p_2 \cdot Q^{(-)}(z_1)) =$$

$$= \exp(\sqrt{2}\, i\, p_2 \cdot Q^{(-)}(z_2)) \exp(\sqrt{2}\, i\, p_1 \cdot Q^{(+)}(z_1)) \cdot$$
$$\cdot \left(1 - \frac{z_1}{z_2}\right)^{-2} \qquad (3.293)$$

Hence the z_2 integral has a double pole at $z_2 = z_1$. Explicitly

$$\frac{V(p_1, z_1)}{z_1} \oint_{z_1} dz_2 \frac{V(p_2, z_2)}{z_2} =$$

$$= \oint_{z_1} dz_2 \frac{z_2}{z_1} \frac{1}{(z_2 - z_1)^2} \exp(i\sqrt{2}\, p_1 \cdot Q^{(-)}(z_1) +$$
$$+ i\sqrt{2}\, p_2\, Q^{(-)}(z_2)) \cdot$$
$$\cdot \exp(i\sqrt{2}\, p_1 \cdot Q^{(0)}(z_1)) \exp(i\sqrt{2}\, p_2 \cdot Q^{(0)}(z_2) \cdot$$
$$\cdot \exp(i\sqrt{2}\, p_1 \cdot Q^{(+)}(z_1) + i\sqrt{2}\, p_2 \cdot Q^{(-)}(z_2)) \qquad (3.294)$$

$$= \frac{\partial}{\partial z_2} [\frac{z_2}{z_1} \exp[i\sqrt{2}\, (p_1 \cdot Q^{(-)}(z_1) + p_2 \cdot Q^{(-)}(z_2))] \cdot$$
$$\cdot \exp(i\sqrt{2}\, [p_1 \cdot Q^{(0)}(z_1) + p_2 \cdot Q^{(0)}(z_2)]) \left(\frac{z_2}{z_1}\right)^{-p_1 \cdot p_2}$$
$$\exp(i\sqrt{2}\, [p_1 \cdot Q^{(+)}(z_1) + p_2 \cdot Q^{(+)}(z_2)])]_{z_1 = z_2} \qquad (3.295)$$

$$= i\sqrt{2}\, p_2 \frac{dQ(z)}{dz} V(p_1 + p_z, z) \qquad (3.296)$$

Since $p_2 \cdot (p_1 + p_2) = 0$ we may regard $p_{2\mu}$ as the polarization vector and write, within an overall normalization,

$$V_\varepsilon(p, z) = \varepsilon \cdot \frac{dQ(z)}{dz} V(p, z) \qquad (3.297)$$

Note that the factor $(z_1/z_2)^{-p_1 \cdot p_2}$ arises from the contraction

$$\left(\frac{z_1}{z_2}\right)^{-p_1 \cdot p_2} = \exp[-p_{1\mu} p_{2\nu} [Q_\mu^{(0)}(z_1), Q_\nu^{(0)}(z_2)]] \qquad (3.298)$$

Now we are ready to construct positive-norm physical states. Let us put $p_{1\mu} = (0,1,0,0,\ldots)$ and introduce auxiliary light-like vectors

$$k_\mu = (\tfrac{1}{2}; -\tfrac{1}{2}, 0, 0, \ldots) \tag{3.299}$$

$$k_{1\mu} = (\tfrac{1}{2}; \tfrac{1}{2}, 0, 0, \ldots.) \tag{3.300}$$

and let the vector momentum be $p_{2\mu} = n\, k_\mu$. Then

$$1 + (p_1 + p_2)^2 = 1 + (p_1 + nk)^2 = n \tag{3.301}$$

Now define

$$A_n{}^i = \frac{1}{\sqrt{2}\,\pi i} \oint \frac{P^i(z)}{z} V(nk, z) \tag{3.302}$$

in which $i = 2, 3, 4, \ldots, (d-1)$ and

$$P_\mu(z) = -iz \frac{d}{dz} Q_\mu(z) \tag{3.303}$$

$$= \sqrt{2}\, P_\mu + i \sum_{n=1}^{\infty} n(a^{(n)+} z^{-n} - a^{(n)} z^n) \tag{3.304}$$

From the properties

$$(P^i(z))^+ = P^i(1/z) \tag{3.305}$$

$$(V(nk, z))^+ = V(-nk, 1/z) \tag{3.306}$$

$$d(1/z) = -dz/z^2 \tag{3.307}$$

we deduce that

$$A_n^{i+} = A_{-n}^i \tag{3.308}$$

Define, in addition,

$$k \cdot A_n = \frac{1}{\sqrt{2}\,\pi i} \oint \frac{k\, P(z)\, V(nk, z)\, dz}{z} \tag{3.309}$$

Note that

$$\frac{d}{dz}(V(nk, z)) = -\sqrt{2}\, n \frac{k\, P(z)\, V(nk, z)}{z} \tag{3.310}$$

and hence

$$k\cdot A_n = \delta_{no} \frac{1}{\sqrt{2}\,\pi i} \oint \frac{dz}{z} \cdot \sqrt{2}\, k\cdot p_1 = \delta_{no} \tag{3.311}$$

Where we have substituted $V(0, z) = 1$, and the explicit form for $p_\mu(z)$ written earlier.

For $p^2 = 0$ $V(p, z)$ transforms, like $Q_\mu(z)$, as a generalized projective scalar, namely

$$[L_n, V(p,z)] = - z^{-n+1} \frac{d}{dz} V(p, z) \quad \text{for } p^2 = 0 \tag{3.312}$$

$$[L_n, Q_\mu(z)] = - z^{-n+1} \frac{d}{dz} Q_\mu(z) \tag{3.313}$$

It follows that

$$[L_n, P_\mu(z)\, V(p, z)] = - z \frac{d}{dz}(z^{-n} P_\mu(z)\, V(p, z)) \tag{3.314}$$

where we used the definition $p_\mu(z) = -iz \frac{d}{dz} Q_\mu(z)$. (3.315)

In particular

$$[L_n, \frac{P_\mu(z)\, V(p, z)}{z}] = - \frac{d}{dz}(z^{-n} P_\mu(z)\, V(p, z)) \tag{3.316}$$

and it follows immediately that

$$[L_n, A_n{}^i] = [L_n, A_n^{i+}] = 0 \tag{3.317}$$

which confirms that states created by $A_n{}^i$ satisfy the physical gauge conditions.

Now consider the important commutator[21]

$$[A_m{}^i, A_n{}^j] = - \frac{1}{2\pi^2} \oint \frac{dz_1}{z_1} \frac{dz_2}{z_2} [P^i(z_1)\, V(mk, z_1), P^j(z_2)\, V(nk, z_2)] \tag{3.318}$$

$$= - \frac{1}{2\pi^2} \oint \frac{dz_1}{z_1} \frac{dz_2}{z_2} [P^i(z_1), P^j(z_2)] \, V(mk, z_1) \cdot$$

$$\cdot \; V(nk, z_2) \tag{3.319}$$

where we have used the fact that the $p^i(z_1)$, $p^j(z_2)$ commute with the $V(mk, z_1)$, $V(nk, z_2)$ and that the two V factors commute with each other. [These simplifications are the reason for considering the massless case].

Put $z_i = e^{i\theta_i}$ then

$$[Q_\mu(z_1 = e^{i\theta_1}), Q_\nu(e^{i\theta_2})] = i\pi \, g_{\mu\nu} \, \varepsilon(\theta_1 - \theta_2) \tag{3.320}$$

and it follows that

$$[P^i(z_1), P^j(z_2)] = - i\pi \, \delta_{ij} \frac{d}{d\theta_2} \delta(\theta_1 - \theta_2) \tag{3.321}$$

Noting that $dz_1/z_1 = id\theta_1$, and that

$$\frac{d}{d\theta_2} V(nk, z_2) = - \sqrt{2} \, i \, k \cdot P(z_2) \, V(nk, z_2) \tag{3.322}$$

we find eventually

$$[A_m{}^i, A_n{}^j] = - \delta_{ij} \, nk \cdot A_{m+n} \tag{3.323}$$

$$= m \, \delta_{ij} \, \delta_{m+n,0} \tag{3.324}$$

To collect the results together, we now have transverse operators $A_n{}^i$ satisfying

$$[L_n, A_n{}^i] = 0 \tag{3.325}$$

$$[A_n{}^i, A_m{}^{j+}] = n \, \delta_{ij} \, \delta_{mn} \tag{3.326}$$

and hence these operators can be used to construct directly linearly-independent positive norm physical states. In space-time dimension d we can so construct $T^{(d-2)}(N)$ states at the level $R = N$. In particular, for $d = d_c$ we can make a complete basis for physical states in this way.

It is instructive to consider the infinite momentum limit of our $A_n^{\ i}$. That is, we act on a ground-state with momentum

$$p_\mu = (E;\ \sqrt{E^2+1},\ 0,\ 0,\ .....) \tag{3.327}$$

and let $E \to \infty$. Now fix

$$1 + (p + \lambda k)^2 = n \tag{3.328}$$

$$2\lambda p\cdot k = n \tag{3.329}$$

and hence

$$\frac{1}{\sqrt{2}\ \pi i} \oint \frac{dz}{z} P^i(z)\ V(\lambda k,\ z) \underset{E\to\infty}{\to}$$

$$\to \frac{1}{\sqrt{2}\ \pi i} \oint \frac{dz}{z^{n+1}} (\sqrt{2}\ P^i + i \sum_{r=1}^{\infty} \sqrt{r}\ (a^{(r)+} z^{-r} - a^{(r)} z^r)) \tag{3.330}$$

$$= \sqrt{2n}\ i\ a_i^{(n)} . \tag{3.331}$$

Thus we obtain just the transverse states created by the original $a^{(n)}$, $a^{(n)+}$. We may say that the $d = d_c$ model appears as a genuine nonrelativistic oscillator in the infinite-momentum limit.

For $d < d_c$, we need further operators involving the longitudinal dimension in order to create all physical states; for full details of the construction we refer the reader to the literature [see Brower, Reference 15]. We may understand intuitively what is happening for $d < d_c$ by considering a simple example. The physical states at the level $R = 2$ are contained in a tensor (in the rest frame)

$$[a_i^{(1)+}\ a_j^{(1)+} - \frac{1}{25}\ \delta_{ij}\ (a^{(1)+}\cdot a^{(1)+})]\ |0\rangle \tag{3.332}$$

where i, j = 1, 2, 3, ... (d-1). For $d = d_c = 26$ this is a single irreducible representation of O(25). For d < 26 it is a linear combination of two irreducible representations of O(d-1); for example for d = 4 it is a linear combination of a spin two and a spin zero. More generally, we may distinguish[28] between the genuine daughters which are present in the critical dimension and trace daughters which occur only in the sub-critical cases. These trace daughters arise simply because one is starting from the O(25) irreducible representations of the model, reducing to a smaller representation space (smaller d) and reducing with respect to the new group O(d-1).

This completes our study of the spectrum of the model, and we now turn to some other useful constructs of the operator formalism.

3.9 TWISTING OPERATOR

It is useful to find an operator[29] $\Omega(\pi)$ which performs the function indicated in Figure 3 (see page 166) namely,

$$\Omega(\pi)\, V(p_1 z_1)\, V(p_2 z_2) \cdots V(p_N z_N)\, |0\rangle$$

$$= V(p_N z_1)\, V(p_{N-1} z_2) \cdots V(p_1 z_N)\, |0\rangle \quad (3.333)$$

with $\pi_\mu = -\sum_{i=1}^{N} p_{i\mu}$.

The appropriate operator is[6,29]

$$\Omega(\pi) = (-1)^R\, e^{-L_-(\pi)} \quad (3.334)$$

as can be checked by using the projective group properties

$$\Omega(\pi)\, V(p_1 z_1)\, V(p_2 z_2) \cdots V(p_N\, z_N)\, |0\rangle$$

$$= (-1)^{L_0+\pi^2}\, V(p_1,\, z_1-1)\, V(p_2,\, z_2-1) \cdots \cdot$$

$$\cdot\, V(p_N,\, z_N-1)\, |0\rangle \qquad (3.335)$$

$$= (-1)^{\pi^2}\, V(p_1,\, 1-z_1)\, V(p_2,\, 1-z_2) \cdots \cdot$$

$$\cdot\, V(p_N,\, 1-z_N)\, |0\rangle \qquad (3.336)$$

$$= (-1)^N\, V(p_N,\, 1-z_N)\, V(p_{N-1},\, 1-z_{N-1}) \cdots \cdot$$

$$\cdot\, V(p_1,\, 1-z_1)\, |0\rangle \qquad (3.337)$$

$$= V(p_N,\, z_1')\, V(p_{N-1},\, z_2') \cdots V(p_1,\, z_N)\, |0\rangle \qquad (3.338)$$

as required. In the final step we simply changed variables to $z_i' = 1 - z_{n-i+1}$, (which has Jacobian $(-1)^N$).

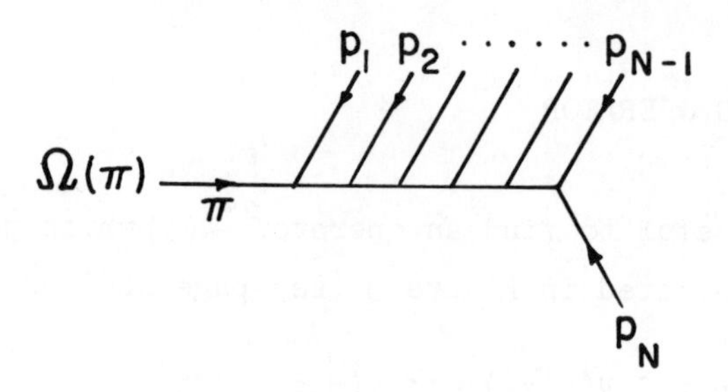

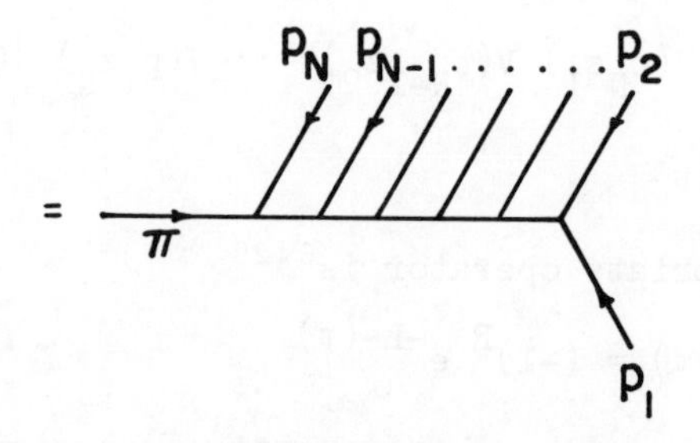

FIGURE 3.3

Twisting Operator

For the earlier steps we have recalled that L_- corresponds to

$$L_1 \leftrightarrow \begin{pmatrix} 0 & 1 \\ 0 & 0 \end{pmatrix} \qquad (3.339)$$

and hence

$$e^{-L_-} V(p, z) e^{L_-} = V(p, z-1) \qquad (3.340)$$

Next we used

$$z^{-L_0} V(p, 1) z^{L_0} = V(p, z) \qquad (3.341)$$

to see that

$$(-1)^{L_0} V(p, z) = V(p, -z) (-1)^{L_0} \qquad (3.342)$$

Finally we used the relationship, derived earlier, that

$$V(p, z) V(p', z') = V(p', z') V(p, z) (-1)^{-2p\cdot p'} \qquad (3.343)$$

which in the present case gives rise to a phase

$$(-1)^{-2\sum_{i\,j} p_i\cdot p_j} = (-1)^{-\pi^2+N} \qquad (3.344)$$

An important property[6] of the twisting operator is the fact that physical states are eigenstates, since

$$= (-1)^R e^{-L_-} |\phi\rangle \qquad (3.345)$$

$$= (-1)^R |\phi\rangle \qquad (3.346)$$

By combining with the propogator, we can arrive at a twisted propagator

$$\tilde{D} = D\Omega \qquad (3.347)$$

which, however, is hermitian only if restricted to the physical subspace. That is, between physical states

$$\langle\phi| D\Omega |\phi'\rangle = \langle\phi| \Omega^+ D |\phi'\rangle \qquad (3.348)$$

To obtain[30,31,32] a hermitian twisted propagator in the full Fock space, it turns out to be useful to define

$$\theta\ (x) = \Omega(1 - x)^{L_0 - L_1} \tag{3.349}$$

whereupon the hermitian twisted propagator is

$$\tilde{D} = \int_0^1 dx\ x^{-1-\alpha(s)+R}\ (1 - x)^{-1+\alpha(o)}\ \theta(x) \tag{3.350}$$

We refer to the literature for details. This hermitian twisted propagator now has the property that it does not propagate spurious states.

It is worth pointing out that the underlying projective group may be embodied into the operator formalism by writing the so-called canonical forms [References 33, 34]. [Reference 34 is also an operator-approach report which, unlike these lectures, concentrates almost exclusively on the O(2,1) subalgebra of the generalized projective algebra.] Notice that

$$x^{L_0} = e^{-p^2 \ell n x}\ x^R \tag{3.351}$$

$$= e^{-p^2 \ell n x}\ :\exp[-\sum_{n=1}^{\infty} n a^{(n)+}\ a^{(n)} (x^n - 1)]: \tag{3.352}$$

where we have used the general formula

$$x^{a+a} = :\exp[a^+ (x - 1)\ a]: \tag{3.353}$$

which is easily checked by taking matrix elements between arbitrary coherent states,

$$<\beta|\ :\sum_{m=0}^{\infty} \frac{1}{m!} (a^+\ (x - 1)\ a)^m\ :\ |\alpha> \tag{3.354}$$

$$= <\beta|\alpha>\ e^{\beta^*\alpha(x-1)} = e^{\beta^*\alpha x} \tag{3.355}$$

$$= <\beta|\ x^{a^+a}\ |\alpha> \tag{3.356}$$

and then noting that the coherent states form an over-complete basis.

More generally we may write the fundamental operators of the theory in the form (canonical form)

$$\mathcal{O}_\Lambda = \exp[-\sum_{n=1}^{\infty} a^{(n)+} A_n] : \exp[-\sum_{m,n=1}^{\infty} a^{(n)+} \cdot$$

$$\cdot (C_{nm} - \delta_{nm}) a^{(m)}] : \cdot$$

$$\cdot \exp[-\sum_{n=1}^{\infty} a^{(n)} B_n] e^{-\phi} \tag{3.357}$$

The matrix C_{mn} may be discovered from the relevant projective transformation

$$z \to z' = \Lambda z = \frac{az + b}{cz + d} \tag{3.358}$$

and it is given by

$$\sum_{m=1}^{\infty} C_{nm} \frac{z^m}{\sqrt{m}} = [(\frac{az + b}{cz + d})^n - (\frac{b}{d})^n] \frac{1}{\sqrt{n}} \tag{3.359}$$

This implies that

$$C_{nm} = \frac{1}{m!} \frac{\sqrt{m}}{\sqrt{n}} [\frac{\partial^m}{\partial z^m} (\frac{az + b}{cz + d})^n] \Bigg|_{z=0} \tag{3.360}$$

For example, for the propagator

$$\begin{pmatrix} a & b \\ c & d \end{pmatrix} = \begin{pmatrix} x & 0 \\ 0 & 1 \end{pmatrix} \tag{3.361}$$

and

$$C_{nm} = \frac{1}{m!} \frac{\sqrt{m}}{\sqrt{n}} \frac{\partial^m}{\partial z^m} x^n \Bigg|_{x=0} \tag{3.362}$$

Of course, this is nothing more than a convenient

prescription, but the group property becomes particularly evident when we multiply two operators

$$\mathcal{O}_{\Lambda} = \mathcal{O}_{\Lambda_1} \mathcal{O}_{\Lambda_2} \tag{3.363}$$

since the product operator is related to the two factors by

$$\phi = \phi_1 + \phi_2 + \sum_{n=1}^{\infty} B_{1n} A_{2n} \tag{3.364}$$

$$A = A_1 + C_1 A_2 \tag{3.365}$$

$$B = B_2 + B_1 C_2 \tag{3.366}$$

$$C = C_1 C_2$$

As an application, consider the twisting operator Ω for which

$$\begin{pmatrix} a & b \\ c & d \end{pmatrix} = \begin{pmatrix} -1 & 1 \\ 0 & 1 \end{pmatrix} \tag{3.368}$$

Hence for this case

$$C_{nm} = (-1)^m \binom{n}{m} \frac{\sqrt{m}}{\sqrt{n}} \tag{3.369}$$

and we may rewrite

$$\Omega(\pi) = (-1)^R e^{-L_-} \tag{3.370}$$

$$= \exp(\pi \cdot \sum_{n=1}^{\infty} \frac{a^{(n)+}}{\sqrt{n}}) * \tag{3.371}$$

$$* \exp(\sum_{m,n=1}^{\infty} [\binom{n}{m} \frac{\sqrt{m}}{\sqrt{n}} (-1)^m - \delta_{mn}] a_n^+ a_m) \tag{3.372}$$

which is the form of $\Omega(\pi)$ originally proposed by Caneschi, Schwimmer and Veneziano.[29]

3.10 MULTIREGGEON VERTEX

Implicit in the set of multiparticle amplitudes A_N for external ground state particles are the amplitudes for excited external particles. The latter may be obtained from the former by factorisation. It is convenient, therefore, to obtain an explicit operator expression from which the amplitudes for arbitrary particles may be directly obtained. The expression we shall use was discovered [Lovelace, 35; see also the modification by Olive, 36] as the culmination of a long series of papers(References 29, 33, 27-44).

The canonical formalism mentioned in the last section may be extended[33,34] to canonical N-vertices of the

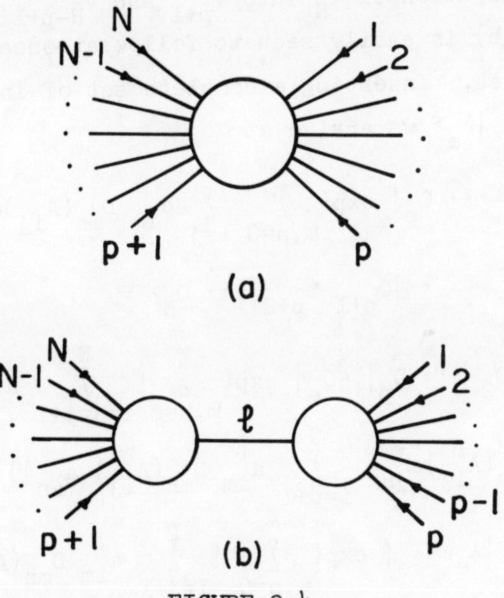

FIGURE 3.4

Canonical N-Vertex

general type (Figure 3.4(a))

$$V_N = \langle 0_1\ 0_2 \cdots 0_p | \exp\{ \sum_{m,n=0}^{\infty} \sum_{i\neq j} \alpha_{im} D_{mn}(A_{ij})\alpha_{jn}\} \cdot$$

$$\cdot |0_{p+1}\ 0_{p+2} \cdots 0_N\rangle \qquad (3.373)$$

Here $\alpha_i = a_i$ if i is a right-pointing leg and $\alpha_i = -a_i^+$ if i is left-pointing. We have introduced separately for each leg i the operators $a_i^{(n)}$, $a_i^{(n)+}$ satisfying

$$[a_{i\mu}^{(n)}, a_{j\nu}^{(m)+}] = -\delta_{ij}\,\delta_{mn}\,g_{\mu\nu} \qquad (3.374)$$

Finally $D_{mn}(A_{ij})$ is the representation matrix, of a projective transformation A_{ij}, which is realized on the Fock space spanned by the $a_{i\mu}^{(n)+}$, $a_{i\nu}^{(n)}$.

Factorisation of V_N into V_{p+1} and V_{N-p+1} according to Figure 3.4(b) is easily seen to follow at once from the group properties. Inserting a complete set of intermediate states $|\lambda_e\rangle$ we arrive at

$$\langle 0_1\ 0_2 \cdots 0_p | \exp\{ \sum_{m,n=0}^{\infty} \sum_{i\neq j} \alpha_{im} D_{mn}(A_{ij})\alpha_{jn}\} \cdot$$

$$\cdot |0_{p+1}\ 0_{p+2} \cdots 0_N\rangle =$$

$$= \langle 0_1\ 0_2 \cdots 0_p | \langle 0_e | \exp\{ \sum_{m,n=o}^{\infty} [\sum_{i\neq j=p+1}^{N} a_{im}^{+} \cdot$$

$$\cdot D_{mn}(A_{ij}^{L})a_{jn}^{+} - \sum_{i=p+1}^{N} a_{im}^{+} D_{mn}(A_{ij}^{L})\, a_{en}]\} \cdot$$

$$\cdot \sum_{\lambda_e} |\lambda_e\rangle\langle\lambda_e| \exp\{ \sum_{m,n=0}^{\infty} [\sum_{i\neq j=1}^{p} a_{im} D_{mn}(A_{ij}^{R})\, a_{jn} -$$

$$- \sum_{i=1}^{p} a_{em}^{+} D_{mn}(A_{ij}^{R})a_{in}]\} |0_e\rangle\, |0_{p+1}\ 0_{p+2} \cdots 0_N\rangle$$

$$(3.375)$$

with the identifications

$$A_{ij} = A^L_{ij} \qquad (p+1) \le i,\ j \le N \tag{3.376}$$

$$= A^R_{ij} \qquad 1 \le i,\ j \le p \tag{3.377}$$

$$= \sum_k A^L_{ik} A^R_{kj} \qquad \text{otherwise} \tag{3.378}$$

For the particular case of the multireggeon vertex, it can be shown[35] that the transformation A_{ij} factorises into

$$A_{ij} = X_i Y_j \tag{3.379}$$

where in terms of Koba-Nielsen z-variables

$$Y_j = \frac{1}{\sqrt{(z_{j-1} - z_{j+1})(z_j - z_{j+1})(z_j - z_{j-1})}} \cdot \begin{pmatrix} -z_{j-1}(z_j - z_{j+1}) & z_j(z_{j-1} - z_{j+1}) \\ -(z_j - z_{j+1}) & (z_{j-1} - z_{j+1}) \end{pmatrix} \tag{3.380}$$

and

$$X_i = \Gamma Y_i^{-1} \tag{3.381}$$

with

$$\Gamma = \begin{pmatrix} 0 & 1 \\ 1 & 0 \end{pmatrix} \tag{3.382}$$

Since

$$\begin{pmatrix} a & b \\ c & d \end{pmatrix}^{-1} = \begin{pmatrix} d & -b \\ -c & a \end{pmatrix} (ad - bc)^{-1} \tag{3.383}$$

it follows that

$$X_i = \frac{1}{\sqrt{(z_{i-1} - z_{i+1})(z_i - z_{i+1})(z_i - z_{i-1})}} \cdot$$

$$\cdot \begin{pmatrix} (z_i - z_{i+1}) & -z_{i-1}(z_i - z_{i+1}) \\ (z_{i-1} - z_{i+1}) & -z_i(z_{i-1} - z_{i+1}) \end{pmatrix} \quad (3.384)$$

and that

$$A_{ij} = a_{ij}[(z_{i-1} - z_{i+1})(z_i - z_{i+1})(z_i - z_{i-1}) \times$$

$$\times \ (z_{j-1} - z_{j+1})(z_j - z_{j+1})(z_j - z_{j-1})]^{-\frac{1}{2}} \quad (3.385)$$

where

$$a_{ij} = \begin{pmatrix} A & B \\ C & D \end{pmatrix} \quad (3.386)$$

$$A = (z_i - z_{i+1})\ (z_j - z_{j+1})(z_{i-1} - z_{j+1}) \quad (3.387)$$

$$B = (z_j - z_{i-1})\ (z_i - z_{i+1})(z_{j-1} - z_{j+1}) \quad (3.388)$$

$$C = (z_i - z_{j-1})\ (z_{i-1} - z_{i+1})(z_j - z_{j+1}) \quad (3.389)$$

$$D = (z_{j-1} - z_{j+1})\ (z_{i-1} - z_{i+1})(z_j - z_i) \quad (3.390)$$

Note, incidentally, that A_{ij} is manifestly projective invariant (consider, as usual, translations, dilations and inversions).

The N-reggeon vertex comprises the appropriate canonical N-vertex together with certain integration measure factors. The full expression is

$$V_N = \int \prod_{i=1}^{N} dz_i \left[\frac{dz_a\ dz_b\ dz_c}{(z_a - z_b)(z_b - z_c)(z_c - z_a)}\right]^{-1}$$

$$\prod_{i=1}^{N} (z_i - z_{i+1})^{\alpha(o)}\ (z_i - z_{i+1})^{-\alpha(o)-1} \ .$$

$$\cdot \langle 0_1\, 0_2 \cdots 0_N |\ \exp\{ \sum_{m,n=0}^{\infty} \sum_{i\neq j} a_{im}\, D_{mn}(X_i Y_j) a_{jn} \} \tag{3.391}$$

It is instructive to take the projection on the ground state particles, because we shall see how important is the role of the projective-spin-zero representation.

We take, as discussed earlier, the $D_{mn}^{(-\varepsilon,+)}$ matrices ($-J = \varepsilon \to 0+$) and calculate the expression

$$V_N \prod_{i=1}^{N} \{ e^{ip_i \cdot Q_i} |0_i\rangle \} \tag{3.392}$$

in which p_i is the momentum (incoming) of the i particle.

In terms of zeroth-mode operators $a^{(s)}$, $a^{(s)+}$ recall that

$$Q_\mu = \frac{1}{\sqrt{2\varepsilon}} (a_\mu^{(\varepsilon)} + a_\mu^{(\varepsilon)+}) \tag{3.393}$$

$$P_\mu = -i \frac{\sqrt{\varepsilon}}{\sqrt{2}} (a_\mu^{(\varepsilon)} - a_\mu^{(\varepsilon)+}) \tag{3.394}$$

$$a_\mu^{(\varepsilon)} = \frac{i}{\sqrt{2\varepsilon}} (P_\mu - i\varepsilon Q_\mu) \tag{3.395}$$

$$a_\mu^{(\varepsilon)+} = -\frac{i}{\sqrt{2\varepsilon}} (P_\mu + i\varepsilon Q_\mu) \tag{3.396}$$

$$[a_\mu^{(\varepsilon)}, a_\mu^{(\varepsilon)+}] = -g_{\mu\nu} \tag{3.397}$$

$$[Q_\mu, P_\nu] = -i\, g_{\mu\nu} \tag{3.398}$$

Therefore only $D_{oo}^{(\varepsilon,+)}(A_{ij})$ gives a contribution to the ground-state projection. Recall from our discussion of the projective group that for a transformation

$$z \to z' = \Lambda z = \frac{az + b}{cz + d} \tag{3.399}$$

the representation matrix is calculable from

$$\frac{\sqrt{\Gamma(m + 2\varepsilon)}}{\sqrt{m!}} (az + b)^m (cz + d)^{-2\varepsilon-m} |\Delta|^\varepsilon =$$

$$= \sum_{n=0}^{\infty} D_{mn}^{(-\varepsilon,+)} (\Lambda) \frac{\sqrt{\Gamma(m + 2\varepsilon)}}{\sqrt{n!}} z^n \tag{3.400}$$

with $\Delta = (ad - bc)$. It follows that

$$D_{oo}^{(-\varepsilon,+)} = (cz + d)^{-2\varepsilon} |\Delta|^\varepsilon \tag{3.401}$$

$$= \exp[- 2\varepsilon \ln(cz + d) + \varepsilon \ln |\Delta|] \tag{3.402}$$

$$= 1 - 2\varepsilon \ln \frac{d}{\sqrt{ad - bc}} + O(\varepsilon^2) \tag{3.403}$$

$$= 1 - \varepsilon \ln \frac{d^2}{(ad - bc)} + O(\varepsilon^2) \tag{3.404}$$

Hence, from our earlier expression for A_{ij} we deduce that

$$D_{oo}^{(-\varepsilon,+)} (A_{ij}) = 1 - \varepsilon \ln\left[\frac{(z_{j-1} - z_{j+1})(z_{i-1} - z_{i+1})(z_j - z_i)(z_i - z_j)}{(z_i - z_{i+1})(z_j - z_{j+1})(z_i - z_{i-1})(z_j - z_{j-1})}\right] + O(\varepsilon^2) \tag{3.405}$$

Now use

$$|0, p_i\rangle = \prod_{i=1}^{N} \{e^{ip_i \cdot Q_i} |0_i\rangle\} = \prod_{i=1}^{N} \{\exp(\frac{p_i \cdot a_i^{(\varepsilon)+}}{\sqrt{2\varepsilon}}) |0_i\rangle\} \tag{3.406}$$

Using coherent state techniques it follows that

$$V_N \prod_{i=1}^{N} \{e^{ip_i \cdot Q_i} |0_i>\} =$$

$$= \int \prod_{i=1}^{N} dz_i \left[\frac{dz_a \; dz_b \; dz_c}{(z_a - z_b)(z_b - z_c)(z_c - z_a)}\right]^{-1} \prod_{i=1}^{N} (z_i - z_{i+2})^{\alpha(o)}$$

$$\prod_{i=1}^{N} (z_i - z_{i+1})^{-\alpha(o)-1} \; M_N(p_i, z_i) \qquad (3.407)$$

where

$$M_N(p_i, z_i) = \lim_{\varepsilon \to 0} <0_1 \; 0_2 \cdots 0_N| \exp(-\sum_{i \neq j} a_i^{(\varepsilon)} \cdot a_j^{(\varepsilon)}) \cdot$$

$$\cdot \exp(\sum_{i \neq j} a^{(\varepsilon)} \cdot a^{(\varepsilon)} \cdot$$

$$\cdot \ell n\left[\frac{(z_{j-1} - z_{j+1})(z_{i-1} - z_{i+1})(z_i - z_j)(z_j - z_i)}{(z_i - z_{i+1})(z_j - z_{j+1})(z_i - z_{i-1})(z_j - z_{j-1})}\right]) \cdot$$

$$\cdot \; |0, p_i> \qquad (3.408)$$

Therefore, dropping an infinite constant

$$\exp(\sum_{i \neq j} \frac{p_i \cdot p_j}{2\varepsilon}) = \exp(\frac{N\alpha(o)}{4\varepsilon}) \qquad (3.409)$$

we find

$$M_N(p_i, z_i) = \prod_{i=j} \cdot$$

$$\cdot \left[\frac{(z_{j-1} - z_{j+1})(z_{i-1} - z_{i+1})(z_j - z_i)(z_i - z_j)}{(z_i - z_{i+1})(z_j - z_{j+1})(z_i - z_{i-1})(z_j - z_{j-1})}\right]^{-\frac{p_i \cdot p_j}{2}} \qquad (3.410)$$

$$= \prod_{i=1}^{N} \left[\frac{z_i - z_{i+2}}{(z_i - z_{i+1})^2}\right]^{-\alpha(o)} \prod_{i<j} |z_i - z_j|^{-2p_i \; p_j} \qquad (3.411)$$

Substituting back this expression for $M_N(p_i, z_i)$ we obtain for the projection of our N-reggeon vertex on to N ground states

$$V_N \prod_{i=1}^{N} \{e^{ip_i \cdot Q_i} |0_i\rangle\} =$$

$$= \int \prod_{i=1}^{N} dz_i \left[\frac{dz_a \, dz_b \, dz_c}{(z_a - z_b)(z_b - z_c)(z_c - z_a)}\right]^{-1} \cdot$$

$$\cdot \prod_{i=1}^{N} (z_i - z_{i+1})^{-1+\alpha(0)} \prod_{i<j} |z_i - z_j|^{-2p_i \cdot p_j} \tag{3.412}$$

which is the familiar N-point function.

In this calculation we should notice how critical it is to have the $J \to 0$ scalar representation in the multi-reggeon vertex, since it gives rise to the factors $[p_i \cdot p_j \ln(z_i - z_j)]$ in the exponent and hence to the basic pole-producing mechanism.

Finally we should note an alternative way[35] to write the matrix element in the N-reggeon vertex. We introduce

$$[C_{m\mu}, C^+_{n\nu}] = - g_{\mu\nu} \, \delta_{mn} \tag{3.413}$$

$$Q^{(i)}_{n\mu} = i \sum_{m=0}^{\infty} \{C^+_{m\mu} D_{mn}(Y_i) + D_{mn}(X_i) \, C_{m\mu}\} \tag{3.414}$$

whereupon we can rewrite

$$\langle 0_1 \, 0_2 \cdots 0_N| \exp\left(\sum_{m,n=0}^{\infty} \sum_{i\neq j} a_i^{\,m} D_{mn}(X_i Y_j) \, a_j^{\,n}\right) =$$

$$= \langle 0_1 \, 0_2 \cdots 0_N| \, \langle 0_c| \prod_{i=1}^{N} : \exp\left(\sum_{k=0}^{\infty} Q^{(i)}_k a_{ik}\right) : |0_c\rangle \tag{3.415}$$

where the normal ordering is with respect to the C-oscillators. This is merely a shorthand which takes advantage of the factorisability in $A_{ij} = X_i Y_j$. This final form was introduced in the original paper of Lovelace.[35] It is easily checked by using coherent states and the Baker-Hausdorf formula, to find that

$$\langle 0_c | \prod_{i=1}^{N} : \exp i\{ \sum_{k=0}^{\infty} \sum_{m=0}^{\infty} [C^{+}_{m\mu} a^{k}_{i\mu} D_{mk} (Y_i) +$$

$$+ D_{km} (X_i) C_{m\mu} a^{k}_{i\mu}]\} : |0_c\rangle =$$

$$= \exp(\sum_{m,n=0}^{\infty} \sum_{i\neq j=1}^{N} a_i^{m} D_{mn} (X_i Y_j) a_{jn}) \qquad (3.415)$$

as required.

To summarise, the multireggeon vertex provides one useful construct in which to study the group properties. Both the factorisation and cyclic symmetry of the theory are manifest. One rather strong result, as we have discussed above, is that the scalar representation of the projective group must be present in order to generate the usual pole producing factor

$$\prod_{i<j} |z_i - z_j|^{-2p_i \cdot p_j} \qquad (3.416)$$

of the multiparticle dual amplitude.

3.11 SUMMARY

By realizing the projective group on the Fock space spanned by harmonic-oscillator operators we are able to write the N-particle amplitudes in a form which exhibits manifestly the two basic properties of cyclic symmetry and factorisability.

An extension to the generalised projective group provides an infinite set of gauge conditions for the physical states in the case of unit intercept. The mechanism by which the gauge operators create null states is similar to that occurring in quantum electrodynamics. The gauge relations remove sufficient states from the Fock space that the remainder have real coupling constants; the proof of this fact involves considerations that depend on the space-time dimensionality.

It is possible to find operators which are written in terms of a Cauchy integral over the vertex function and which create linearly independent positive-norm physical states.

In constructing the twisting operator and the multireggeon vertex the projective group properties are again the guiding principle. In the general vertex the necessity of the projective-scalar representation, in order to produce resonance poles, becomes evident.

REFERENCES

1. V. Bargmann, Ann. Math. 48, 568 (1947).
2. A. O. Barut and C. Fronsdal, Proc. Roy. Soc. A287, 532 (1965).
3. S. Fubini and G. Veneziano, Nuovo Cimento 67A, 29 (1970).
4. L. Clavelli and P. Ramond, Phys. Rev. D2, 973 (1970).
5. L. Clavelli and P. Ramond, Phys. Rev. D3, 988 (1971).
6. F. Gliozzi, Lettere al Nuovo Cimento 2, 846 (1969).
7. C. B. Thorn, Phys. Rev. D1, 1693 (1970).
8. C. B. Chiu, S. Matsuda and C. Rebbi, Phys. Rev. Letters 23, 1526 (1969).
9. C. B. Chiu, S. Matsuda and C. Rebbi, Nuovo Cimento 67A, 437 (1970).
10. M. A. Virasoro, Phys. Rev. D1, 2933 (1970).
11. L. Clavelli and J. H. Weis, as quoted in P. Ramond, NAL preprint THY 15(1971).
12. P. Ramond, Nuovo Cimento 4A, 544 (1971).
13. S. Fubini and G. Veneziano, Ann. Phys. 63, 12 (1971).
14. E. Del Guidice and P. DiVecchia, Nuovo Cimento 70A, 579 (1970).
15. R. C. Brower, Phys. Rev. D6, 1655 (1972).
16. P. Goddard and C. B. Thorn, Phys. Letters 40B, 235 (1972).
17. F. Galzerati, F. Gliozzi, R. Musto and F. Nicodemi, Letter al Nuovo Cimento 4, 991 (1970).
18. P. Campagna, E. Napolitano, S. Sciuto and S. Fubini, Nuovo Cimento 2A, 911 (1971).
19. R. C. Brower and C. B. Thorn, Nucl. Phys. B31, 163 (1971).

20. E. Del Guidice, P. D. Vecchia and S. Fubini, Ann. Phys. 70, 378 (1972).
21. R. C. Brower and P. Goddard, Nucl. Phys. B40, 437 (1972).
22. P. H. Frampton and H. B. Nielsen, Nucl. Phys. B45, 318 (1972).
23. Our analysis in the present subsection (3.5) differs from that of the original references. The main result is, of course, well known.
24. Y. Nambu, Proceedings of the International Conference on Symmetries and Quark Models, Detroit 1969. Edited by R. Chand, Gordon and Breach (1970).
25. J. J. Sakurai, Advanced Quantum Mechanics, Addison-Wesley (1967).
26. W. Heitler, The Quantum Theory of Radiation, Oxford University Press (1954).
27. J. Schwinger, Selected Papers on Quantum Electrodynamics, Dover Publications (1958).
28. P. H. Frampton, Physics Letters 41B, 364 (1972).
29. L. Caneschi, A. Schwimmer and G. Veneziano, Physics Letters 38B, 251 (1969).
30. D. Amati, M. LeBellac and D. I. Olive, Nuovo Cimento 66A, 815 (1970).
31. D. Amati, M. LeBellac and D. I. Olive, Nuovo Cimento 66A, 831 (1970).
32. V. Alessandrini, D. Amati, M. LeBellac and D. I. Olive, Physics Letters 32B, 285 (1970).
33. J. D. Collop, Nuovo Cimento 1A, 217 (1971).
34. V. Alessandrini, D. Amati, M. LeBellac and D. I. Olive, Physics Reports 1C, 269 (1971).
35. C. Lovelace, Physics Letters 32B, 703 (1970).

36. D. I. Olive, Nuovo Cimento 3A, 399 (1971).
37. S. Sciuto, Lettere al Nuovo Cimento 2, 411 (1969).
38. L. Caneschi and A. Schwimmer, Lettere al Nuovo Cimento 3, 213 (1970).
39. I. T. Drummond, Nuovo Cimento 67A, 71 (1970).
40. G. Carbone and S. Sciuto, Lettere al Nuovo Cimento 3, 246 (1970).
41. D. J. Collop and I. T. Drummond, Nuovo Cimento 69A, 261 (1970).
42. J. M. Kosterlitz and D. A. Wray, Lettere al Nuovo Cimento 3, 491 (1970).
43. L. P. Yu, Phys. Rev. D2, 1010 (1970).
44. L. P. Yu, Phys. Rev. D2, 2256 (1970).

4

INTERNAL SYMMETRY

4.1 INTRODUCTION

We now consider in turn the incorporation of the three internal symmetries relevant to hadron physics, namely those associated with conservation of isospin, strangeness, and baryon number.

The treatment of isospin is simplest. By introducing a multiplicative factor one can preserve the fundamental properties of cyclic symmetry and factorisability of the multiparticle dual amplitude. At the same time the observed exchange degeneracy and absence of exotics is introduced in a natural way. The multiplicative factor for the SU(2) (isospin) case is a trace of Pauli matrices.

To include strangeness, we may generalise the multiplicative factor by extending the Pauli matrices to the Gell-Mann λ - matrices of SU(3). This does not necessarily imply that there is SU(3) mass degeneracy in the Born Amplitude.

Next we briefly analyse the factorisation of unequal intercepts in the N-point function, and a solution is found where the intercepts depend quadratically on the internal quantum numbers of the external particles.

Baryon number cannot be dealt with by a multiplicative approach because the pole structure is different when baryons are present. One method is to use the rubber string picture. The mesonic dual amplitudes may be derived from calculational rules where a meson is pictured either as a linear rubber string with a quark at one end ($\sigma = 0$) and an antiquark at the other end ($\sigma = \pi$), or equivalently as a circular string ($0 \leq \sigma \leq 2\pi$) with quarks at these points. Baryons may then be imagined as circular strings with quarks at $\sigma = 0$, $2\pi/3$ and $4\pi/3$; this leads to the pole structure expected from Harari-Rosner quark diagrams, and to generalisations to exotic mesons and baryons.

A more rigorous treatment of the rubber string for free mesons reveals that the elimination of the ghost states, through the generalised projective gauge conditions, may be re-interpreted as being a direct consequence of the general covariance of the classical action of the string under arbitrary re-parametrisations of the two-dimensional world-sheet mapped out by the string as it propagates in space-time.

4.2 MULTIPLICATIVE INTERNAL SYMMETRY FACTOR

We begin with a discussion of isospin invariance, which we may regard as an exact symmetry of the strong interactions. That is, we assume that the mass differences between different members of the same isospin multiplet are entirely electromagnetic in origin.

When all the external mesons are isoscalars we write the full S-matrix as a sum over inequivalent cyclic permutations. Each particular configuration contains poles only in the planar channels (i.e. channels corresponding to groups of adjacent external particles in the planar diagram). The principal properties of B_N that we shall need are that it is cyclic symmetric and that it satisfies factorisation; for example at a three-particle ground-state pole in B_6 one knows that

$$\lim_{\alpha_{123} \to 0} \alpha_{123} \, B(p_1 p_2 p_3 p_4 p_5 p_6) =$$

$$= B_4(s_{12}, s_{23}) \, B_4(s_{45}, s_{56}). \qquad (4.1)$$

We do not need to know the detailed form of B_N. We must introduce isospin in such a way that the following requirements are satisfied:

i) factorisation and bootstrap consistency

ii) cyclic symmetry, if the external particles are identical

iii) absence of poles with isospin > 1 (this is the input of an experimental result).

The solution[1] is to multiply each term in the sum over permutations by a simple isospin multiplicative factor. The factorisation has to be proved but this turns out to be remarkably simple.

Let us begin with the case where all N external particles have isospin T = 1 with isospin labels a_i

(i = 1, 2, ..., N; a_i = 1, 2, 3). Then the proposal is to write

$$T_N^{a_1 a_2 \cdots a_n} = \sum \frac{1}{2} \mathrm{Tr}\,(\tau_{a_1} \tau_{a_2} \cdots \tau_{a_N}) \times$$

$$\times\, B_N(p_1 p_2 \cdots p_N) \tag{4.2}$$

where the sum is over all $\frac{1}{2}$ (N-1)! inequivalent permutations. The τ_{a_i} are Pauli matrices. The cyclic symmetry (ii) follows at once from the cyclic property of the trace. The absence of exotics is ensured by the fact that the products of 2 × 2 matrices are again 2 × 2 matrices and hence can be expanded as linear combinations of isoscalar and isovector. Factorisation follows from the fundamental identity for the Pauli matrices

$$\frac{1}{2} \mathrm{Tr}(\tau_{a_1} \tau_{a_2} \tau_{a_3} \cdots\cdots \tau_{a_N}) =$$

$$= \sum_{k=0,1,2,3} \frac{1}{2} \mathrm{Tr}(\tau_{a_1} \tau_{a_2} \cdots \tau_{a_m} \tau_k) \times$$

$$\times \frac{1}{2} \mathrm{Tr}(\tau_k \tau_{a_{m+1}} \cdots \tau_{a_N}) \tag{4.3}$$

where we have written

$$\tau_0 = \begin{pmatrix} 1 & 0 \\ 0 & 1 \end{pmatrix} \quad \tau_1 = \begin{pmatrix} 0 & 1 \\ 1 & 0 \end{pmatrix} \quad \tau_2 = \begin{pmatrix} 0 & -i \\ i & 0 \end{pmatrix}$$

$$\tau_3 = \begin{pmatrix} 1 & 0 \\ 0 & -1 \end{pmatrix} \tag{4.4}$$

To prove the factorisation property we need use only the fact that

$$\tau_i \tau_j = -\tau_j \tau_i = i\tau_k \tag{4.5}$$

and

$$\tau_i^{\ 2} = \tau_0 \tag{4.6}$$

to re-write

$$\tau_L = (\tau_{a_i} \ \tau_{a_2} \cdots \tau_{a_M}) \tag{4.7}$$

$$\tau_R = (\tau_{a_{M+1}} \ \tau_{a_{M+2}} \cdots \tau_{a_N}) \tag{4.8}$$

and thence

$$\tau_L = \alpha_1^{\ L} \tau_1 + \alpha_2^{\ L} \tau_2 + \alpha_3^{\ L} \tau_3 + \alpha_0^{\ L} \tau_0 \tag{4.9}$$

$$\tau_R = \alpha_1^{\ R} \tau_1 + \alpha_2^{\ R} \tau_2 + \alpha_3^{\ R} \tau_3 + \alpha_0^{\ R} \tau_0 \tag{4.10}$$

whereupon

$$\frac{1}{2} \mathrm{Tr}(\tau_{a_1} \ \tau_{a_2} \cdots \tau_{a_N}) = \frac{1}{2} \mathrm{Tr}(\tau_L \ \tau_R) \tag{4.11}$$

$$= \alpha_1^{\ L} \alpha_1^{\ R} + \alpha_2^{\ L} \alpha_2^{\ R} + \alpha_3^{\ L} \alpha_3^{\ R} + \alpha_0^{\ L} \alpha_0^{\ R} \tag{4.12}$$

$$= \sum_{k=0,1,2,3} \frac{1}{2} \mathrm{Tr}(\tau_L \ \tau_k) \frac{1}{2} \mathrm{Tr}(\tau_k \ \tau_R) \tag{4.13}$$

where we observed that

$$\mathrm{Tr}(\tau_i \ \tau_j) = 2 \ \delta_{ij} \tag{4.14}$$

This procedure has the immediate consequence of exchange degenerate isocalar - isovector pairs of trajectories since we may write

$$\tau_a \tau_b = \delta_{ab} + i \ \varepsilon_{abx} \ \tau_x \tag{4.15}$$

After summing over permutations one finds that the T = 0 and T = 1 have opposite signatures. (This is obvious for a two-particle channel, for example).

The multiplicative approach for isospin may be regarded as very satisfactory. The only note of consternation (which might ultimately prove serious) is that it leads to π-η mass degeneracy.

To incorporate strange mesons we can extend the matrices to 3×3 matrices corresponding to the broken SU(3) symmetry of Gell-Mann-Ne'emann. [References 2, 3 reprinted in reference 4]. In fact for the Gell-Mann matrices[2]

$$\lambda_1 = \begin{pmatrix} 0 & 1 & 0 \\ 1 & 0 & 0 \\ 0 & 0 & 0 \end{pmatrix} \quad \lambda_2 = \begin{pmatrix} 0 & -i & 0 \\ i & 0 & 0 \\ 0 & 0 & 0 \end{pmatrix} \quad \lambda_3 = \begin{pmatrix} 1 & 0 & 0 \\ 0 & -1 & 0 \\ 0 & 0 & 0 \end{pmatrix}$$

$$\lambda_4 = \begin{pmatrix} 0 & 0 & 1 \\ 0 & 0 & 0 \\ 1 & 0 & 0 \end{pmatrix} \quad \lambda_5 = \begin{pmatrix} 0 & 0 & -i \\ 0 & 0 & 0 \\ i & 0 & 0 \end{pmatrix} \quad \lambda_6 = \begin{pmatrix} 0 & 0 & 0 \\ 0 & 0 & 1 \\ 0 & 1 & 0 \end{pmatrix}$$

$$\lambda_7 = \begin{pmatrix} 0 & 0 & 0 \\ 0 & 0 & -i \\ 0 & i & 0 \end{pmatrix} \quad \lambda_8 = \frac{1}{\sqrt{3}}\begin{pmatrix} 1 & 0 & 0 \\ 0 & 1 & 0 \\ 0 & 0 & -2 \end{pmatrix}$$

$$\lambda_0 = \frac{2}{3}\begin{pmatrix} 1 & 0 & 0 \\ 0 & 1 & 0 \\ 0 & 0 & 1 \end{pmatrix} \tag{4.16}$$

we find once again

$$\mathrm{Tr}(\lambda_i \lambda_j) = 2\,\delta_{ij} \tag{4.17}$$

which ensures the factorisability. Furthermore only singlet and octet representations will be present, thus ensuring absence of unwanted (exotic) particles with $T > 1$ or $|S| > 1$. The assignment of, say, the pseudoscalar mesons will be according to the scheme

$$\begin{pmatrix} \frac{1}{\sqrt{6}}\eta - \frac{1}{\sqrt{2}}\pi^o & \pi^- & K^o \\ \pi^+ & \frac{1}{\sqrt{6}}\eta + \frac{1}{\sqrt{2}}\pi^o & K^+ \\ \overline{K^o} & K^- & -\frac{\sqrt{2}}{\sqrt{3}}\eta \end{pmatrix} \tag{4.18}$$

Concerning the multiplicative internal symmetry factors it should be noted that

(i) Although SU(2) mass degeneracy (isospin) is essential, it is not necessary to assume SU(3) mass degeneracy. Indeed it is unlikely that we should begin with SU(3) mass degeneracy in the Born amplitude because it is so badly broken in Nature. It is, however, fortunately consistent with the multiplicative approach that the strange ($S = \pm 1$) octet members have a different Regge intercept than the non-strange ($S = 0$) members, since by strangeness conservation they always occur in different planar channels. The only note of consternation is that it seems difficult to avoid the (undesirable) equality of π and η masses.

(ii) The uniqueness of the form of the multiplicative factor, assuming the requirements listed above, has been demonstrated for SU(2) by Tornqvist[5] and for SU(3) by several authors. [References 6-8].

(iii) The approach described here is readily re-interpreted in terms of the quark model, and in particular in terms of the planar Harari-Rosner quark diagrams which we described much earlier.

4.3 FACTORISATION OF UNEQUAL INTERCEPTS

One possible way to introduce internal symmetry, which we now discuss briefly, is to make the intercepts depend on the internal quantum numbers of the external particles. We shall be content simply to indicate how restrictive the requirement of factorisation is on the form of such a dependence; the question of ghosts will not be treated, but it seems certain that for the conventional generalized Euler B function ghosts will be present for all cases unless at least one intercept is set equal to one.

We have written the N-point function in the form

$$B_N = \int_0^1 du_{1j}\, u_{1j}^{-\alpha_{1j}-1} \prod_{2\le i<j\le(N-1)} \cdot$$

$$\cdot\, (1 - u_{1i} u_{1i+1} \cdots\cdots u_{1j-1})^{-2p_i\cdot p_j - C_{ij}} \tag{4.19}$$

in which

$$C_{ij} = \alpha_{ij}(o) + \alpha_{i+1,j-1}(o) -$$

$$- \alpha_{i+1,j}(o) - \alpha_{i,j-1}(o) \tag{4.20}$$

For example, if the intercept $\alpha_n(o)$ depends only on the number n of external particles coupling to the trajectory this gives

$$C_{ij} = + \alpha_{j-i+1}(o) - 2\alpha_{j-i}(o) + \alpha_{j-i-1}(o) \tag{4.21}$$

More generally we would like to write

$$-2p_i\cdot p_j - C_{ij} = -2\hat{p}_i\cdot\hat{p}_j \tag{4.22}$$

where $\hat{p}_i$ is a (4 + D) dimensional vector. In such a case we may factorise simply by adding D extra dimensions to the harmonic operators.

Let us assume that [References 9-11]

$$\alpha_{ij}(o) = B + \sum_{k=i}^{j} d_k + \Big(\sum_{k=i}^{j} e_k\Big)^2 \tag{4.23}$$

$$= B + D_{ij} + E_{ij} \qquad (4.24)$$

where d_k, e_k depend on the internal quantum numbers of particle k.

[For example, we may take d_k, e_k proportional to the total quark numbers ($\nu = N_q + N_{\bar{q}}$), in which case the dependence must be at least quadratic to avoid stable high-mass exotics since the scalar states would have masses satisfying

$$M_\nu - M_{\nu-1} \underset{\nu\to\infty}{\to} 0 \qquad (4.25)$$

if $\alpha(o)$ depends only linearly on ν but this difference $(M_\nu - M_{\nu-1})$ can remain finite for $\nu \to \infty$ in the case of $\alpha(o)$ depending quadratically on ν].

Inserting our ansatz for $\alpha_{ij}(o)$ one finds

$$C_{ij} = 2e_i e_j \qquad (4.26)$$

Now certain consistency conditions must be met. Consider first that the number of external particles N be held fixed. Then there is the requirement that $\alpha_{ij}(o) = \alpha_{j-1,N}(o)$. In particular

$$\sum_{k=1}^{N-1} d_k + \left(\sum_{k=1}^{N-1} e_k\right)^2 = d_N + e_N^2 \qquad (4.27)$$

It follows that

$$d_N = -\sqrt{E_{1,N}}\, e_N + \tfrac{1}{2}(D_{1,N} + E_{1,N}) \qquad (4.28)$$

More generally we use

$$D_{1,j} = D_{1,N} - D_{j+1,N} \qquad (4.29)$$

$$E_{1,j} = E_{1,N} - 2\sqrt{E_{1,N}\, E_{j+1,N}} + E_{j+1,N} \qquad (4.30)$$

to find that the requirement

$$\alpha_{1j}(o) = D_{ij} + E_{1j} = \alpha_{j+1,N}(o) = D_{j+1,N} + E_{j+1,N} \tag{4.31}$$

implies

$$D_{1,N} - 2D_{j+1,N} + E_{1,N} - 2\sqrt{E_{1,N}\, E_{j+1,N}} = 0 \tag{4.32}$$

Combining the results we see that

$$D_{j+1,N-1} = -\sqrt{E_{1,N}\, E_{j+1,N-1}} \tag{4.33}$$

for all j. The most general solution of this is to take

$$d_k = -\sqrt{E_{1,N}}\, e_k \qquad 2 \le k \le (N-1) \tag{4.34}$$

Similarly we find

$$d_1 = \frac{1}{2} D_{1,N} + \frac{1}{2} E_{1,N} - \sqrt{E_{1,N}}\, e_1 \tag{4.35}$$

This leads to

$$\alpha_{ij} = B + D_{ij} + E_{ij} \tag{4.36}$$

$$= B - \sqrt{E_{ij}}\,(\sqrt{E_{1,N}} - \sqrt{E_{ij}}) +$$

$$+ \frac{1}{2}(D_{1,N} + E_{1,N})\,(\delta_{1i} + \delta_{jN}) \tag{4.37}$$

Because of the arbitrariness in the cycle labelling consistency now requires

$$D_{1,N} + E_{1,N} = 0 \tag{4.38}$$

whereupon

$$\alpha_{ij}(o) = B + \left(\sum_{k=i}^{j} e_k\right)^2 - \sqrt{E_{1,N}} \sum_{k=i}^{j} e_k \tag{4.39}$$

is a consistent possibility for fixed N.

When we vary N, however, one soon finds contradictions unless the quantity e_k is conserved, that is

$$\sqrt{E_{1,N}} = \sum_{k=1}^{N} e_k = 0 \tag{4.40}$$

We should add the remarks:

(i) Suppose that we take all external particles to be identical. Then we must have, say, $d_k = e_k = \pm 1$, and consistency is possible only for N even.

(ii) We may factorize the extra piece $2e_i e_j$ in $2\hat{p}_i \cdot \hat{p}_j$ by adding one additional (fifth) dimension. This increases the density of hadronic states $\rho(m)$ as a function of mass m. Recall that

$$\lim_{m\to\infty} \left[\frac{m}{\ell n\ \rho(m)}\right] = T_H(d) = \frac{\sqrt{6}}{2\pi} \frac{1}{\sqrt{\alpha' d}} \tag{4.41}$$

where T_H is the Hagedorn temperature and $d = (4 + D)$ is the dimensionality of oscillators. Putting $\alpha' = 1\ \mathrm{GeV}^{-2}$ one finds $T_H(4) = 195$ MeV, $T_H(5) = 174$ MeV. Both are close to that suggested by Hagedorn's fits to transverse momentum distributions; if one takes the numerology very seriously the extra dimension improves the agreement. [Reference 9].

4.4 THE RUBBER STRING MODEL

The operator formalism is suggestive of a quantum system with an infinite number of degrees of freedom, for example black-body radiation inside a cavity or a one-dimensional elastic continuum. It is by now traditional to discuss a massless rubber string. [References 12-16].

The discussion of the string, that is given in the present section, will be at such a level that we use the string simply to set up calculational rules (analogous to Feynman rules) that reproduce the generalised Euler B function. We shall here be non-rigorous about the method of embedding the string in Minkowski space-time; in a later section we shall give a more detailed treatment, using Riemannian geometry, of the two-dimensional world-sheet mapped out in space-time by the propogation of a one-dimensional string. The price paid for the higher rigour in the second discussion will be that, although we obtain a beautiful understanding of the level structure, we shall not deal at all with the (non-local) interactions. In the present less rigorous first discussion we shall include interactions and will already gain some insight into how our third and last internal symmetry, baryon-number, might be incorporated.

Consider a set of N mass points arranged along a one-dimensional line, and with nearest-neighbour harmonic-oscillator interactions. The classical energy will now be in the form of a sum of kinetic and potential terms

$$E = \frac{1}{2} \sum_{i=1}^{N} \dot{X}_\mu^{(i)} \dot{X}_\mu^{(i)} + \frac{1}{2} \sum_{i=1}^{N} |X_\mu^{(i)} - X_\mu^{(i-1)}|^2 \tag{4.42}$$

where $X_\mu^{(i)}$ is the position of the i th particle. Taking a continuum limit to a string of length π gives

$$E = \frac{1}{2} \int_0^\pi d\sigma (\dot{X}_\mu(\theta)\, \dot{X}_\mu(\theta) + \frac{\partial X_\mu}{\partial \sigma} \frac{\partial X_\mu}{\partial \sigma}) \tag{4.43}$$

The relevant equation of motion is the wave equation

$$\ddot{X}_\mu(\sigma) - \frac{\partial^2}{\partial \sigma^2} X_\mu(\sigma) = 0 \tag{4.44}$$

and the solution of this can be expanded in terms of normal modes by

$$X_\mu(\sigma) = \sum_{n=1}^{\infty} \frac{\sqrt{2}}{\sqrt{n}} (a_\mu^{(n)} + a_\mu^{(n)+}) \cos n\sigma \tag{4.45}$$

with $[a_\mu^{(m)}, a_\nu^{(n)+}] = - \delta_{mn} g_{\mu\nu}$ and where we have selected those solutions which respect the boundary condition

$$\frac{\partial X_\mu(\sigma)}{\partial \sigma} = 0 \quad \text{at } \sigma = 0, \pi \tag{4.46}$$

corresponding to the vanishing tension at the ends of the massless string.

We can now use this field $X_\mu(\sigma)$ to set up calculational rules, based on diagrams, that faithfully reproduce the generalised Euler B functions.

The calculational rules for the N-meson Born amplitude are [Reference 16]

(1) We take two of the external lines (chosen arbitrarily and interpret them as the incoming and outgoing rubber string ($\tau = -\infty$ and $\tau = +\infty$) which maps out the region $0 \leq \sigma \leq \pi$ and $-\infty \leq \tau \leq +\infty$ in the complex $W = (\tau + i\sigma)$ plane (see Figure 4.1).

(2) The remaining (N - 2) scalars are emitted and absorbed from the two sides $\sigma = 0$ and $\sigma = \pi$ of the basic infinite strip; one scalar must be at a fixed τ-value, say $\tau = 0$ but the remaining (N - 3) lines must be integrated over all possible τ - orderings which respect the cyclic ordering of the external lines.

(3A) For each emitted or absorbed line we write a factor $T_0(p)$ or $T_\pi(p)$ corresponding to $\sigma = 0$ or $\sigma = \pi$ respectively in

$$T_\sigma(p) = \exp(ip_\mu\, X_\mu(\sigma)) \tag{4.47}$$

(3B) Internal propogation of the rubber string is described by a factor (for unit intercept)

$$P(s = p_\mu p_\mu) = \frac{1}{R - s - 1} = i \int_0^\infty d\tau\; e^{i(L_0-1)\tau} \tag{4.48}$$

where p_μ is the total four momentum.

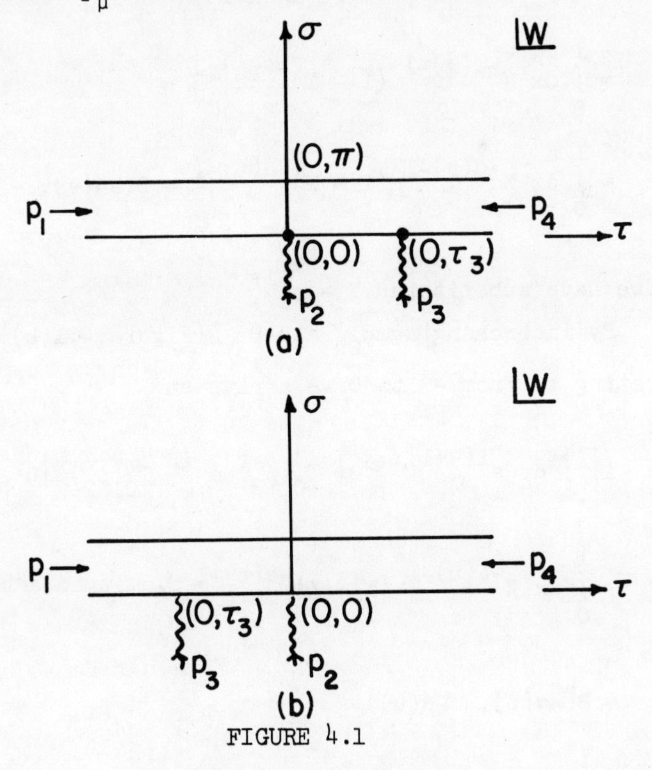

FIGURE 4.1

The st and tu Diagrams

Consider first the four-meson Born amplitude, in the configuration corresponding to Figure 4.1(a). We have the amplitude

$$i \int_0^\infty d\tau \; e^{-i(s+1)\tau} \langle 0| \; T_0(p_2) \; e^{iR} \; T_0(p_3) \; |0\rangle =$$

$$= \int_0^1 dx \; X^{-s-2} \langle 0| \exp(i\sqrt{2} \sum_{n=1}^{\infty} \frac{p_2 \cdot a^{(n)}}{\sqrt{n}}) \; X^R \; \cdot$$

$$\cdot \exp(i\sqrt{2} \sum_{n=1}^{\infty} \frac{p_3 \cdot a^{(n)+}}{\sqrt{n}}) |0\rangle \qquad (4.49)$$

$$= \int_0^1 dx \; X^{-\alpha(s)-1} \; (1 - X)^{-2p_2 \cdot p_3} \qquad (4.50)$$

$$= \int_0^1 dx \; X^{-\alpha(s)-1} \; (1 - X)^{-\alpha(t)-1} = B(-\alpha(s), \; -\alpha(t)) \qquad (4.51)$$

where we have substituted $X = e^{-i\tau}$.

By interchanging $P_{2\mu}$ and $P_{3\mu}$ (Figure 4.1(b)) and integrating τ_3 from $-\infty$ to 0 we arrive at

$$i \int_{-\infty}^0 d\tau \; e^{i(u+1)\tau} \langle 0| \; T_0(p_3) \; e^{-iR} \; T_0(p_2) \; |0\rangle =$$

$$= \int_0^1 dx \; X^{-\alpha(u)-1} \; (1 - X)^{-\alpha(t)-1} \qquad (4.52)$$

$$= B(-\alpha(t), \; -\alpha(u)) \qquad (4.53)$$

More interesting, since the direct connection to the operator factorisation is slighly less immediate, is to consider the $B(-\alpha(u), -\alpha(s))$ term which arises from the sum of two time orderings as depicted in Figure 4.2.

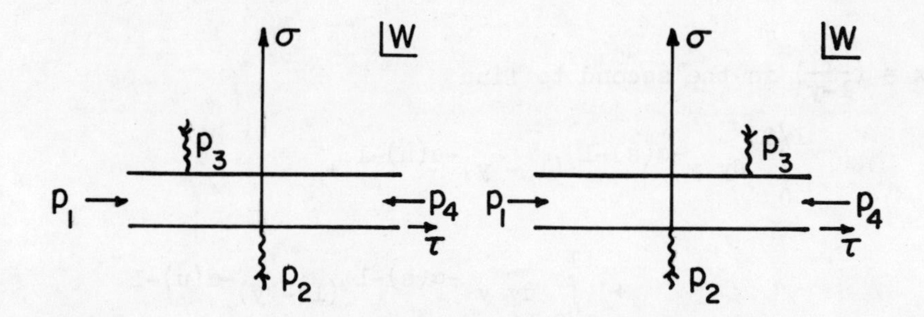

FIGURE 4.2

The us Diagram

The amplitude for this permutation is then

$$i \int_0^\infty d\tau \, [e^{-i\tau(u+1)} \langle 0| \, T_\pi(p_3) \, e^{iR\tau} \, T_0(p_2) \, |0\rangle +$$

$$+ e^{i\tau(s+1)} \langle 0| \, T_0(p_2) \, e^{iR\tau} \, T_\pi(p_3) |0\rangle] =$$

$$= \int_0^1 dX \, [x^{-u-2} \langle 0| \exp(i\sqrt{2} \sum_{n=1}^\infty \frac{p_3 \cdot a^{(n)} (-1)^n}{\sqrt{n}}) \, x^R \, \cdot$$

$$\cdot \exp(i\sqrt{2} \sum_{n=1}^\infty \frac{p_2 \cdot a^{(n)+}}{\sqrt{n}}) \, |0\rangle +$$

$$+ x^{-s-2} \langle 0| \exp(i\sqrt{2} \sum_{n=1}^\infty \frac{p_2 \cdot a^{(n)}}{\sqrt{n}}) \, x^R \, \cdot$$

$$\cdot \exp(i\sqrt{2} \sum_{n=1}^\infty \frac{p_3 \cdot a^{(n)}}{\sqrt{n}} (-1)^n) \, |0\rangle \qquad (4.54)$$

$$= \int_0^1 dx \, [x^{-\alpha(u)-1} (1 + x)^{-\alpha(t)-1} +$$

$$+ x^{-\alpha(s)-1} \, (1 + x)^{-\alpha(t)-1}] \qquad (4.55)$$

Now we combine these two terms by using $-\alpha(t) - 1 = \alpha(s) + \alpha(u)$ and substituting $x = (\frac{1-y}{y})$ in the first,

$x = (\frac{y}{1-y})$ in the second to find

$$\int_0^{1/2} dy\, y^{-\alpha(s)-1} (1 - y)^{-\alpha(u)-1} +$$

$$+ \int_{1/2}^{1} dy\, y^{-\alpha(s)-1} (1 - y)^{-\alpha(u)-1}$$

$$= B(-\alpha(u), -\alpha(s)). \tag{4.56}$$

Now that we understand that the two τ-orderings fit together in this way we may write formally a full vertex

$$T(p) = T_0(p) + T_\pi(p) \tag{4.57}$$

and then obtain the sum over all permutations by writing

$$\langle 0|\ (T(p_2)\ P(s)\ T(p_3) + T(p_3)\ P(u)\ T(p_2))\ |0\rangle =$$

$$= 2[B(-\alpha(s), -\alpha(t)) + B(-\alpha(t), -\alpha(u)) +$$

$$B(-\alpha(u), -\alpha(s))] \tag{4.58}$$

where the eight contributing diagrams are indicated in Figure 4.3.

Does this procedure continue to work in the multi-particle extensions? The answer is yes, it does. We shall give some details of the N = 5 case only, since it already gives a non-trivial combination over τ-order Let us keep particles 1 and 5 as the ends of the rubber strip; then there are 48 possible distributions of the remaining three lines (analogous to the 8 of Figure 4.3).

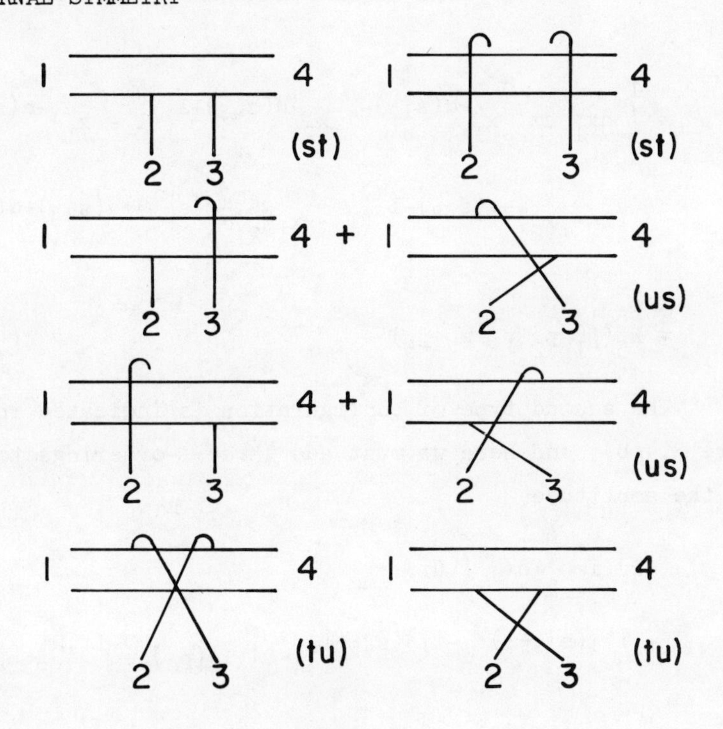

FIGURE 4.3

Eight Inequivalent Four-Meson Diagrams

These fall into two basic types, the first of which is exemplified by Figure 4.4 (a), whose amplitude is

$$(-i)\ i \int_{-\infty}^{0} d\tau_1\ e^{-i\tau_1(s_{12}+1)} \int_{0}^{\infty} d\tau_2\ e^{-i\tau_2(s_{45}+1)}$$

$$\langle 0|\ T_0(p_2)\ e^{i\tau_1 R}\ T_0(p_3)\ e^{i\tau_2 R}\ T_0(p_4)\ |0\rangle =$$

$$= \int_0^1 dx_1\ dx_2\ x_1^{-s_{12}-1}\ x_2^{-s_{45}-1} \langle 0|\ e^{i\sqrt{2}\ p_2 \sum \frac{a^{(n)}}{\sqrt{n}}} \cdot$$

$$\cdot\ x_1^{R}\ e^{i\sqrt{2}\ p_3 \sum \frac{a^{(n)+}}{\sqrt{n}}}\ e^{i\sqrt{2}\ p_3 \sum \frac{a^{(n)}}{\sqrt{n}}} \cdot$$

$$\cdot\ x_2^{R}\ e^{i\sqrt{2}\ p_4 \sum \frac{a^{(n)+}}{\sqrt{n}}}\ |0\rangle \qquad (4.59)$$

$$= \int_0^1 dx_1\, dx_2\, x_1^{-\alpha(s_{12})-1}\, x_2^{-\alpha(s_{45})-1}\, (1-x_1)^{-\alpha(s_{23})-1} \cdot$$

$$\cdot (1-x_2)^{-\alpha(s_{34})-1}\, (1-x_1x_2)^{-\alpha(s_{51})+\alpha(s_{23})+\alpha(s_{34})} \tag{4.60}$$

$$= B_5(p_1\, p_2\, p_3\, p_4\, p_5) \tag{4.61}$$

The second type of configuration is indicated in Figure 3.4(b), and here we must add three τ-orderings to give the amplitude

$$(-i)\, i \int_{-\infty}^{0} d\tau_1 \int_0^{\infty} d\tau_2 \cdot$$

$$\cdot [e^{i\tau_1(s_{12}+1)}\, e^{-i\tau_2(s_{45}+1)} \langle 0|\, T_\pi(p_2)\, e^{-i\tau_1 R} \cdot$$

$$\cdot T_0(p_3)\, e^{i\tau_2 R}\, T_0(p_4)\, |0\rangle +$$

$$+ e^{i\tau_1(s_{13}+1)}\, e^{-i\tau_2(s_{45}+1)} \langle 0|\, T_0(p_3)\, e^{-i\tau_1 R} \cdot$$

$$\cdot T_\pi(p_2)\, e^{i\tau_2 R}\, T_0(p_4)\, |0\rangle +$$

$$+ e^{i\tau_1(s_{13}+1)}\, e^{-i\tau_2(s_{23}+1)} \langle 0|\, T_0(p_3)\, e^{-i\tau_1 R} \cdot$$

$$\cdot T_0(p_4)\, e^{i\tau_2 R}\, T_\pi(p_2)\, |0\rangle] =$$

$$= \int_0^1 dx_1\, dx_2\, [x_1^{-\alpha_{12}-1}\, x_2^{-\alpha_{45}-1} (1+x_1)^{-2p_2\cdot p_3} \cdot$$

$$\cdot (1-x_2)^{-2p_3\cdot p_4}\, (1+x_1x_2)^{-2p_3\cdot p_4} +$$

$$+ x_1^{-\alpha_{13}-1}\, x_2^{-\alpha_{45}-1}\, (1+x_1)^{-2p_2\cdot p_3}\, (1+x_2)^{-2p_2\cdot p_4} \cdot$$

$$\cdot (1-x_1x_2)^{-2p_3\cdot p_4} +$$

$$+ x_1^{-\alpha_{13}-1} x_2^{-\alpha_{25}-1}(1 - x_1)^{-2p_3 \cdot p_4} (1 + x_2)^{-2p_2 \cdot p_4} \cdot$$

$$\cdot (1 + x_1 x_2)^{-2p_2 \cdot p_3}] \qquad (4.62)$$

Now we use (for unit intercept)

$$-2p_2 \cdot p_3 = -\alpha_{45} + \alpha_{12} + \alpha_{13} \qquad (4.63)$$

$$-2p_3 \cdot p_4 = -\alpha_{34} - 1 \qquad (4.64)$$

$$-2p_2 \cdot p_4 = -\alpha_{13} + \alpha_{25} + \alpha_{45}. \qquad (4.65)$$

and we make the changes of variables denoted by

$$\left.\begin{aligned} u_{12} &= \frac{x_1}{1 + x_1} && (4.66a) \\ u_{45} &= \frac{x_2(1 + x_1)}{1 + x_1 x_2} && (4.66b) \end{aligned}\right\}$$

$$\left.\begin{aligned} u_{12} &= (1 + x_1)^{-1} && (4.67a) \\ u_{45} &= \frac{x_2(1 + x_1)}{(1 + x_2)} && (4.67b) \end{aligned}\right\}$$

$$\left.\begin{aligned} u_{12} &= (1 + x_1 x_2)^{-1} && (4.68a) \\ u_{45} &= \frac{1 + x_1 x_2}{1 + x_2} && (4.68b) \end{aligned}\right\}$$

respectively in the three terms of the amplitude. After a little algebra one now finds that the three pieces fit together precisely to give the full integration region in

$$\int_0^1 du_{12} \int_0^1 du_{45}\, u_{12}^{-\alpha_{12}-1} (1 - u_{12})^{-\alpha_{13}-1} (1 - u_{45})^{-\alpha_{34}-1}$$

$$u_{45}^{-\alpha_{45}-1} (1 - u_{12}u_{45})^{-\alpha_{52}+\alpha_{13}+\alpha_{34}} \qquad (4.69)$$

$$= B_5(p_2\ p_1\ p_3\ p_4\ p_5) \tag{4.70}$$

as required. The three terms give respectively the three domains of integration

$$0 \le u_{12} \le \frac{1}{2}\ ;\ 0 \le u_{45} \le 1 \tag{4.71}$$

$$\frac{1}{2} \le u_{12} \le 1\ ;\ 0 \le u_{45} \le 1\ ;\ u_{12}u_{45} \le \frac{1}{2} \tag{4.72}$$

$$\frac{1}{2} \le u_{12} \le 1\ ;\ \frac{1}{2} \le u_{45} \le 1\ ;\ u_{12}u_{45} \ge \frac{1}{2} \tag{4.73}$$

which combine nicely into the unit square $0 \le u_{12}, u_{45} \le 1$.

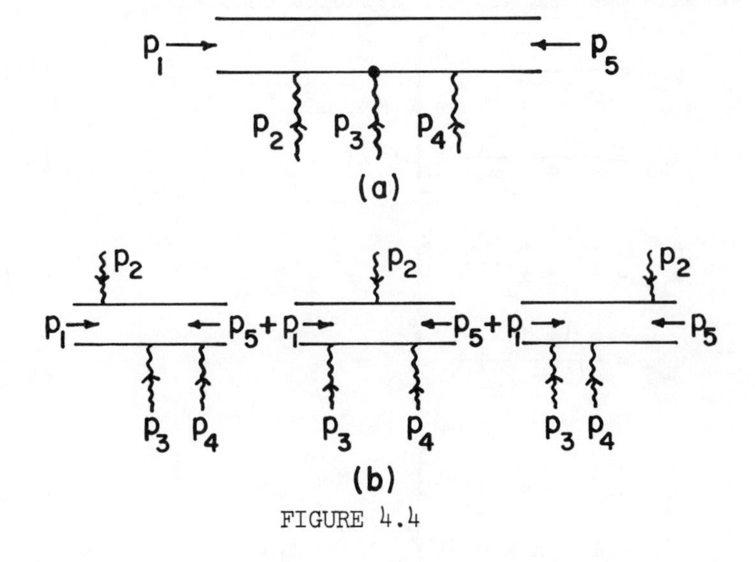

FIGURE 4.4

Five-Point Functions

By repeated application of the two basic calculations (Figure 4.4(a) and Figure 4.4(b)) one finds that the 48 string diagrams add together to give each of the 12 inequivalent permutations added with equal coefficients (the coefficients are all equal to 2).

A similar result presumably obtains for the generalisation to N external particles. (See Reference 16, where such a conjecture is made).

We should now add some general remarks.

(i) As mentioned at the beginning, these rubber string ideas as presented so far are no more than calculational rules, and we have not directly confronted the problem of how the string is embedded in Minkowski space-time. We give a more satisfactory treatment of this point later.

(ii) We have treated two external lines asymmetrically, and have not needed to exhibit any rubber string character for the remaining (N-2) ground state particles. The final result, however, was always fully cyclic symmetric and clearly independent of the particular choice of external state to treat preferentially.

(iii) Everywhere we used unit intercept $\alpha(o) = 1$, since the calculational rules are most simple for this case. One can, if necessary, modify the rules in a somewhat ad hoc manner (analogous to adding a fifth dimension) to deal with the cases $\alpha(o) \neq 1$ [for details, see Frye, Reference 17].

(iv) The way in which the different τ-orderings combine neatly together becomes less surprising when we note that the boundary of the infinite strip is nothing more than a re-mapping of the Koba-Nielsen circle. To be explicit, suppose we consider the unit circle in a Z-plane with $Z_1 = 1$, $Z_{n-1} = i$, $Z_N = -1$ and $Z_i = e^{i\theta_i}$ $(2 \leq i \leq N-2,\ 0 \leq Z_i \leq \frac{\pi}{2})$ together with its mapping into an x-plane with $x_1 = 0$, $x_{N-1} = 1$ and $x_N = \infty$ $(0 \leq x_i \leq 1$ (real); $2 \leq i \leq N-2)$. Now we map these into a W-plane strip with $W_1 = -\infty$, $W_{N-1} = +\infty$, $W_N = i\pi$. The appropriate transformations are found to be

$$\left.\begin{aligned} x &= \frac{i(1 - Z)}{(1 + Z)} && (4.74a) \\ Z &= \frac{i - x}{i + x} && (4.74b) \end{aligned}\right\}$$

$$\left.\begin{aligned} W &= \ln\left(\frac{x}{1 - x}\right) && (4.75a) \\ x &= \frac{e^W}{1 + e^W} && (4.75b) \end{aligned}\right\}$$

$$\left.\begin{aligned} Z &= \frac{1 + e^W(1 + i)}{1 + e^W(1 - i)} && (4.76a) \\ W &= \ln\left(\frac{i(1 - Z)}{1 - i) + Z(1 + i)}\right) && (4.76b) \end{aligned}\right\}$$

With these transformations in mind, we can understand that the tortuous summations over τ-orderings are no more than the integration of Koba-Nielsen variables on the unit circle (or real axis) suitably respecting the cyclic ordering of the external particles.

(v) We may make duality transformations directly on the strip. For example in Figure 4.5(a) we may transform from the first to the second configurations by the strip-preserving transformation.

$$W' = \ln(1 - e^W) \qquad (4.77)$$

It is, of course, very tempting to identify these string diagrams directly with Harari-Rosner diagrams. One edge of the strip is now a quark, and the other is an antiquark, as indicated by the arrows in Figure 4.5. The duality transformation we have just made is then interpreted as proceeding from the s-channel representation to the t-channel

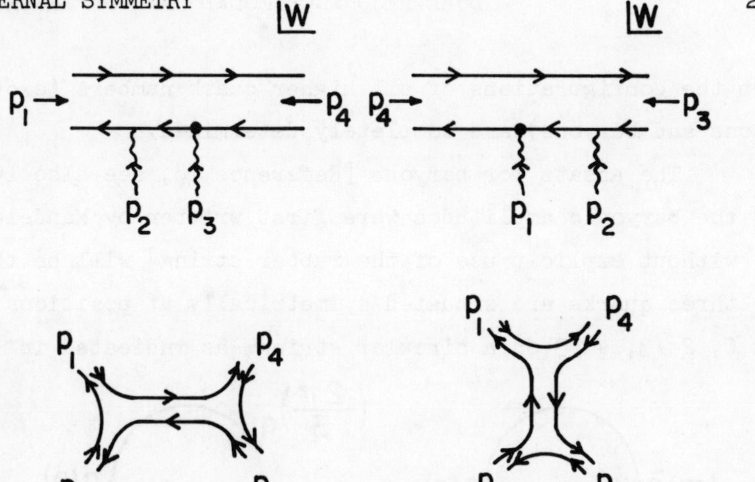

FIGURE 4.5

Duality Transformation

representation, equal by duality, as indicated in Figure 4.5(b).

4.5 BARYONS AND EXOTICS

When we have recognized a correspondence between the strip diagrams and the Harari-Rosner diagrams, a natural question to ask is: can we exploit this connection to arrive at an amplitude for meson-baryon scattering? Of course, at the present level we ignore the presence of half-integer spins, and consider spinless ground-state baryons. Baryon number will be incorporated only in the sense that baryons contain three "quarks", while mesons contain only two.

We shall show that once an ansatz has been made for the configuration of quarks on the rubber band for baryons, then the configurations of all higher quark numbers (exotic mesons and baryons) are completely determined.

The ansatz for baryons [Reference 18, see also 19, 20; the baryonic amplitudes were first written by Mandelstam, 21, without explicit use of the rubber string] will be that the three quarks are situated symmetrically at positions $\sigma = 0, 2\pi/3, 4\pi/3$ on a circular string, as indicated in

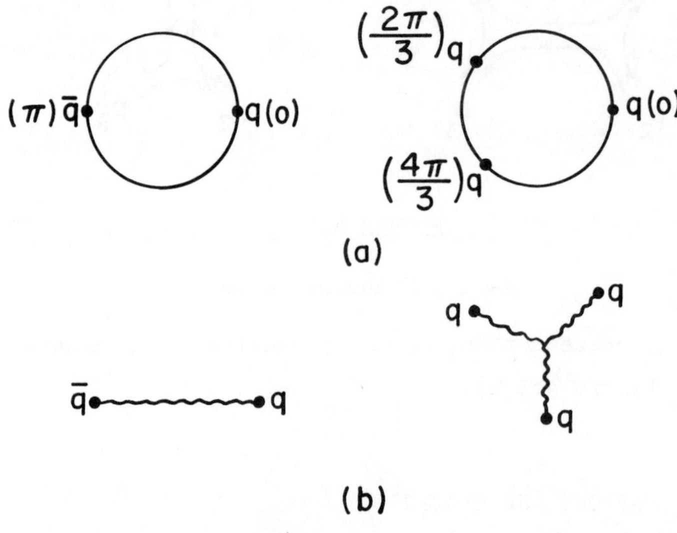

FIGURE 4.6

Meson and Baryon Models

Figure 4.6(a). We should emphasise that this is only an assumption, taken as one of the simplest generalisations of the mesonic case.

We first consider the diagram of Figure 4.7(a), for meson-baryon scattering. The baryon strip carries quarks at $\sigma = 0, 2\pi/3, 4\pi/3$. We use a field containing both cosine and sine modes of the form

$$x_\mu(\sigma,\tau) = \sum_{n=1}^{\infty} \frac{\sqrt{2}}{\sqrt{n}} [a_\mu^{(n)} e^{-n\tau} + a_\mu^{(n)+} e^{n\tau}) \cos n\sigma + + (b_\mu^{(n)} e^{-n\tau} + b_\mu^{(n)+} e^{n\tau}) \sin n\sigma] \quad (4.78)$$

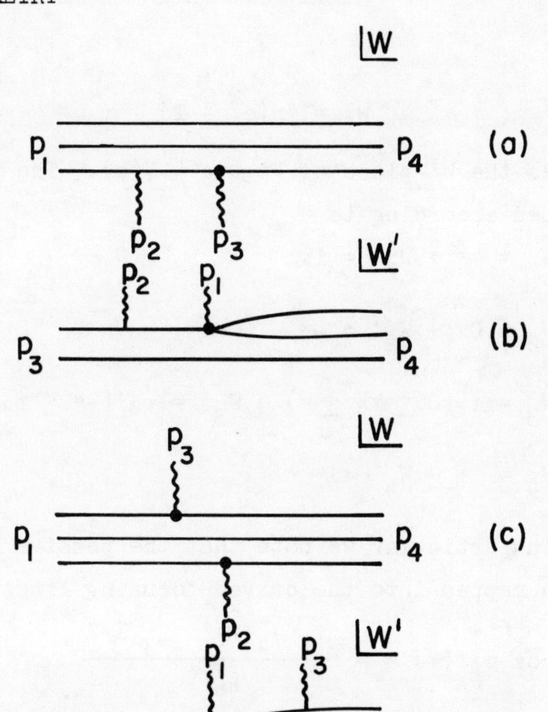

FIGURE 4.7

Baryon Emission Vertex

and then the amplitude is given by

$$\int_0^\infty d\tau_3 \langle 0|\, e^{ip_2 \cdot X(0,0)}\, e^{ip_3 \cdot X(0,\tau_3)}\, |0\rangle =$$

$$= B(-\alpha_s, -\alpha_t) \qquad (4.79)$$

That is, we obtain the Euler B function. This is because the calculation is identical to that for the mesonic diagram of Figure 4.1(a).

We may now look for the vertex to describe baryon emission,[19] by making the strip-preserving transformation.

$$W' = \ell n(e^W - 1) \tag{4.80}$$

This gives the W'-plane of Figure 4.7(b). The points have been mapped according to

$$W_1 = -\infty \rightarrow W_1' = i\pi \tag{4.81}$$

$$W_2 = 0 \rightarrow W_2' = -\infty \tag{4.82}$$

$$W_3 = \tau_3(0 \leq \tau_3 \leq \infty) \rightarrow W_3' = \tau_3'(-\infty \leq \tau_3' \leq \infty) \tag{4.83}$$

$$W_4 = \infty \rightarrow W_4' = \infty \tag{4.84}$$

In particular we note that the passive quark lines have been mapped into the curved focusing lines $C_\pm$ given by

$$\cos \sigma_\pm'(\tau) = -\frac{3 + (4e^{2\tau'} - 3)^{1/2}}{4e^{\tau'}} \tag{4.85}$$

selected by

$$\sigma_\pm'(\infty) = \pi \pm \frac{1}{3}\pi \tag{4.86}$$

for the original lines $\sigma_\pm = \pi \pm \frac{1}{3}\pi$ respectively.
To check this write

$$W_\pm = \pi i \pm \frac{i\pi}{3} + \tau \tag{4.87}$$

whereupon

$$W_\pm' = \ell n[-\frac{1}{2} \mp \frac{i\sqrt{3}}{2}) e^\tau - 1] \tag{4.88}$$

$$e^{\tau_\pm' + i\sigma_\pm'} = e^\tau(-\frac{1}{2} \mp i\frac{\sqrt{3}}{2}) - 1 \tag{4.89}$$

$$e^{2\tau_\pm'} = 1 + e^\tau + e^{2\tau} \tag{4.90}$$

$$e^\tau = -\frac{1}{2} \pm \frac{1}{2}\sqrt{4e^{2\tau_\pm'} - 3} \tag{4.91}$$

$$\cos \sigma_{\pm}' = e^{-\tau_{\pm}'} \left(-\frac{1}{2} e^{\tau} - 1\right) \tag{4.92}$$

$$= -\left[\frac{3 \pm \sqrt{4e^{2\tau_{\pm}'} - 3}}{4e^{\tau_{\pm}'}}\right] \tag{4.93}$$

as required.

To ensure that the calculation of the diagram Figure 4.7(b) gives the same result, namely $B(-\alpha_s, -\alpha_t)$, as Figure 4.7(a) we deduce that the baryon vertex is given by

$$e^{ip_1 \cdot x(\sigma,\tau)} \tag{4.94}$$

precisely as for the mesonic case. This deduction will enable us to proceed to more complicated processes with more baryonic external lines.

First we must discuss the other permutations of the meson-baryon scattering case. The tu-term (1324) is obtained simply by interchanging $p_2 \leftrightarrow p_3$ in the st-diagram (1234) and clearly gives $B(-\alpha_t, -\alpha_u)$, again an Euler B function. The us-term, Figure 4.7(c), is more interesting, since in this case baryons in the u-channel build up baryons in the s-channel by duality, and we expect the new (sine) modes in the baryon to be excited. Indeed the calculation gives now

$$\int_{-\infty}^{0} d\tau_3 <0| \, e^{ip_3 \cdot X(\theta,\tau_3)} \, e^{ip_2 \cdot X(0,0)} \, |0> \; +$$

$$+ \int_{0}^{\infty} d\tau_3 <0| \, e^{ip_2 \cdot X(0,0)} \, e^{ip_3 \cdot X(\theta,\tau_3)} \, |0> \tag{4.95}$$

with $\theta = 4\pi/3$. We will, however, leave θ free to see how the general four-point function depends on the angular separation of the active quarks (for $\theta = \pi$ we should regain the Euler B function, for example.)

The amplitude becomes

$$\int_0^1 dx\, x^{-\alpha_u - 1} \langle 0|\, \exp(\sqrt{2}\, ip_3 \cdot \sum_{n=1}^{\infty} \frac{a^{(n)}}{\sqrt{n}} \cos n\theta)\, x^R$$

$$\exp(\sqrt{2}\, ip_2 \cdot \sum_{n=1}^{\infty} \frac{a^{(n)+}}{\sqrt{n}})\, |0\rangle + (\alpha_s \leftrightarrow \alpha_u)$$

$$= \int_0^1 dx\, x^{-\alpha_u - 1} \exp[p_2 \cdot p_3 \sum_{n=1}^{\infty} \frac{x^n}{n}(e^{in\theta} - e^{-in\theta})] +$$

$$+ (\alpha_s \leftrightarrow \alpha_u) \tag{4.96}$$

$$= \int_0^1 dx\, x^{-\alpha_u - 1} (1 - 2x\cos\theta + x^2)^{-\frac{1}{2}(1+\alpha_t)} +$$

$$+ (\alpha_s \leftrightarrow \alpha_u) \tag{4.97}$$

Now make the change of variables

$$u = (1 - 2x\cos\theta + x^2)^{-\frac{1}{2}} \tag{4.98}$$

$$v = xu \tag{4.99}$$

and use $\alpha_s + \alpha_t + \alpha_u = -1$ (for unit intercepts) to rewrite the amplitude

$$\int_0^{\frac{1}{2\sin\theta}} \frac{du}{v - u\cos\theta}\, u^{-\alpha_s - 1}\, v^{-\alpha_u - 1} +$$

$$+ \int_{\frac{1}{2\sin\theta}}^1 \frac{dv}{u - v\cos\theta}\, v^{-\alpha_s - 1}\, u^{-\alpha_u - 1} =$$

$$= \int_0^1 \frac{du}{v - u\cos\theta} u^{-\alpha_s - 1} v^{-\alpha_u - 1} \tag{4.100}$$

$$= 2 \int_0^1 du \int_0^1 dv\, u^{-\alpha_s - 1} v^{-\alpha_u - 1} \delta(u^2 + v^2 - 2uv\cos\theta - 1) \tag{4.101}$$

where we have noticed that

$$u^2 + v^2 - 2uv\cos\theta = 1 \tag{4.102}$$

If we now make the change of variable

$$u = Z[Z^2 + (1 - Z)^2 - 2Z(1 - Z)\cos\theta]^{1/2} \tag{4.103}$$

we may re-write the amplitude

$$\int_0^1 dZ\, Z^{-\alpha_s - 1} (1 - Z)^{-\alpha_u - 1} [1 - 4Z(1 - Z)\cos^2\tfrac{\theta}{2}]^{\frac{\alpha_s + \alpha_u}{2}} \tag{4.104}$$

so that in addition to the usual branch points of the integrand at $Z = 0, 1, \infty$ there are new branch points situated at

$$Z = \frac{1}{2} \pm \frac{1}{2} i \left(\sec^2 \frac{\theta}{2} - 1\right)^{\frac{1}{2}} \tag{4.105}$$

In particular, for the specific case in which we are interested the meson-baryon amplitude, with $\theta = 4\pi/3$, becomes

$$\int_0^1 dx\, x^{-\alpha_s - 1} (1 - x)^{-\alpha_u - 1} (1 - x + x^2)^{\frac{\alpha_s + \alpha_u}{2}} \tag{4.106}$$

This amplitude was first written by Mandelstam[21], without the use of the rubber string picture.

To check the consistency of the approach we may re-calculate the (us)-diagram by using the configuration of Figure 4.7(d) (see page 209). This is obtained from Figure 4.7(c) by the strip preserving duality transformation

$$W' = \ell n(e^{W} - 1) \tag{4.107}$$

The interesting feature now is that in the W'-plane one meson (with momentum $p_{3\mu}$) is emitted from a focusing curve C_+. The appropriate amplitude is

$$i \int_0^\infty d\tau_3 \left[1 + \left(\frac{d\sigma_+'(\tau_3')}{d\tau_3'}\right)^2\right]^{1/2}$$

$$\langle 0| e^{ip_1 \cdot x(0,\pi)} e^{ip_3 \cdot x(\tau_3', \sigma_+'(\tau_3'))} |0\rangle \tag{4.108}$$

with $\sigma_+'(\tau_3)$ given by the explicit equation for C_+ derived above.

The vacuum expectation value becomes (with $x = e^{-\tau_3}$)

$$\langle 0| e^{i\sqrt{2}\, p_1 \sum \frac{a^{(n)}(-1)^n}{\sqrt{n}}} x^R e^{i\sqrt{2}\, p_3 \sum \frac{a^{(n)+}}{\sqrt{n}} \cos n\sigma_+} |0\rangle$$

$$= \exp[p_1 \cdot p_3 \sum_{n=1}^{\infty} \frac{x^n}{n} (-1)^n (e^{in\sigma_+} + e^{-in\sigma_+})] \tag{4.109}$$

$$= [(1 + x\, e^{i\sigma_+})(1 + x\, e^{-i\sigma_+})]^{p_1 \cdot p_3} \tag{4.110}$$

$$= (1 + 2x \cos \sigma_+ + x\,)^{p_1 \cdot p_3} \tag{4.111}$$

$$= (1 - \frac{1}{2} x^2 - \frac{x}{2}\sqrt{4 - 3x^2}\,)^{p_1 \cdot p_3} \tag{4.112}$$

$$= (-\frac{1}{2} x + \frac{1}{2} \sqrt{4 - 3x^2})^{2p_1 \cdot p_3} \tag{4.113}$$

where in the last step we simply notice that

$$(-\frac{1}{2} x + \frac{1}{2} \sqrt{4 - 3x^2})^2 = 1 - \frac{1}{2} x^2 - \frac{x}{2} \sqrt{4 - 3x^2} \tag{4.114}$$

To calculate the arc length note that

$$\sin \sigma_+' = \frac{\sqrt{3}}{4} (e^{-\tau_3'} \pm e^{-\tau_3'} \sqrt{4e^{2\tau_3'} - 3}\,) \tag{4.115}$$

and hence

$$\cos\sigma_+' \frac{d\sigma_+'}{d\tau_3'} = \frac{\sqrt{3}}{4\sqrt{4e^{2\tau_3'} - 3}} \cdot$$

$$\cdot \left[3e^{-\tau_3'} \pm e^{-\tau_3'} \sqrt{4e^{2\tau_3'} - 3} \right] \quad (4.116)$$

$$= \frac{\sqrt{3}}{\sqrt{4e^{2\tau_3'} - 3}} \cos\sigma_+' \quad (4.117)$$

whereupon

$$\left[1 + \left(\frac{d\sigma_+'}{d\tau_3'}\right)^2\right]^{1/2} = \frac{2}{\sqrt{4 - 3x}} \quad (4.118)$$

Collecting results the amplitude is

$$2\int_0^1 \frac{dx}{\sqrt{4 - 3x^2}} x^{-\alpha_s - 1} \left(-\frac{1}{2}x + \frac{1}{2}\sqrt{4 - 3x^2}\right)^{-\alpha_u - 1} \quad (4.119)$$

Putting

$y = (-\frac{1}{2}x + \frac{1}{2}\sqrt{4 - 3x^2})$ we notice that

$$x^2 + y^2 + xy = 1 \quad (4.120)$$

$$x + 2y = \sqrt{4 - 3x^2} \quad (4.121)$$

and hence the original (us)-amplitude may be regained, namely

$$2\int_0^1 dx \int_0^1 dy\, x^{-\alpha_s - 1} y^{-\alpha_u - 1} \delta(x^2 + y^2 + xy - 1) \quad (4.122)$$

This confirms that the pictorial idea of two quarks focusing to a point vertex is consistent at the four-point level.

Knowledge of the baryon emission vertex enables us to generalize further to include more baryon external lines, whereupon (as is well known) exotic intermediate states become essential. Here we can directly investigate how the extra quarks of such states should be situated on the rubber string. [Reference 22].

The amplitude for baryon-antibaryon scattering is indicated by Figure 4.8(a) in the W-plane. By making the

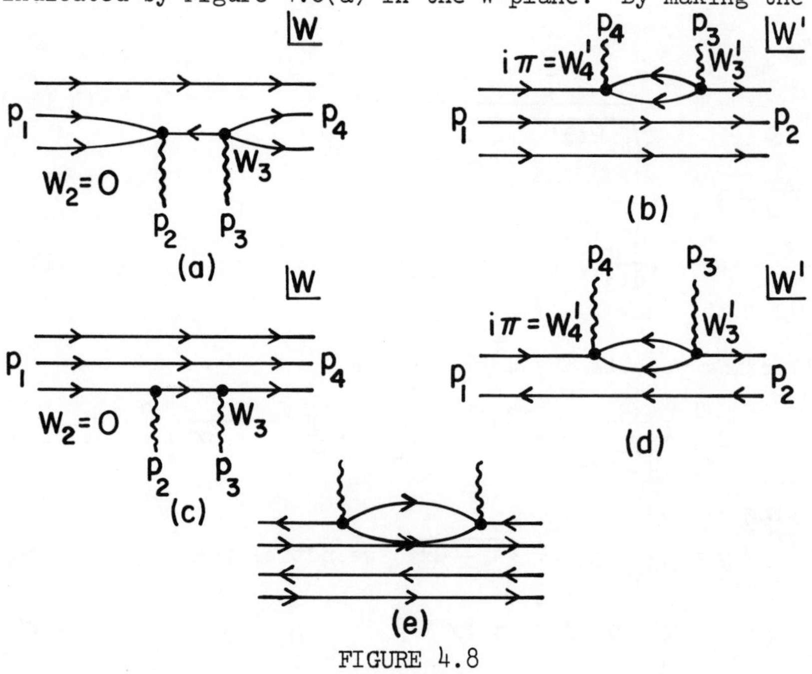

FIGURE 4.8

Baryons and Exotics

strip-preserving duality transformation

$$W' = \ln(e^{-W} - 1)^{-1} \tag{4.123}$$

we arrive at the configuration of Figure 4.8(b) which exhibits an exotic meson in the t-channel. The t-channel pole arises from $(\tau_3' - \tau_4') \to \infty$ corresponding to $(\tau_3 - \tau_2) \to 0$ and so we are interested in the asymmetric positions of the two quarks and two antiquarks in the intermediate state. It is not necessary to make an an explicit calculation (the precise equations are, however, contained in reference 22). The required information can be obtained by comparison to Figure 4.8(c), 4.8(d) for meson-baryon scattering where a similar quark "loop" occurs. The comparison reveals that the quarks are at $\sigma = 0, 2\pi/3$ and the antiquarks at $\sigma = \pi, 5\pi/3$. By scattering these exotic mesons, now as external legs, off of baryons (Figure 4.8(e)) we arrive at exotic baryon intermediate states with four quarks at $\sigma = 0$ (two), $2\pi/3$, $4\pi/3$ plus an antiquark at $\sigma = \pi$.

Following such an iterative approach one finds that the following rule (c f. Figure 4.9(a)).

All quarks are placed at the vertices of one equilateral triangle (T_1) of a regular starred hexagon inscribed into the circular rubber band, and all antiquarks are placed on the other equilateral triangle (T_2). There are the restrictions that for each $q(\bar{q})$ at a vertex of $T_1(T_2)$ there must appear <u>either</u> one $q(\bar{q})$ each at the other two vertices of $T_1(T_2)$ <u>or</u> one $\bar{q}(q)$ at the diametrically opposed vertex of $T_2(T_1)$.

Examples for total quark numbers four and five are given in Figures 4.9(b) and 4.9(c) respectively. (Page 218).

We should add two remarks:

(a) Once we have agreed upon the quark configuration $\sigma = 0, 2\pi/3, 4\pi/3$ for baryons, the above Star of David rule followed. Nevertheless it is quite possible to envisage other generalisations from two to three quarks such as that indicated schematically in Figure 4.6(b) (see page 208).

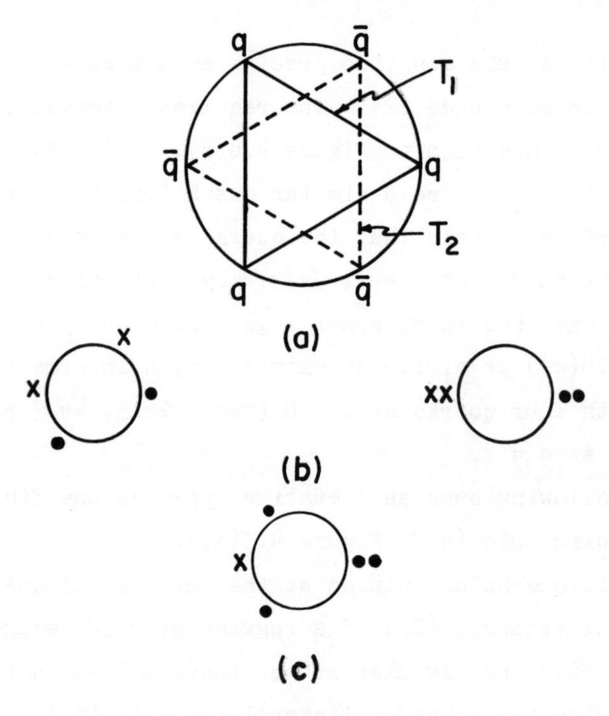

FIGURE 4.9

Star Of David Rule

(b) It is not possible to use a simple multiplicative factor for baryon number similar to that which we have discussed earlier for isospin and hypercharge. To see this point, consider the non-planar tree

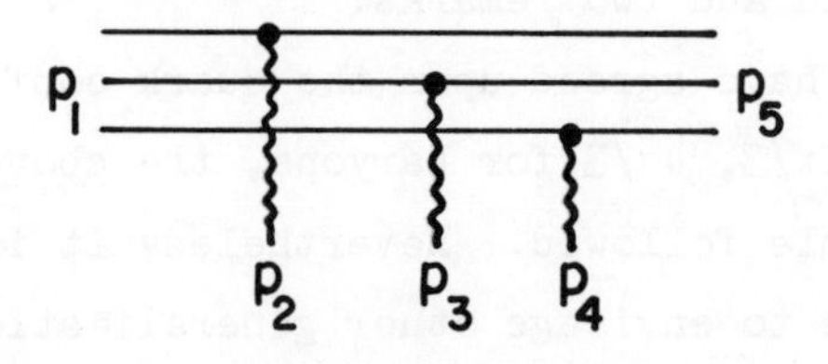

FIGURE 4.10

Non-planar Tree Amplitude

diagram of Figure 4.10 (which must be present for consistent factorisation); this is a five point function with poles in six non-planar, rather than five planar, channels and clearly therefore cannot be adequately described by the mesonic B_5. Instead we must make a more essential modification as we have done above.

4.6 MORE ON THE RUBBER STRING

We shall now give a more mathematically rigorous treatment of the rubber string model. In particular we shall be concerned with how the string is embedded into Minkowski space-time. This procedure might ultimately lead to a much simpler reformulation of the dual models. For the moment, it only provides us with a useful, different viewpoint of the spectrum of physical states.

The underlying idea [of Nambu, Reference 23; also References 24-28] is to take as the action of the freely propogating rubber string the area of the world sheet mapped out in space time. Defining two coordinates $\xi_1 = \sigma$, $\xi_2 = \tau$ to parametrise the sheet we write

$$S = \frac{1}{2\pi}\int_0^\pi d\sigma \int_{\tau_i}^{\tau_f} d\tau \sqrt{-g} \qquad (4.124)$$

with

$$g = ||g_{ab}|| \qquad (4.125)$$

$$g_{ab} = g_{\mu\nu}\frac{\partial x^\mu}{\partial \xi_a}\frac{\partial x^\nu}{\partial \xi_b} \qquad (4.126)$$

and $x^\mu(\sigma, \tau)$ is the space-time position regarded as a field on variables σ, τ.

We introduce the notation

$$\dot{x}_\mu = \frac{\partial x_\mu}{\partial \tau} \tag{4.127}$$

$$x_\mu{}' = \frac{\partial x_\mu}{\partial \sigma} \tag{4.128}$$

whereupon

$$S = \int_0^\pi d\sigma \int_{\tau_i}^{\tau_f} d\tau \, L \tag{4.129}$$

$$= \frac{1}{2\pi} \int_0^\pi d\sigma \int_{\tau_i}^{\tau_f} d\tau \, \sqrt{(\dot{x}\cdot x')^2 - \dot{x}^2 x'^2} \tag{4.130}$$

To see that this is the area mapped out by the world sheet consider the infinitesimal

$$dx \;\; = \dot{x}^\mu \, d\tau + x'^\mu \, d\sigma \tag{4.131}$$

The corresponding area element is

$$dA = [-\,(dx^\mu \wedge dx^\nu)\,(dx^\mu \wedge dx^\nu)]^{1/2} \tag{4.132}$$

where

$$dx^\mu \wedge dx^\nu = d\sigma \;\, d\tau[\dot{x}^\mu x'^\nu - \dot{x}^\nu x'^\mu] \tag{4.133}$$

and hence

$$dA = d\sigma \;\, d\tau[(\dot{x}\cdot x')^2 - \dot{x}^2 x'^2]^{1/2} \tag{4.134}$$

as required.

This action satisfies

i) Poincare invariance.

ii) General covariance under arbitrary redefinition of the coordinates σ, τ. This follows because the action has been constructed with $\sqrt{-g}$ the inner metric of the sheet. It is also physically clear because the area cannot depend on the choice of parametrisation. This general covariance will be shown to be the origin of the generalized projective gauge conditions. Clearly the form of the action is closely analogous to the use of $\sqrt{-\bar{g}}$, $\bar{g} = |g^{\mu\nu}|$ in general relativity (e.g. Reference 29); note that probably no clearer exposition of the latter exists than the original by Einstein [Reference 30 translated and reprinted in Reference 31] in what is perhaps the most celebrated article contributed yet to theoretical physics since Newton. From the point of view of general covariance in the case of the two-dimensional sheet, we might equally consider using the outer metrics

$$b^{\alpha}_{ab} = g_{\mu\nu} \frac{\partial \hat{n}_{\alpha}^{\ \mu} \ \partial x^{\nu}}{\partial \xi^{a} \ \partial \xi^{b}} \tag{4.135}$$

where $\hat{n}_{\alpha}^{\ \mu}$ ($\alpha = 1, 2$) are unit vectors normal to the sheet. However, the action we have written is the simplest possibility and it possesses the third that iii) it gives rise to equations of motion of second order

Concerning the general covariance (ii), it is here very useful to bear in mind the simpler situation of a point particle propogating in space-time. To make a manifestly covariant description there we may introduce a four component space-time position $x_{\mu}(\tau)$ regarded as a field on τ and then introduce an action

$$S = \int_{\tau_i}^{\tau_f} d\tau \ \sqrt{\dot{x}^2} \tag{4.136}$$

which is the length of the world-line. This action is covariant under general coordinate changes $\tau \to f(\tau)$; physically this corresponds to the fact that only three independent dynamical variables exist since for example $x_i(\tau = x_0 = t)$, $i = 1, 2, 3$, completely specifies the shape of the world-line.

To find the equations of motion and the boundary conditions we apply Hamilton's principle of stationary action, keeping fixed the initial and final $\tau = \tau_i, \tau_f$ configurations of string. This gives

$$0 = \delta S = \int d\sigma\, d\tau\, \delta L(\dot{x}_\mu, x_\mu{}') \tag{4.137}$$

$$= \int d\sigma \int d\tau \left[\frac{\partial L}{\partial \dot{x}_\mu} \frac{\partial}{\partial \tau}(\delta x_\mu) + \frac{\partial L}{\partial x_\mu{}'} \frac{\partial}{\partial \sigma}(\delta x_\mu)\right] \tag{4.138}$$

$$= \int d\tau \left(\delta x_\mu \frac{\partial L}{\partial x_\mu{}'}\right)\Bigg|_{\sigma=0}^{\sigma=\pi} -$$

$$- \int d\tau \int d\sigma \left(\frac{\partial}{\partial \tau} \frac{\partial L}{\partial \dot{x}_\mu} + \frac{\partial}{\partial \sigma} \frac{\partial L}{\partial x_\mu{}'}\right) \tag{4.139}$$

Hence the equations of motion are

$$\frac{\partial}{\partial \tau}\left(\frac{\partial L}{\partial \dot{x}_\mu}\right) + \frac{\partial}{\partial \sigma}\left(\frac{\partial L}{\partial x_\mu{}'}\right) = 0 \tag{4.140}$$

and the boundary conditions are

$$\frac{\partial L}{\partial x_\mu{}'} = 0 \qquad \text{at } \sigma = 0, \pi \tag{4.141}$$

The equations of motion are, more explicitly,

$$0 = \frac{\partial}{\partial \tau}\left(\frac{(\dot{x}\cdot x')x_\mu{}' - (x')^2 \dot{x}_\mu}{\sqrt{(\dot{x}\cdot x')^2 - \dot{x}^2 x'^2}}\right) +$$

$$+ \frac{\partial}{\partial \sigma}\left(\frac{(\dot{x}\cdot x')\dot{x}_\mu - (\dot{x}^2)x_\mu{}'}{\sqrt{(\dot{x}\cdot x')^2 - \dot{x}^2 x'^2}}\right) \tag{4.142}$$

Conjugate momentum densities are defined by

$$P^{\mu} = \frac{\partial L}{\partial \dot{x}_{\mu}} \tag{4.143}$$

and

$$P_{\sigma}^{\ \mu} = \frac{\partial L}{\partial x_{\mu}'} \tag{4.144}$$

the total momentum of the string being given by

$$P^{\mu} = \int_{0}^{\pi} P^{\mu} \, d\sigma \tag{4.145}$$

Similarly we can define densities for the generators of the homogeneous Lorentz group by

$$M^{\mu\nu} = x^{\mu} P^{\nu} - x^{\nu} P^{\mu} \tag{4.146}$$

$$M_{\sigma}^{\mu\nu} = x^{\mu} P_{\sigma}^{\ \nu} - x^{\nu} P_{\sigma}^{\ \mu} \tag{4.147}$$

The total Lorentz generator for the string is

$$M^{\mu\nu} = \int_{0}^{\pi} d\sigma \, M^{\mu\nu} \tag{4.148}$$

The equations of motion show that these densities are all locally conserved, namely

$$\frac{\partial}{\partial \tau} P^{\mu} + \frac{\partial}{\partial \sigma} P_{\sigma}^{\ \mu} = 0 \tag{4.149}$$

$$\frac{\partial}{\partial \tau} M^{\mu\nu} + \frac{\partial}{\partial \sigma} M_{\sigma}^{\mu\nu} = 0 \tag{4.150}$$

and the boundary conditions yield

$$\left.\begin{aligned} P_{\sigma}^{\ \mu} &= 0 \qquad (4.151a) \\ M_{\sigma}^{\mu\nu} &= 0 \qquad (4.151b) \end{aligned}\right\} \text{ at } \sigma = 0, \pi$$

The general covariance enables us to select at our convenience a particular choice of σ, τ parametrisation, i.e. a particular gauge, in order that we simplify (linearise) the Lagrangian density and the equations of motion. We therefore choose a coordinate system in which the inner metric g^{ab} is both diagonal, and traceless (i.e. scale-invariant) by demanding

$$g_{12} = g_{21} = 0 \tag{4.152}$$

$$g_{11} + g_{22} = 0 \tag{4.153}$$

That is,

$$\dot{x} \cdot x' = 0. \tag{4.154}$$

$$\dot{x}^2 + x'^2 = 0 \tag{4.155}$$

This linearises the Lagrangian density to

$$L = \frac{1}{2\pi} \sqrt{-g} = \frac{1}{4\pi} (\dot{x}^2 - x'^2) \tag{4.156}$$

The equations of motion become

$$\ddot{x}_\mu - x_\mu'' = 0 \tag{4.157}$$

while the boundary conditions become $x_\mu' = 0$ at $\sigma = 0,\pi$.

The Poincaré densities are now

$$P^\mu = \frac{1}{2\pi} \dot{x}_\mu \tag{4.158}$$

$$P_\sigma{}^\mu = -\frac{1}{2\pi} x_\mu' \tag{4.159}$$

$$M^{\mu\nu} = \frac{1}{2\pi} (x_\mu \dot{x}_\nu - x_\nu \dot{x}_\mu) \tag{4.160}$$

$$M_\sigma^{\mu\nu} = \frac{1}{2} (x_\mu x_\nu' - x_\nu x_\mu') \tag{4.161}$$

As mentioned already, we expect constraints between

the components of P_μ and x_μ and that therefore we shall not regard them as independent dynamical variables in setting up the classical canonical formalism. Indeed with our particular choice of gauge these constraints read

$$x' \cdot \mathcal{P} = 0 \tag{4.162}$$

$$\mathcal{P}^2 + \frac{1}{(2\pi)^2}(x')^2 = 0 \tag{4.163}$$

Our procedure will be to eliminate certain components from the equations of motion, by using these gauge conditions.

The particular identification of σ, τ is a matter only of our convenience. For example, in the point particle case mentioned above, it seems very natural to identify the coordinate τ with x_0 = time, although this is not essential. In the present case there is a slight calculational advantage in using null-plane variables, since there the gauge constraints separate rather neatly, but again this is not essential.

For any d-vector A_μ (μ = 0, 1, 2, 3, 4, ..., (d-2), (d-1)) we denote

$$A_\pm = \frac{1}{\sqrt{2}}(A_0 \pm A_L) \tag{4.164}$$

with A_L (L for longitudinal) = A_{d-1}. The remaining transverse components will be denoted by

$$\underset{\sim}{A_i} \quad (i = 1, 2, \ldots, d-2) \tag{4.165}$$

To be specific we will identify τ with the null-plane component x^+ by

$$x^+ = 2\,P^+\,\tau \tag{4.166}$$

where P^+, a component of the total four-momentum P^μ, is inserted for convenience only (we shall see shortly that P^+ is a constant of the motion). The identification of σ will be such that $\sigma(0 \leq \sigma \leq \pi)$ is π times that fraction of the total string momentum along the τ-direction carried by the portion between $\sigma = 0$ and $\sigma = \sigma$. This definition implies that the momentum density component $\mathcal{P}^+$ is constant along the string and is given by

$$P^+ = \mathcal{P}^+/\pi \tag{4.167}$$

We now outline the principal steps and results in this null-plane language. (see Goddard et al., Reference 28].

Using the fact that now $x_+{}' = 0$ the gauge constraints becomes

$$\frac{x'_- \cdot \mathcal{P}^+}{\pi} = \underset{\sim}{x}'_- \cdot \underset{\sim}{\mathcal{P}} \tag{4.168}$$

and

$$\underset{\sim}{\mathcal{P}}^2 + \frac{\underset{\sim}{x}'^2}{(2\pi)^2} = \frac{2P_+P_-}{\pi} \tag{4.169}$$

We now choose to eliminate P_- and x_+ from the equations of motion. We have immediately

$$P_- = \frac{\pi}{2P_+}(\underset{\sim}{\mathcal{P}}^2 + \frac{\underset{\sim}{x}'^2}{(2\pi)^2}) \tag{4.170}$$

Let us introduce the center of mass coordinate

$$q_-(\tau) = \frac{1}{\sqrt{2}\pi}\int_0^\pi d\sigma\, x_-(\sigma, \tau) \qquad \text{(definition)} \tag{4.171}$$

then x_- is eliminated through

$$x_- = \sqrt{2}\, q^-(\tau) + \frac{\pi}{P^+}\int_0^\pi d\sigma' \left(\frac{\sigma'}{\pi} - \theta(\sigma' - \sigma)\right) \underset{\sim}{x}' \cdot \underset{\sim}{\mathcal{P}} \tag{4.172}$$

To see this write

$$x_-{}' = \frac{\pi}{P^+}\underset{\sim}{x}' \cdot \underset{\sim}{\mathcal{P}} \tag{4.173}$$

$$x_-(\sigma, \tau) = x_-(\pi,\tau) - \int_0^\pi d\sigma' \frac{\pi}{P^+}\underset{\sim}{x}' \cdot \underset{\sim}{\mathcal{P}}\, \theta(\sigma' - \sigma) \tag{4.174}$$

Now re-write

$$\sqrt{2}\,\pi\, q_-(\tau) = \int_0^\pi d\sigma'\; x_-(\sigma',\tau) = \sigma' x_-(\sigma',\tau)\Big|_0^\pi -$$

$$- \int_0^\pi d\sigma'\; \sigma' x_-'(\sigma',\tau) \tag{4.175}$$

$$= \pi\, x_-(\pi,\tau) - \int_0^\pi d\sigma'\; \sigma'\; x_-'(\sigma',\tau) \tag{4.176}$$

to arrive at the final form since

$$x_-(\pi,\tau) = \sqrt{2}\, q_-(\tau) + \frac{1}{\pi}\int_0^\pi d\sigma'\; \sigma'\; x_-'(\sigma',\tau) \tag{4.177}$$

To write the Hamiltonian note that

$$P\cdot x = 2P_+P_-\,\tau + P_+x_- - \underset{\sim}{P}\cdot\underset{\sim}{x} \tag{4.178}$$

so that for translations in τ the generator is

$$H = 2P_+P_- = \pi\int_0^\pi d\sigma\;(\underset{\sim}{P}^2 + \underset{\sim}{x}'^2/(2\pi)^2) \tag{4.179}$$

We are now ready to write the classical equations of motion for the independent dynamical variables $(\underset{\sim}{x},\underset{\sim}{\tau})$. These follow at once from

$$\frac{\partial}{\partial\tau}\left(\frac{\partial L}{\partial\dot{x}_\mu}\right) + \frac{\partial}{\partial\sigma}\left(\frac{\partial L}{\partial x_\mu'}\right) = 0 \tag{4.180}$$

with $L = \frac{1}{2}(\dot{x}^2 - x'^2)$ to be

$$\dot{\underset{\sim}{P}} = \frac{1}{2\pi}\,\underset{\sim}{x}'' \tag{4.181}$$

and

$$\dot{\underset{\sim}{x}} = 2\pi\,\underset{\sim}{P} \tag{4.182}$$

The equations demand that the Poisson brackets are

$$\{x^i, x^j\} = \{P^i, P^j\} = 0 \tag{4.183}$$

$$\{x^i(\sigma), P^j(\sigma')\} = \delta^{ij}\ \delta(\sigma - \sigma') \tag{4.184}$$

by using the form for H given earlier and requiring that

$$\dot{\underset{\sim}{x}} = \{\underline{x}, H\} \tag{4.185}$$

$$\dot{\underset{\sim}{P}} = \{\underset{\sim}{P}, H\} \tag{4.186}$$

as the Hamilton equations of motion.

In addition we know that P_+ = constant, and that

$$\dot{q}_- = \frac{1}{\sqrt{2}\, P_+} H \tag{4.187}$$

Hence

$$\dot{q}_- = \frac{1}{\sqrt{2}\, P_+} H\tau + q_{-0} \tag{4.188}$$

The following Poisson brackets are appropriate to q_{-0} and P_+

$$\{\sqrt{2}\, q_{0-}, P_+\} = -1 \tag{4.189}$$

and

$$\{P_+, \underset{\sim}{x}\} = \{P_+, \underset{\sim}{P}\} = \{q_{0-}, \underset{\sim}{x}\} = \{q_{0-}, \underset{\sim}{P}\} = 0 \tag{4.190}$$

Before making the transition to quantum mechanics, it is convenient to avoid the use of the continuous coordinates σ and τ by making the Fourier expansion into normal modes according to

$$x_\mu(\sigma, \tau) = \sqrt{2}\, [q_0{}^\mu + \alpha_0{}^\mu \tau + \sum_{\substack{n=-\infty \\ n\neq 0}}^{\infty} \frac{\alpha_n{}^\mu}{n} \cos n\sigma\, e^{-in\tau}] \tag{4.191}$$

This exploits the wave equation and boundary conditions satisfied by $x_\mu(\sigma, \tau)$. Of course, we expect that the $\alpha_n{}^\mu$

will be identifiable (after quantisation) with the harmonic-oscillator-like operators of the operator formalism.

The independent variables are now the constants of the motion $\underset{\sim}{\alpha}_n$ and $\underset{\sim}{q}_0$. The plus components all vanish:

$$q_{0+} = \alpha_n^+ = 0 \quad (n \neq 0) \tag{4.192}$$

except for

$$\alpha_0^+ = \sqrt{2}\, P_+ \tag{4.193}$$

On the other hand the minus components can be re-expressed in terms of the transverse components by

$$\alpha_n^- = \frac{1}{\alpha_0^+} L_n \tag{4.194}$$

where

$$L_n = \frac{1}{2} \sum_{\substack{m=-\infty \\ m\neq 0}}^{\infty} \underset{\sim}{\alpha}_{-n} \cdot \underset{\sim}{\alpha}_{n+m} \tag{4.195}$$

To derive this expression note that

$$\underset{\sim}{x}' = -\, i \sqrt{2} \sum_{\substack{n=-\infty \\ n\neq 0}}^{\infty} \alpha_n \sin n\sigma\, e^{-in\tau} \tag{4.196}$$

$$\underset{\sim}{P} = \dot{x}/2\pi = \frac{\sqrt{2}}{2\pi} [\underset{\sim}{\alpha}_0 + \sum_{\substack{n=-\infty \\ n\neq 0}}^{\infty} \underset{\sim}{\alpha}_n \cos n\sigma\, e^{-in\tau}] \tag{4.197}$$

Now use the relationship

$$\sqrt{2}\, i \sum_{\substack{n=-\infty \\ n\neq 0}}^{\infty} \frac{\alpha_n^-}{n} \cos n\sigma\, e^{-in\tau} =$$

$$= \frac{\pi}{P^+} \int_0^{\pi} d\sigma' (\frac{\sigma'}{\pi} - \theta(\sigma' - \sigma))\, \underset{\sim}{x}' \cdot \underset{\sim}{P} \tag{4.198}$$

and project out α_n^- by operation with $\int_{-\infty}^{\infty} d\tau \int_0^{\pi} d\sigma\, e^{in\tau} \cos n\sigma$ to deduce that

$$\alpha_n^- = \frac{1}{2\sqrt{2}P^+} \sum_{\substack{m=-\infty \\ m\neq 0}}^{\infty} \underset{\sim}{\alpha}_{-n} \cdot \alpha_{n+m} = \frac{1}{\alpha_0^+} L_n \tag{4.199}$$

as required. Here we have used

$$\int_0^{\pi} d\sigma \frac{\sigma'}{\pi} \sin A\sigma' \cos B\sigma' - \int_0^{\pi} d\sigma' \sin A\sigma' \cos B\sigma' =$$

$$= -\frac{1}{2}\left[\frac{\cos(A+B)\sigma}{A+B} - \frac{\cos(A-B)\sigma}{A-B}\right] \tag{4.200}$$

The Poisson brackets of the independent variables in normal modes are

$$\{\alpha_n^i, \alpha_m^j\} = -i\, n\, \delta_{n+m,o}\, \delta^{ij} \tag{4.201}$$

$$\{q_0^i, \alpha_0^j\} = \delta^{ij} \tag{4.202}$$

$$\{q_0^i, q_0^j\} = 0 \tag{4.203}$$

$$\{q_0^-, \alpha^+\} = -1 \tag{4.204}$$

$$\{q_0^-, \alpha_n^i\} = \{q_0^-, q_0^i\} = \{\alpha_0^+, \alpha_n^i\} = \{\alpha_0^+, q_0^i\} = 0 \tag{4.205}$$

For the dependent modes one finds the generalised projective algebra (for Poisson brackets)

$$\{L_n, L_m\} = -i\,(n-m)\, L_{n+m} \tag{4.206}$$

$$\{L_n, \alpha_m^i\} = i\, m\, \alpha_{m+n}^i \tag{4.207}$$

$$\{q_0^i, L_m\} = \alpha_m^i \tag{4.208}$$

We take the Hamiltonian to be L_0. It is given by

$$H = L_0 = 2P_+P_- = \underset{\sim}{P}^2 + \sum_{n=1}^{\infty} \alpha_n \cdot \alpha_n^* \tag{4.209}$$

The total momentum is

$$P^\mu = \frac{1}{2\pi} \int_0^\pi d\sigma\ \dot{x}^\mu = \frac{1}{\sqrt{2}} \alpha_0{}^\mu \tag{4.210}$$

and the Lorentz generators are

$$M^{\mu\nu} = \int_0^\pi d\sigma\ (P^\nu x^\mu - P^\mu x^\nu) \tag{4.211}$$

$$= \frac{1}{2\pi} \int_0^\pi d\sigma\ (\dot{x}^\nu x^\mu - \dot{x}^\mu x^\nu) \tag{4.212}$$

$$= \frac{1}{\pi} \int_0^\pi d\sigma\ (q_0{}^\mu + a_0{}^\mu \tau + i \sum_{n \neq 0} \frac{\alpha_n{}^\mu}{n} \cos n\sigma\ e^{-in\tau}) \cdot$$

$$\cdot (\alpha_0{}^\nu + \sum_{n' \neq 0} \alpha_n{}^\nu \cos n'\sigma\ e^{-in'\tau}) -$$

$$- (\mu \leftrightarrow \nu) \tag{4.213}$$

$$= (q_0{}^\mu \alpha_0{}^\nu - q_0{}^\nu \alpha_0{}^\mu) + i \sum_{n \neq 0} (\frac{\alpha_n{}^\mu \alpha_{-n}{}^\nu - \alpha_n{}^\nu \alpha_{-n}{}^\mu}{n}) \tag{4.214}$$

We are now ready to proceed to the quantum mechanics of our massless relativistic string. We accomplish this by imposing the Dirac quantum conditions (see Chapter 4 of Dirac's book, reference 32) i.e. by replacing

$$i\ \{\text{Poisson bracket}\} \to [\text{commutator}] \tag{4.215}$$

This gives immediately the canonical commutation relations for the independent dynamical variables.

The crucial point is that in expressing the dependent variables in terms of the independent ones, there now may occur a serious ambiguity in ordering the (non-commutative) operators. In the present case this happens for L_0 in α_0^-. We define

$$\alpha_0^- = \frac{1}{a_0^+}[L_0 - \alpha(0)] \tag{4.216}$$

where

$$L_0 = \sum_{n=1}^{\infty} \underset{\sim}{\alpha}_n^+ \cdot \underset{\sim}{\alpha}_n + \underset{\sim}{P}^2 \tag{4.217}$$

is normal ordered. The interpretation of $\alpha(0)$ is, as our choice of symbol indicates, the leading trajectory intercept. To see this note that the invariant squared mass is

$$m^2 = 2P_+P_- - \underset{\sim}{P}^2 = \sum_{n=1}^{\infty} \underset{\sim}{\alpha}_n^+ \cdot \underset{\sim}{\alpha}_n - \alpha(0) \tag{4.218}$$

so that $(-\alpha(0))$ is the squared mass of the ground state. We can already see by counting that since there are only transverse states we will run into trouble with covariance unless the first excited vector state is massless (i.e. $\alpha(0) = 1$). We shall, however, display this in a slightly more general manner by writing the Lorentz generators $M^{\mu\nu}$. The canonical components of $M^{\mu\nu}$ are

$$M^{ij} = q_0{}^i \alpha_0{}^j - q_0{}^j \alpha_0{}^i - i \sum_{n=1}^{\infty} \left(\frac{\alpha^i_{-n}\alpha^j_n - \alpha^j_{-n}\alpha^i_n}{n}\right) \tag{4.219}$$

$$M^{i+} = - M^{+i} = q_0{}^i \alpha_0{}^+ \tag{4.220}$$

$$M^{+-} = - \frac{1}{2}(q_0{}^- \alpha_0{}^+ + \alpha_0{}^+ q_0{}^-) \tag{4.221}$$

But the components $M^{i-} = M^{-i}$ contain the non-canonical variables L_n which satisfy (as we know from our earlier discussion of the operator formalism).

$$[L_n, L_m] = (n - m)L_{n+m} + \left(\frac{d - 2}{12}\right) n(n^2 - 1)\, \delta_{n+m,0} \tag{4.222}$$

where d = dimension of space-time. The expression for M^{i-} is

$$M^{i-} = A^{i-} + B^{i-} \tag{4.223}$$

where

$$A^{i-} = \frac{q_0^i}{\alpha_0^+}(L_0 - \alpha(0)) - q_0^- \alpha_0^i \tag{4.224}$$

$$B^{i-} = -\frac{i}{\alpha_0^+} \sum_{n=1}^{\infty} \left(\frac{\alpha_{-n}^i L_n - L_{-n}\alpha_n^i}{n}\right) \tag{4.225}$$

Now the commutator $[M^{i-}, M^{j-}]$ should vanish according to the Lorentz algebra

$$[M^{\mu\nu}, M^{\rho\sigma}] = g^{\mu\rho} M^{\nu\sigma} + g^{\nu\sigma} M^{\mu\rho} - g^{\mu\sigma} M^{\nu\rho} - g^{\nu\rho} M^{\mu\sigma} \tag{4.226}$$

There is, however, a contribution proportional to (d - 2) in $[B^{i-}, B^{j-}]$ of the form [using [AB, CD] = [A,C]BD + A[B,C]D + C[A,D]B + CA[B,D]]

$$-\frac{1}{\alpha_0^{+2}}\left(\frac{d-2}{12}\right) \sum_{n=1}^{\infty} \frac{n(n^2 - 1)}{n^2} [\alpha_{-n}^i \alpha_n^j - \alpha_{-n}^j \alpha_n^i] \tag{4.227}$$

arising from the anomaly in $[L_n, L_{-n}]$. Also there is a term in $[A^{i-}, B^{j-}]$ proportional to $\alpha(o)$ namely

$$+\frac{2}{\alpha_0^{+2}} \sum_{n=1}^{\infty} \frac{\alpha(0)}{n} (\alpha_{-n}^i \alpha_n^j - \alpha_{-n}^j \alpha_n^i) \tag{4.228}$$

Computation of the full commutator gives the answer

$$[M^{i-}, M^{j-}] = -\frac{2}{\alpha_0^{+2}} \sum_{n=1}^{\infty} \left[\left(1 - \frac{d-2}{24}\right)n + \left(\frac{d-2}{24} - \alpha(0)\right)\frac{1}{n}\right] (\alpha_{-n}^i \alpha_n^j - \alpha_n^i \alpha_{-n}^j) \tag{4.229}$$

which vanishes only if $\alpha(0) = 1$ and $d = 26$. This confirms the completeness of the transverse states under these conditions, as was demonstrated earlier; the use of the null-plane quantization here is most closely related to the infinite momentum limit discussed there. The present treatment gives us some more insight into the anomaly term of the generalised projective algebra: the anomaly term is

(i) a purely quantum mechanical effect and

(ii) a result of the infinite number of degrees of freedom which make normal ordering essential to avoid divergent matrix elements.

The most important general result of this subsection is that the decoupling of spurious states of the dual model is a direct consequence of the general covariance of the classical action of the rubber string. This observation might lead not only to a reformulation of existing models but eventually to the building of improved ghost-free models with more realistic resonance spectra. [See References 33-36].

4.7 SUMMARY

We have shown how a simple multiplicative approach to internal symmetry can work well for isospin and for strangeness. Both cyclic symmetry and factorisability can be maintained, together with the absence of unwanted states with exotic quantum numbers. Although the λ - matrices of SU(3) were used to accomodate strangeness, it was not necessary to have exact SU(3) mass degeneracy in the Born amplitude. This last fact is important because SU(3) symmetry is strongly broken in Nature and one is anxious not to attribute any large (symmetry breaking) effects to the perturbatively unitary corrections.

A method of introducing baryons, while overlooking the existence of half-integer spin, is to picture them as circular rubber strings with three quarks placed symmetrically on the circumference. This leads to non-planar tree diagrams having an essentially different pole structure from the mesonic case. Such an ansatz for baryons, together with the already-known configuration for mesons, dictates by projective invariance the configurations of all exotic hadrons with total quark number greater than three.

The observation that links the ghost-eliminating gauges of the unit intercept dual model to the general covariance of the classical action of a rubber string is an interesting one. Indeed, one may hope on the basis of this idea that he can perhaps simplify the formulation of the theory from the postulation of a complicated set of on-mass-shell tree amplitudes to the postulation of a limited number of axioms in the language of the string.

REFERENCES

1. J. E. Paton and Chan Hong-Mo, Nucl. Phys. B10, 516 (1969).
2. M. Gell-Mann, California Institute of Technology Synchrotron Laboratory Report CTSL-20(1961).
3. Y. Ne'emann, Nuclear Phys. 26, 222 (1961).
4. M. Gell-Mann and Y. Ne'emann, The Eightfold Way, Benjamin (1964).
5. N. A. Tornqvist, Nuclear Phys. B26, 104 (1971).
6. I. Gonzales Mestres, Lettere al Nuovo Cimento 4, 1207 (1970).
7. I. Bars, Nucl. Phys. B31, 15 (1971).
8. I. Bars, Nucl. Phys. B31, 29 (1971).
9. P. Olesen, Nucl. Phys. B18, 459 (1970).
10. P. Olesen, Nucl. Phys. B19, 589 (1970).
11. S. Mandelstam, Phys. Rev. D6, 1734 (1970).
12. Y. Nambu, Proceedings of the International Conference on Symmetries and Quark Models, edited by R. Chand. Gordan and Breach (1970) p. 269.
13. H. B. Nielsen, in High Energy Physics, proceedings of the Fifteenth International Conference on High Energy Physics, Kiev, 1970, edited by V. Shelest, (Naukova, Dunika, Kiev, U.S.S.R., 1972).
14. L. Susskind, Phys. Rev. Letters 23, 545 (1969).
15. L. Susskind, Phys. Rev. D1, 1182 (1970).
16. L. Susskind, Nuovo Cimento 69A, 457 (1970).
17. G. Frye, Phys. Rev. D1, 1194 (1970).
18. G. Frye, C. W. Lee and L. Susskind, Nuovo Cimento 69A, 497 (1970).

19. P. H. Frampton and P. G. O. Freund, Nucl. Phys. B24, 453 (1970).

20. E. Corrigan, Ph.D. thesis, Cambridge, England. Chapter two of this thesis reports on unpublished work by E. Corrigan, C. Montonen and D. I. Olive.

21. S. Mandelstam, Phys. Rev. D1, 1720 (1970).

22. S. Ellis, P. H. Frampton, P. G. O. Freund and D. Gordon, Nucl. Phys. B24, 465 (1970).

23. Y. Nambu, lecture notes prepared for the Summer Institute of the Niels Bohr Institute (SINBI), (1970).

24. L. N. Chang and F. Mansouri, Phys. Rev. D5, 2535 (1972).

25. F. Mansouri and Y. Nambu, Phys. Letters 39B, 375 (1972).

26. T. Goto, Prog. Theor. Phys. 46, 1560 (1971).

27. Y. Hara, Prog. Theor. Phys. 46, 1549 (1971).

28. P. Goddard, J. Goldstone, C. Rebbi and C. B. Thorn, Nucl. Phys. B56, 109 (1973).

29. S. Weinberg, Gravitation and Cosmology, Wiley (1972).

30. A. Einstein, Annalen der Physik 49, 769 (1916).

31. The Principle of Relativity with notes by A. Sommerfeld, translated by W. Perrett and G. B. Jeffery, Dover (1923) page 109.

32. P. A. M. Dirac, The Principles of Quantum Mechanics, Oxford University Press. Fourth Edition (1958).

33. H. B. Nielsen and P. Olesen, Nucl. Phys. B61, 45 (1973).

34. H. C. Tze, Nuov. Cim. Lett. 7, 401 (1973).

35. P. Olesen, Phys. Lett. 50B, 255 (1974).

36. J. L. Gervais and B. Sakita, Phys. Rev. Letters 30, 716 (1973).

5

SPIN

5.1 INTRODUCTION

Here we study the introduction of extra spin degrees of freedom into the model. First we discuss a multiplicative approach to spin, analogous to that used successfully for isospin, and show that it leads to both parity doubling and to spin ghosts.

An approach to half-integer spin of applying a correspondence principle to the Dirac equation leads to a dual theory of free fermions with sufficient gauges to eliminate spin ghosts, and involving both commuting and anticommuting harmonic-oscillator-like operators.

Using similar sets of oscillators, a multipion amplitude is constructed with an enlarged gauge algebra and with different families of trajectories for even and odd G-parity. The gauges are adequate to remove all negative probability states. Although the trajectories have unphysical intercepts (a unit intercept rho trajectory persists) the spectrum of low-lying states is qualitatively similar to that observed for strangeness-zero mesons.

The pion model and the fermion model can be unified to give a consistent set of Born amplitudes with external fermion legs; this shows that there is no <u>essential</u> difficulty to introducing half-integer spins into dual theories.

Consideration of simple bilinear realisations of the generalised projective algebra reveals that only two such realisations are possible, corresponding to generalised projective spin $S = 0$ and commuting operators, and $S = -\frac{1}{2}$ and anticommuting operators. Combining these additively gives a group-theoretical derivation of the dual pion model within the operator formalism.

Finally, a non-planar extension of the dual pion model, analogous to the Shapiro-Virasoro model, is discussed.

5.2 MULTIPLICATIVE SPIN FACTOR

In view of the success in dealing with an exact internal symmetry, namely isospin, in a simple multiplicative fashion, it is natural to attempt to introduce half-integer spin through a similar method. We shall therefore begin our discussion of spin by describing such a direct approach, as proposed by Mandelstam[1] and by Bardakci and Halpern[2], since the difficulties which arise here will strongly motivate the later more successful developments.

We write the amplitude for the planar quark diagram of Figure 5.1 as

$$A_{2n} = \Gamma_{2n} B_{2n} \tag{5.1}$$

where

$$\Gamma_{2n} = \bar{u}(2n)\, u(1)\, \bar{u}(2)\, u(3) \cdots\cdots \bar{u}(2n-2)\, u(2n-1) \tag{5.2}$$

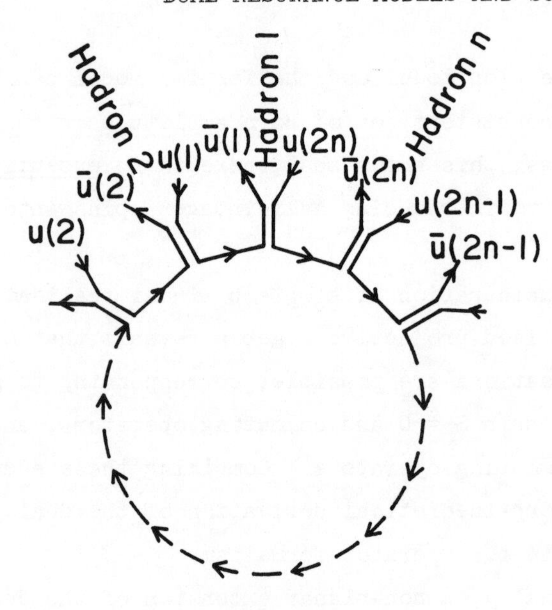

FIGURE 5.1

Multiplicative Spin Factor

where the u(i) are Dirac spinors describing the external spin 1/2 quarks. Next we make a Fierz transformation

$$u(i)\bar{u}(j) = \frac{1}{4}(S_{ji}1 + P_{ji}\gamma^5 + V^{\mu}_{ji}\gamma^{\mu} - A^{\mu}_{ji}\gamma^5\gamma_{\mu} + \frac{1}{2}\Sigma^{\mu\nu}_{ji}\alpha_{\mu\nu}) \tag{5.3}$$

in which

$$S_{ji} = \bar{u}(j)u(i) \tag{5.4}$$

$$P_{ji} = \bar{u}(j)\gamma^5 u(i) \tag{5.5}$$

$$V^{\mu}_{ji} = \bar{u}(j)\gamma^{\mu}u(i) \tag{5.6}$$

$$A^{\mu}_{ji} = \bar{u}(j)\gamma^5\gamma^{\mu}u(i) \tag{5.7}$$

$$\Sigma^{\mu\nu}_{ji} = \bar{u}(j)\sigma_{\mu\nu}u(i) \tag{5.8}$$

with

$$\sigma_{\mu\nu} = \frac{i}{2}[\gamma_\mu, \gamma_\nu] = -\sigma_{\nu\mu} \tag{5.9}$$

These quantities are 4 × 4 Dirac matrices and we may now re-write, inserting the spinor indices

$$\Gamma_{2n} = (\bar{u}(2n))_a \; (u(1)\bar{u}(2))_{ab} \; (u(3)\bar{u}(4))_{bc} \cdots .$$

$$(u(2n-3)\bar{u}(2n-1))_{yz} \; u(2n-1)_z \tag{5.10}$$

$$= (u(1)\bar{u}(2))_{ab} \; (u(3)\bar{u}(4))_{bc} \; \cdots \; (u(2n-1)\bar{u}(2n))_{za} \tag{5.11}$$

$$= \frac{1}{4}\,\mathrm{Tr}[u(1)\bar{u}(2),\; u(3)\bar{u}(4),\; \ldots\; ,\; u(2n-1)\bar{u}(2n)] \tag{5.12}$$

$$= \frac{1}{4}\,\mathrm{Tr}[S_{(1)}1 + P_{(1)}\gamma_5 + V^{\mu}_{(1)}\gamma_\mu - A^{\mu}_{(1)}\gamma_5\gamma_\mu +$$

$$+ \frac{1}{2}\,\Sigma^{\mu\nu}_{(1)}\sigma_{\mu\nu},\; \ldots.,\; S_{(n)}1 + P_{(n)}\gamma_5 + V^{\mu}_{(n)}\gamma_\mu -$$

$$- A^{\mu}_{(n)}\gamma_5\gamma_\mu + \frac{1}{2}\,\Sigma^{\mu\nu}_{(n)}\sigma_{\mu\nu}] \tag{5.13}$$

where we have adopted the notation

$$S_{(i)} = \bar{u}(2i)\; u(2i-1) \tag{5.14}$$

$$P_{(i)} = \bar{u}(2i)\; \gamma_5\; u(2i-1) \tag{5.15}$$

$$V^{\mu}_{(i)} = \bar{u}(2i)\; \gamma_\mu\; u(2i-1) \tag{5.16}$$

$$A^{\mu}_{(i)} = u(2i)\; \gamma^5\gamma^\mu\; u(2i-1) \tag{5.17}$$

$$\Sigma^{\mu\nu}_{(i)} = \bar{u}(2i)\; \sigma_{\mu\nu}\; u(2i-1) \tag{5.18}$$

to identify with the n hadronic $q\bar{q}$ channels of the diagram. [Notice that the strictly analogous treatment of isospin, starting with 2n two-component spinors, would lead to the usual isospin trace factor.]

This spin factor, written as a trace, is manifestly cyclic symmetric. To discuss the factorisation on an internal hadron state we must recognise that for any two Dirac matrics A and B there is the completeness relation

$$\mathrm{Tr}(AB) = \mathrm{Tr}(A)\,\mathrm{Tr}(B) + \mathrm{Tr}(A\gamma_5)\,\mathrm{Tr}(\gamma_5 B) + \\ + \mathrm{Tr}(A\gamma_\mu)\,\mathrm{Tr}(\gamma_\mu B) - \mathrm{Tr}(A\gamma_5\gamma_\mu)\,\mathrm{Tr}(\gamma_5\gamma_\mu B) + \\ + \frac{1}{2}\,\mathrm{Tr}(A\sigma_{\mu\nu})\,\mathrm{Tr}(\sigma_{\mu\nu}B) \quad (5.19)$$

Taking the momenta of the two subsystems as $q_A{}^\mu = -q_B{}^\mu$ we now decompose these terms into pure spin states.

Defining

$$\gamma_A{}^\mu = \gamma^\mu - \frac{\not{q}_A\, q_A{}^\mu}{(q_A{}^2)} \quad (5.20)$$

$$\gamma_B{}^\mu = \gamma^\mu - \frac{\not{q}_B\, q_B{}^\mu}{(q_B{}^2)} \quad (5.21)$$

and using

$$\sigma_{\mu\nu} = \frac{1}{q^2}\,(q_\mu \sigma_{\lambda\nu} q_\lambda - \sigma_{\mu\lambda} q_\lambda q_\nu + i\,\varepsilon_{\mu\nu\lambda\kappa} q_\lambda \gamma_5 \sigma_{\kappa\rho} q_\rho) \quad (5.22)$$

for $q^\mu = q_A{}^\mu$ or $q_B{}^\mu$ we find eventually that (using repeatedly $q_A{}^\mu = -\,q_B{}^\mu$)

$$\mathrm{Tr}(AB) = \mathrm{Tr}(A)\,\mathrm{Tr}(B) + \mathrm{Tr}(A\gamma_5)\,\mathrm{Tr}(\gamma_5 B) \\ + \mathrm{Tr}(A\gamma_A{}^\mu)\,\mathrm{Tr}(\gamma_B{}^\mu B) - \mathrm{Tr}\,\Big(\frac{A\,\not{q}_A}{\sqrt{q_A{}^2}}\Big)\,\mathrm{Tr}\,\Big(\frac{\not{q}_B B}{\sqrt{q_B{}^2}}\Big) \\ - \mathrm{Tr}\Big(\frac{A\gamma_5\gamma_\mu{}^A}{\sqrt{q_A{}^2}}\Big)\,\mathrm{Tr}\Big(\frac{\gamma_5\gamma_\mu{}^B B}{\sqrt{q_B{}^2}}\Big)$$

$$+ \mathrm{Tr}\left(\frac{A\gamma_5 \not{q}_A}{\sqrt{q_A^2}}\right) \mathrm{Tr}\left(\frac{\gamma_5 \not{q}_B B}{\sqrt{q_B^2}}\right)$$

$$- \mathrm{Tr}\left(\frac{A\sigma_{\mu\nu} q_A{}^2}{\sqrt{q_A^2}}\right) \mathrm{Tr}\left(\frac{\sigma^{\mu\nu} q_\lambda{}^B B}{\sqrt{q_B^2}}\right)$$

$$- \mathrm{Tr}\left(\frac{A\gamma_5 \sigma_{\mu\nu} q_A{}^2}{\sqrt{q_A^2}}\right) \mathrm{Tr}\left(\frac{\gamma_5 \sigma^{\mu\nu} q_\lambda{}^B B}{\sqrt{q_B^2}}\right) \tag{5.23}$$

To arrive at this result we have used

$$\mathrm{Tr}(A\gamma^\mu)\, \mathrm{Tr}(\gamma^\mu B) = \mathrm{Tr}\left(A\gamma_A{}^\mu + \frac{A\not{q}_A q_A{}^\mu}{q_A{}^2}\right) \cdot$$

$$\cdot\, \mathrm{Tr}\left(\gamma_B{}^\mu B + \frac{\not{q}_B q_B{}^\mu B}{q_B{}^2}\right) \tag{5.24}$$

$$= \mathrm{Tr}(A\gamma_A{}^\mu)\, \mathrm{Tr}(\gamma_B{}^\mu B) - \mathrm{Tr}\left(\frac{A\not{q}_A}{\sqrt{q_A^2}}\right) \mathrm{Tr}\left(\frac{\not{q}_B B}{\sqrt{q_B^2}}\right) \tag{5.25}$$

since $q_A{}^\mu \gamma_B{}^\mu = 0$, $q_A{}^\mu = -q_B{}^\mu$. Similarly we treat the axial vector piece and finally for the tensor piece we use

$$\frac{1}{2} \mathrm{tr}(A\sigma_{\mu\nu})\, \mathrm{Tr}(\sigma_{\mu\nu} B) = \frac{1}{2q_A{}^2 q_B{}^2} [\mathrm{Tr}(A q_\mu{}^A \sigma_{\lambda\nu} q_\lambda{}^A) \cdot$$

$$\cdot\, \mathrm{Tr}(q_\mu{}^B \sigma_{\kappa\nu} q_\kappa{}^B B) + \mathrm{Tr}(A\, q_\nu{}^A\, \sigma_{\lambda\mu}\, q_\mu{}^A) \cdot$$

$$\cdot\, \mathrm{Tr}(q_\nu{}^B\, \sigma_{\kappa\mu}\, q_\mu{}^B\, B) + 2(g_{\lambda\lambda'} g_{\kappa\kappa'} - g_{\lambda\kappa'} g_{\kappa\lambda'}) \cdot$$

$$\cdot\, \mathrm{Tr}(A q_\lambda{}^A\, \gamma_5 \sigma_{\kappa\rho} q_\rho{}^A)\, \mathrm{Tr}(q_{\lambda'}{}^B \gamma_5 \sigma_{\kappa'\rho'} q_{\rho'}{}^B B)] \tag{5.26}$$

$$= -\, \mathrm{Tr}\left(\frac{A\sigma_{\mu\nu} q_A{}^\nu}{\sqrt{q_A^2}}\right) \mathrm{Tr}\left(\frac{\sigma_{\mu\lambda} q_B{}^\lambda B}{\sqrt{q_B^2}}\right)$$

$$- \mathrm{Tr}\left(\frac{A\gamma_5\sigma_{\mu\nu}q_A{}^{\nu}}{\sqrt{q_A{}^2}}\right) \mathrm{Tr}\left(\frac{\gamma_5\sigma_{\mu\lambda}q_B{}^{\lambda}B}{\sqrt{q_B{}^2}}\right) \tag{5.27}$$

since

$$\varepsilon_{\mu\nu\kappa\lambda}\,\varepsilon_{\mu\nu\kappa'\lambda'} = -2(g_{\kappa\kappa'}g_{\lambda\lambda'} - g_{\kappa\lambda'}g_{\lambda\kappa'}). \tag{5.28}$$

This shows clearly that eight independent spin states are present corresponding to two each of $J^P = 0^+, 0^-, 1^+, 1^-$. Superimposing this result on the known spectrum of the B_{2n} function, we see that all levels exhibit parity-doubling.

Next we study the hermiticity of the couplings, by comparison with the usual Feynman rules.[3] We are concerned with the quantities Γ satisfying

$$\gamma_0\Gamma^+\gamma_0 = \Gamma \tag{5.29}$$

Since for these quantities one knows that

$$\bar{u}(q_2)\ \Gamma\ u(q_1) = [\bar{u}(q_1)\ \Gamma\ u(q_2)]^+ \tag{5.30}$$

as required for hermiticity. With the usual definitions and conventions

$$\gamma^0 = \begin{pmatrix} 1 & 0 \\ 0 & -1 \end{pmatrix} \tag{5.31}$$

$$\underline{\gamma} = \begin{pmatrix} 0 & \underline{\sigma} \\ -\underline{\sigma} & 0 \end{pmatrix} \tag{5.32}$$

$$\gamma_5 = i\gamma_0\gamma_1\gamma_2\gamma_3 = \begin{pmatrix} 0 & 1 \\ 1 & 0 \end{pmatrix} \tag{5.33}$$

then the following are hermitian couplings

$$\Gamma = 1,\ i\gamma^5,\ \gamma^{\mu},\ i\gamma^5\gamma^{\mu},\ \sigma_{\mu\nu},\ i\gamma^5\sigma_{\mu\nu} \tag{5.34}$$

Taking into account the factors $+i$ and $-ig^{\mu\nu}$ associated with the spin zero and spin one propagators, we find the appropriate vertex factors to be

$$J^P = 0^+ \qquad -i\ ;\ \not{q}/\sqrt{q^2} \tag{5.35}$$

$$J^P = 0^- \qquad \gamma_5 \; ; \; \gamma_5 \not{q}/\sqrt{q^2} \tag{5.36}$$

$$J^P = 1^- \qquad -i\gamma_\mu \; ; \; \sigma_{\mu\nu} q^\nu/\sqrt{q^2} \tag{5.37}$$

$$J^P = 1^+ \qquad -i\gamma_5\gamma_\mu \; ; \; -i\gamma_5\sigma_{\mu\nu} q^\nu/\sqrt{q^2} \tag{5.38}$$

For the second set of couplings, we have noted that $q_\mu \to -q_\mu$ under the hermitian conjugation.

Comparison of these correct couplings with those occurring in the completeness relation reveals that precisely half the states, namely the scalars and axial vectors, are ghosts with non-hermitian coupling. These are called spin ghosts and in the later developments we shall overcome this problem (although not the other problem of parity-doubling).

Why do we obtain ghosts with a spin multiplicative factor and not with the SU_2 or SU_3 internal symmetry factors? The reason is that the spin group associated with leaving invariant the Lorentz scalar

$$u^+\gamma^0 u \qquad \gamma^0 = \begin{pmatrix} +1 & & & \\ & +1 & & \\ & & -1 & \\ & & & -1 \end{pmatrix} \tag{5.39}$$

is a non-campact group U(2,2) and any finite-dimensional representation must be non-unitary. The groups SU_2 and SU_3 are compact and finite dimensional UIR are possible.

From another viewpoint, the negative signs in γ^0 correspond to the negative energy solutions of the Dirac equation, and are connected to the well-known re-interpretation in terms of the positive-energy holes in the Dirac sea.

We can of course, combine the two types of multiplicative factor; combining the spin factor with SU_2, SU_3 gives rise to the larger invariances contained in

$$U(4,4) \supset U(2,2) \otimes SU_2 \tag{5.40}$$

and

$$U(6,6) \supset U(2,2) \otimes SU_3 \tag{5.41}$$

respectively. [c f. Salam, Delbourgo and Strathdee, Reference 4; reprinted in Reference 5]. Half of the states are spin-ghosts which are theoretically unacceptable. The parity-doubling, although not objectional on purely theoretical grounds, is physically unacceptable since it is not observed.

Concerning this parity-doubling problem there is a more general discussion[6] that can be made as follows. Let us multiply a dual amplitude B_N by a factor dependent on the external momenta

$$A_N = F_N(p_i)\, B_N \tag{5.42}$$

Consider the case of N = 6 pseudoscalars and the factorisation at the internal three-particle ground-state pseudoscalar. We know that B_6 factorises correctly and hence we must have

$$F_6 = P(s_{12}, s_{23})\, P(s_{45}, s_{56}) + \alpha_{123}\, \tilde{F}_6^{(123)} \tag{5.43}$$

$$= P(s_{23}, s_{34})\, P(s_{56}, s_{61}) + \alpha_{234}\, \tilde{F}_6^{(234)} \tag{5.44}$$

where $s_{ij} = (p_i + p_j)^2$, $\alpha_{ijk} = \alpha((p_i + p_j + p_k)^2)$. Now put $\alpha_{123} = \alpha_{234} = 0$ and realise that all six energies $s_{i,i+1}$ (i = 1, 2, 3, 4, 5, 6) are independent variables to see that only a trivial solution P = constant survives. Hence we cannot modify B_N by a non-trivial multiplicative factor <u>and</u> keep only a pseudoscalar in the 3π channel. By extending the argument (Reference 6) it can be shown that both 1^+ <u>and</u> 1^- must be present, and hence parity-doubling is inevitable.

The general conclusion of this subsection is that any simple multiplicative spin factor with the generalised Euler B function is unsuccessful and certainly cannot lead to a physically-acceptable Born amplitude. A less trivial introduction of spin becomes essential and we shall devote the rest of our general discussion of spin to such an attempt, the Ramond-Neveu-Schwarz (RNS) theory, which has high internal consistency. It will succeed in eliminating spin ghosts although not parity doubling; it will not describe the real world but it does provide a very stimulating development.

5.3 FREE FERMIONS (Ramond)

To introduce a dual theory for free fermions we first present a different viewpoint[7] on the bosonic case. We introduce a hamiltonian

$$H = \sum_{n=0}^{\infty} \frac{1}{2} (p_\mu^{(n)} p_\mu^{(n)} + n^2 q_\mu^{(n)} q_\mu^{(n)}) \tag{5.45}$$

at each point x_μ, p_μ of Lorentz phase space, with the commutators

$$[q_\mu^{(m)}, p_\nu^{(n)}] = - i g_{\mu\nu} \delta_{mn} \tag{5.46}$$

$$[q_\mu^{(m)}, q_\nu^{(n)}] = [p_\mu^{(m)}, p_\nu^{(n)}] = 0 \tag{5.47}$$

Defining generalised momentum and position operators through

$$P_\mu = \sqrt{2} \sum_{n=0}^{\infty} p_\mu^{(n)} \tag{5.48}$$

$$Q_\mu = \frac{1}{\sqrt{2}} \sum_{n=0}^{\infty} q_\mu^{(n)} \tag{5.49}$$

we introduce a τ-variable through the Heisenberg equations

$$P_\mu(\tau) = e^{i\tau H} P_\mu e^{-i\tau H} \tag{5.50}$$

$$Q_\mu(\tau) = e^{i\tau H} Q_\mu e^{-i\tau H} \tag{5.51}$$

Defining

$$a_\mu^{(n)+} = \frac{1}{\sqrt{2}} [\sqrt{n}\, q_\mu^{(n)} - \frac{i}{\sqrt{n}} p_\mu^{(n)}] \tag{5.52}$$

$$a_\mu^{(n)} = \frac{1}{\sqrt{2}} [\sqrt{n}\, q_\mu^{(n)} + \frac{i}{\sqrt{n}} p_\mu^{(n)}] \tag{5.53}$$

$$p_\mu^{(n)} = i \frac{\sqrt{n}}{\sqrt{2}} (a_\mu^{(n)+} - a_\mu^{(n)}) \tag{5.54}$$

$$q_\mu^{(n)} = \frac{1}{\sqrt{2n}} (a_\mu^{(n)+} + a_\mu^{(n)}) \tag{5.55}$$

$$[a_\mu^{(n)}, a_\nu^{(m)+}] = -g_{\mu\nu}\, \delta_{mn} \tag{5.56}$$

then

$$p^{(n)} \cdot p^{(n)} + n^2 q^{(n)+} \cdot q^{(n)} = 2n a^{(n)+} \cdot a^{(n)} + 4n \tag{5.57}$$

and

$$P_\mu(\tau) = \sqrt{2}\, p_\mu^{(0)} + i \sum_{n=1}^{\infty} \sqrt{n}\, e^{i\tau n a^{(n)+} a^{(n)}} \cdot$$

$$\cdot\, (a_\mu^{(n)+} - a_\mu^{(n)})\, e^{-i\tau n a^{(n)+} a^{(n)}} \tag{5.58}$$

$$= \sqrt{2}\, p_\mu^{(0)} - i \sum_{n=1}^{\infty} \sqrt{n}\, [a_\mu^{(n)}\, e^{-in\tau} - a_\mu^{(n)+}\, e^{in\tau}] \tag{5.59}$$

We may consider τ as a cyclic variable, and physical quantities are taken to be averages over $-\pi \leq \tau \leq +\pi$. With the notation

$$\langle A(\tau) \rangle = \frac{1}{2\pi} \int_{-\pi}^{\pi} d\tau\, A(\tau) \tag{5.60}$$

we see that

$$\frac{1}{\sqrt{2}} \langle P_\mu(\tau) \rangle = p_\mu^{(0)} \tag{5.61}$$

$$\sqrt{2}\, \langle Q_\mu(\tau) \rangle = q_\mu^{(0)} \tag{5.62}$$

For products of operators, we adopt a prescription (correspondence principle) that a product of averages is replaced by an average over the product, suitably normal ordered. Thus

$$p_\mu p_\mu = \frac{1}{2} \langle P_\mu(\tau)\rangle\langle P_\mu(\tau)\rangle \tag{5.63}$$

$$\to \frac{1}{2} \langle : P^2(\tau) :\rangle \tag{5.64}$$

With such a correspondence principle the Klein-Gordon equation becomes

$$\left(\frac{1}{2} \langle :P^2(\tau):\rangle - m^2\right) |\phi\rangle = 0 \tag{5.65}$$

or

$$\left(p^2 + \sum_{n=1}^{\infty} n a^{(n)+} \cdot a^{(n)} - m^2\right) |\phi\rangle = 0 \tag{5.66}$$

that is

$$(L_0 + m^2) |\phi\rangle = 0 \tag{5.67}$$

which is the familiar equation for the spectrum of dual models.

We can go further to identify the gauge conditions. To eliminate ghosts we need subsidiary conditions of the form

$$p_\mu \, a_\mu^{(n)} |\phi\rangle = 0 \tag{5.68}$$

that is

$$\langle P_\mu(\tau)\rangle\langle e^{in\tau} P_\mu(\tau)\rangle |\phi\rangle = 0 \tag{5.69}$$

Through the correspondence principle this becomes

$$\langle e^{in\tau} : P(\tau)^2 :\rangle |\phi\rangle = 0 \tag{5.70}$$

and we know from earlier discussions that this is equivalent to the Virasoro gauges

$$\Big[i \sqrt{2n}\, p\cdot a^{(n)} - \sum_{m=1}^{\infty} \sqrt{m(m+n)}\, a^{(m)+}\cdot a^{(m+n)} + \frac{1}{2} \sum_{m=1}^{n-1} \sqrt{m(n-m)}\, a^{(m)}\cdot a^{(n-m)}\Big] |\phi\rangle = 0 \tag{5.71}$$

These results suggest that we should attempt a similar treatment of the Dirac equation[8]. To do so we need a generalisation $\Gamma_\mu(\tau)$ of the Dirac gamma matrix γ_μ. We require that $\Gamma_\mu(\tau)$ satisfy the following three requirements

$$\text{i)} \quad \langle \Gamma_\mu(\tau) \rangle = \gamma_\mu \tag{5.72}$$

$$\text{ii)} \quad \gamma_0 \Gamma_\mu(z)^+ \gamma_0 = \Gamma_\mu(z) \tag{5.73}$$

$$\text{iii)} \quad \{\Gamma_\mu(\tau), \Gamma_\nu(\tau')\}_+ = 2\, g_{\mu\nu}\, \delta(\tfrac{1}{2\pi}(\tau - \tau')) \tag{5.74}$$

These properties will generalise the well-known

$$\gamma_0 \gamma_\mu^+ \gamma_0 = \gamma_\mu \tag{5.75}$$

$$\{\gamma_\mu, \gamma_\nu\}_+ = 2 g_{\mu\nu} \tag{5.76}$$

Concerning the argument in the delta function of $\{\Gamma_\mu(\tau), \Gamma_\nu(\tau')\}_+$ we may motivate it by noticing that

$$[Q_\mu(\tau), P_\nu(\tau')] = -i\, g_{\mu\nu}\, \delta(\tfrac{1}{2\pi}(\tau - \tau')) \tag{5.77}$$

To satisfy the three requirements we find

$$\Gamma_\mu(\tau) = \gamma_\mu + i\sqrt{2}\, \gamma_5 \sum_{n=1}^{\infty} (b_\mu^{(n)+} e^{in\tau} + b_\mu^{(n)} e^{-in\tau}) \tag{5.78}$$

where

$$\{b_\mu^{(m)}, b_\nu^{(n)+}\}_+ = -g_{\mu\nu}\, \delta_{mn} \tag{5.79}$$

Property (i) is immediately satisfied; property (ii) follows from

$$\gamma^0 \gamma_\mu^+ \gamma^0 = \gamma_\mu \tag{5.80}$$

$$\gamma^0 \gamma_5^+ \gamma^0 = \gamma_5 \tag{5.81}$$

and the anticommutator (iii) is

$$\{\Gamma_\mu(\tau),\ \Gamma_\nu(\tau')\}_+ = \{\gamma_\mu,\gamma_\nu\}_+ -2 \sum_{n=1}^{\infty} [\{\gamma_5\, b_\mu^{(n)+}, \cdot$$

$$\cdot\ \gamma_5 b_\nu^{(n)}\}_+ \, e^{in(\tau-\tau')} +$$

$$+ \{\gamma_5\, b_\mu^{(n)}, \gamma_5\, b_\nu^{(n)+}\}_+ \, e^{-in(\tau-\tau')} \tag{5.82}$$

$$= 2g_{\mu\nu} \sum_{n=-\infty}^{\infty} e^{in(\tau-\tau')} \tag{5.83}$$

$$= 2g_{\mu\nu}\, \delta(\frac{1}{2\pi}\,(\tau - \tau')) \tag{5.84}$$

as required.

We are now ready to apply the correspondence principle to the Dirac equation through the steps

$$(\gamma\cdot p - m)\ |\phi\rangle = 0 \tag{5.85}$$

$$(\frac{1}{\sqrt{2}} \langle\Gamma_\mu(\tau)\rangle\langle P_\mu(\tau)\rangle - m)\ |\phi\rangle = 0 \tag{5.86}$$

$$\to (\frac{1}{\sqrt{2}} \langle\Gamma_\mu(\tau)\cdot P_\mu(\tau)\rangle - m)\ |\phi\rangle = 0 \tag{5.87}$$

Hence

$$[\gamma\cdot p - m - \gamma_5 \sum_{n=1}^{\infty} \sqrt{n}\ (a^{(n)+}\cdot b^{(n)} - a^{(n)}\cdot b^{(n)+})]\,|\phi\rangle = 0$$

By squaring the dual fermion equation, we can arrive eventually at a form exhibiting linear Regge trajectories, as follows. Start from

$$(\frac{1}{2} \langle\Gamma\cdot P\rangle\langle\Gamma\cdot P\rangle - m^2)\ |\phi\rangle = 0 \tag{5.89}$$

Now since

$$\frac{1}{2} <\Gamma\cdot P><\Gamma\cdot P> = \frac{1}{2} << (-\Gamma_{\nu}(\tau')\Gamma_{\mu}(\tau) +$$

$$+ 2g_{\mu\nu}\delta(\frac{1}{2\pi}(\tau - \tau'))) (P_{\nu}(\tau') P_{\mu}(\tau) -$$

$$- ig_{\mu\nu}\frac{d}{d\tau} \delta(\frac{1}{2\pi}(\tau - \tau')))>> \qquad (5.90)$$

$$= - \frac{1}{2} <\Gamma\cdot P><\Gamma\cdot P> + <P^2> - \frac{i}{2} <\Gamma\cdot\dot{\Gamma}> \qquad (5.91)$$

with

$$\dot{\Gamma}(\tau) = \frac{d}{d\tau} \Gamma(\tau) \qquad (5.92)$$

$$= - \sqrt{2}\, \gamma_5 \sum_{n=1}^{\infty} n(b_{\mu}^{(n)+} e^{in\tau} - b_{\mu}^{(n)} e^{-in\tau}) \qquad (5.93)$$

and since

$$- \frac{1}{4} i <\Gamma\cdot\dot{\Gamma}>= \sum_{n=1}^{\infty} n\, b_{\mu}^{(m)+} b_{\mu}^{(n)} \qquad (5.94)$$

we arrive at

$$(\frac{1}{2} <\Gamma\cdot P><\Gamma\cdot P> - m^2)|\phi> = (\frac{1}{2} <P^2> - \frac{1}{4} i<\Gamma\cdot\dot{\Gamma}> - m^2)|\phi> \qquad (5.95)$$

$$= [p^2 - m^2 + \sum_{n=1}^{\infty} n(a^{(n)+}\cdot a^{(n)} + b^{(n)+}\cdot b^{(n)})]|\phi> = 0 \qquad (5.96)$$

from which the general nature of the spectrum is manifest: namely, linear trajectories.

For consistency we require the term $- \frac{i}{4} <\Gamma\cdot\dot{\Gamma}>$ to generate τ translations also. By using

$$[AB, C]_- = A\{B, C\}_+ - \{A, C\}_+ B \qquad (5.97)$$

we can check that

$$[- \frac{1}{4} i <\Gamma\cdot\dot{\Gamma}>, \Gamma_{\mu}(\tau)] = - \frac{1}{4} i<\Gamma_{\mu}(\tau')\{\dot{\Gamma}_{\nu}(\tau'), \Gamma_{\mu}(\tau)\}_+$$

$$- \{\Gamma_{\mu}(\tau'), \Gamma_{\mu}(\tau)\}_+ \dot{\Gamma}_{\nu}(\tau')> \qquad (5.98)$$

$$= -\frac{1}{4} i \langle \Gamma_\mu(\tau)\ 2 \frac{d}{d\tau} \delta(\frac{1}{2\pi}(\tau - \tau'))$$

$$- 2\dot{\Gamma}_\mu(\tau)\ \delta(\frac{1}{2\pi}(\tau - \tau')) \tag{5.99}$$

$$= i \frac{d}{d\tau} \Gamma_\mu(\tau) \tag{5.100}$$

as expected.

Putting $Z = e^{-i\tau}$, we introduce the operators

$$L_n^{\Gamma} = + \frac{1}{4} i \langle Z^{-n} : \Gamma(Z)\cdot\dot{\Gamma}(Z) :\rangle \tag{5.101}$$

$$= \frac{1}{2} \sum_{m=1}^{\infty} (n + 2m)\ b^{(m)+}\cdot b^{(m+n)} -$$

$$- \frac{1}{4} \sum_{m=1}^{n-1} (n - 2m)\ b^{(m)+}\cdot b^{(n-m)} +$$

$$+ \frac{\sqrt{2}}{4} i\ n\ \gamma\cdot b^{(n)}\gamma^5 \tag{5.102}$$

Aside from possible anomaly terms at m + n = 0 we can find by careful use of

$$[AB, CD] = A\{B, C\}_+D - \{A, C\}_+BD + CA\{B, D\}_+ -$$

$$- C\{A, D\}_+ B \tag{5.103}$$

that these operators satisfy the generalised projective algebra

$$[L_m^{\Gamma}, L_n^{\Gamma}] = (m - n)\ L_{m+n}^{\Gamma}\ (+ \text{ anomaly term}) \tag{5.104}$$

The anomaly for n = -m arises from two terms: firstly the quadratic in $b^{(n)}$ piece gives rise to

$$\frac{d}{16} \sum_{m=1}^{m-1} 2(m - 2n)^2 = \frac{d}{24}(m^3 - 3m^2 + 2m) \tag{5.105}$$

Secondly there is the term

$$[\frac{\sqrt{2}\ im\ \gamma\cdot b^{(m)}\gamma^5}{4}, \frac{\sqrt{2}\ im\ \gamma\cdot b^{(m)+}\gamma^5}{4}] \tag{5.106}$$

which contains $(m^2 d/8)$. This gives altogether

$$[L_m^{\Gamma}, L_n^{\Gamma}] = (m - n) L_{m+n}^{\Gamma} + \frac{d}{24} m(m^2 + 2) \delta_{m+n,0} \tag{5.107}$$

suggesting that we re-define

$$L_0^{\Gamma'} = L_0^{\Gamma} + \frac{d}{16}, \; L_n^{\Gamma'} = L_n^{\Gamma} \; (n \neq 0) \tag{5.108}$$

whereupon

$$[L_m^{\Gamma'}, L_n^{\Gamma'}] = (m - n) L_{m+n}^{\Gamma'} + \frac{d}{24} m(m^2 - 1)\delta_{m+n,0} \tag{5.109}$$

Combining these operators into

$$L_n^f = L_n^{\Gamma'} + L_n^a \tag{5.110}$$

$$= -\frac{1}{2} \langle Z^{-n} : (P^2 - \frac{i}{2} \Gamma\cdot\dot{\Gamma}) :\rangle + \frac{d}{16} \delta_{n0} \tag{5.111}$$

we arrive at

$$[L_m^f, L_n^f] = (m - n) L_{m+n}^f + \frac{d}{8} m(m^2 - 1) \delta_{m+n,0} \tag{5.112}$$

by bearing in mind the anomaly $\frac{d}{12} m(m^2 - 1) \delta_{m+n,0}$ of the L_n^a algebra.

The gauge condition

$$L_n^f |\phi\rangle = 0 \tag{5.113}$$

is compatible with the spectrum equation

$$(L_0^f + m^2 - \frac{d}{16}) |\phi\rangle - 0 \tag{5.114}$$

The crucial point is that we may introduce further gauges, to eliminate spin ghosts, in the form

$$F_n = \frac{1}{\sqrt{2}} \langle Z^{-n} : \Gamma(z)_\mu P_\mu(z) :\rangle \tag{5.115}$$

$$= \gamma \sum_{m=1}^{\infty} \sqrt{m}\,(a_m \cdot b^+_{m-n} - a^+_m \cdot b_{m+n})$$

$$+ i\sqrt{2}\, p_\mu \gamma_5 b_\mu^{(n)} - i\,\sqrt{n/2}\;\gamma \cdot a^{(n)} \tag{5.116}$$

which satisfy

$$[L_m^f, F_n] = (\frac{m}{2} - n)\, F_{m+n} \tag{5.117}$$

$$\{F_m, F_n\}_+ = -\,2\,L^f_{m+n} - \frac{d}{2}\,(m^2 - \frac{1}{4})\,\delta_{m+n,0} \tag{5.118}$$

To arrive at the $[L_m^f, F_n]$ commutator we use

$$[P^2(z), P_\mu(z')] = 4\pi i\, P_\mu(z)\,\frac{d}{d\tau}\,(\tau - \tau') \tag{5.119}$$

$$[\Gamma(z)\cdot\dot{\Gamma}(z), \Gamma_\mu(z')] = [4\pi\;\Gamma_\mu(z)\,\frac{d}{d\tau}\,\delta(\tau - \tau') -$$

$$-\,4\pi\;\dot{\Gamma}_\mu(z)\;\delta(\tau - \tau')] \tag{5.120}$$

To arrive at the anticommutator $\{F_m, F_n\}_+$ we first use

$$\{P(z)\cdot\Gamma(z), P(z')\cdot\Gamma(z')\}_+ = 2\pi i\;\Gamma(z)\cdot\Gamma(z)\;\delta'(\tau - \tau') +$$

$$+\,4\pi P(z)\cdot P(z)\;\delta(\tau - \tau') \tag{5.121}$$

and then to obtain the anomaly term, we find two pieces, namely

$$-d \sum_{p=1}^{m-1} p = -\,\frac{d}{2}\,(m^2 - m) \tag{5.122}$$

and

$$-\,\frac{m}{2}\,\gamma_\mu\gamma_\mu = -\,\frac{dm}{2} \tag{5.123}$$

coming from the quadratic term and from the $\gamma\cdot a^{(n)}$ term respectively. The sum gives $(-\,\frac{d}{2}\,m^2)\;\delta_{m+n,0}$ and hence the required result when we use $L_0^{\Gamma'} = (L_0^{\Gamma} + d/16)$.

To summarise the gauge algebra, it is (for the

Ramond theory)

$$[L_m^f, L_n^f] = (m - n) L_{m+n}^f + \frac{d}{8} m(m^2 - 1)\delta_{m+n,0} \quad (5.124)$$

$$[L_m^f, F_n] = (\frac{m}{2} - n) F_{m+n} \quad (5.125)$$

$$\{F_m, F_n\} = - 2L_{m+n}^f - \frac{d}{2} (m - \frac{1}{4})\delta_{m+n,0} \quad (5.126)$$

The F_n gauges now provide further subsidiary conditions since we know that

$$0 = F_n (F_0 - m) |\phi\rangle \quad (5.127)$$

$$= [2L_n^f - (F_0 + m) F_n] |\phi\rangle \quad (5.128)$$

and therefore

$$F_n|\phi\rangle = 0 \quad (5.129)$$

It is amusing that these are the extension, through the correspondence principle, of the well-known Rarita-Schwinger subsidiary condition in the forms

$$p\cdot b^{(n)} |\phi\rangle = 0 \quad (5.130)$$

$$\gamma\cdot a^{(n)} |\phi\rangle = 0 \quad (5.131)$$

We shall discuss the spectrum later, but some general features are

i) The parent trajectory is doubly degenerate since we may take either of the forms, for the parent states

$$(a_\mu^{(1)+})^J |0\rangle \quad (5.132)$$

$$(a_\mu^{(1)+})^{J-1} (b_\nu^{(1)+}) |0\rangle \quad (5.133)$$

ii) All levels but the ground-state are parity-doubled.

5.4 DUAL PION MODEL (Neveu and Schwarz)

We now introduce a bosonic multiparticle amplitude, due to Neveu and Schwarz[9], that involves anticommuting operators similar, but not identical, to those introduced in the previous section. For the moment we regard this development as completely separate from that for fermions, but at a later stage we shall couple the two together in the full RNS theory of bosons and fermions.

We introduce operators satisfying*

$$\{b_\mu^{(r)}, b_\nu^{(s)+}\}_+ = - g_{\mu\nu}\delta_{rs} \tag{5.134}$$

where r, s $= \frac{1}{2}, \frac{3}{2}, \frac{5}{2}, \ldots$ are now half-odd integers. Also we introduce a field

$$H_\mu(\tau) = \sum_{r=-\infty}^{\infty} b_\mu^{(r)} e^{-ir\tau} \tag{5.135}$$

whereupon

$$\{H_\mu(\tau), H_\nu(\tau')\}_+ = - 2\pi g_{\mu\nu} \delta(\tau - \tau') \tag{5.136}$$

Now we define

$$L_n^b = - \frac{1}{2} i \langle e^{in\tau} : \dot{H}(\tau)\cdot H(\tau) :\rangle \tag{5.137}$$

where

$$\dot{H}_\mu(\tau) = \frac{d}{d\tau} H_\mu(\tau) \tag{5.138}$$

It follows that

---

*We adopt consistently the convention throughout that indices m, n on $b_\mu^{(m)+}$, $b_\mu^{(n)+}$, have integer values appropriate to the fermion sector, whereas indices r, s on $b_\mu^{(r)+}$, $b_\mu^{(s)+}$ have half-odd-integer values appropriate to the boson sector.

$$L_n^{\ b} = -\frac{1}{2} \sum_{r=1/2}^{\infty} (n + 2r)\, b_r^{\ +} \cdot b_{n+r}$$

$$- \frac{1}{4} \sum_{r=1/2}^{n-1/2} (n - 2r)\, b_r \cdot b_{n-r} \qquad (5.139)$$

and these operators satisfy the algebra

$$[L_m^{\ b}, L_n^{\ b}] = (m - n)\, L_{m+n}^{b} + \frac{d}{24} m(m^2 - 1)\, \delta_{m+n,0} \qquad (5.140)$$

The anomaly term arises from the piece

$$\frac{1}{16} \sum_{r=1/2}^{m-1/2} \sum_{r=1/2}^{n-1/2} (m - 2r)\,(m - 2s)\, [b_r\, b_{m-r},\, b_{n-s}^{+}\, b_s^{\ +}] \qquad (5.141)$$

and using

$$[AB, CD] = A\{BC\}_+ D - \{AC\}_+ BD + CA\{BD\}_+ - C\{AD\}_+ B \qquad (5.142)$$

we find, after normal ordering the extra term

$$\frac{d}{8} \sum_{r=1/2}^{m-1/2} (m - 2r)^2 = \frac{d}{24} m(m^2 - 1) \qquad (5.143)$$

as required.

Adding*

$$L_n = L_n^{\ a} + L_n^{\ b} \qquad (5.144)$$

then gives

$$[L_m, L_n] = (m - n)\, L_{m+n} + \frac{d}{8} m(m^2 - 1)\, \delta_{m+n,0} \qquad (5.145)$$

---

*For the full gauge operator (commuting part plus anticommuting part) we write consistently L_n with no superscript for the boson sector, to distinguish from $L_n^{\ \Gamma}$ for the fermion sector. For the anticommuting part alone we use $L_n^{\ b}$ for bosons, $L_n^{\ \Gamma}$ for fermions; the commuting part $L_n^{\ a}$ is the same for both sectors.

In terms of the vertex $V_0(p, z)$ defined for the generalised Veneziano model through

$$V_0(p, z) = : \exp(\sqrt{2}\, ip\cdot Q(z)) : \qquad (5.146)$$

we define the Neveu-Schwarz vertex

$$V(p, z) = p\cdot H(z)\, V_0(p\ z) \qquad (5.147)$$

Now we need the commutator

$$[L_n, H_\mu(z)] = -\frac{1}{2}\, i < Z'^{-n}[H_{\mu'}(z')\, \dot{H}_\mu(z'), H_\mu(z)]> \qquad (5.148)$$

$$= -\frac{1}{2}\, i < Z'^{-n}\{H_\mu(z')\, -2\pi \frac{d}{d\tau'}\, \delta(\tau - \tau') -$$

$$- \dot{H}_\mu(z')\cdot 2\pi\, \delta(\tau - \tau') \qquad (5.149)$$

$$= -\frac{1}{2}\, nz^{-n}\, H_\mu(z) - iz^{-n} \frac{d}{d\tau}\, H(z) \qquad (5.150)$$

Now

$$z = e^{-i\tau} \qquad (5.151)$$

$$z\frac{d}{dz} = i\frac{d}{d\tau} \qquad (5.152)$$

$$\therefore\ [L_n \cdot H_\mu(z)] = -z^{-n}(z\frac{d}{dz} - \frac{n}{2})\, H_\mu(z) \qquad (5.153)$$

Combining this result with

$$[L_n, V_0(p, z)] = -\, z^{-n}(z\frac{d}{dz} + np^2)\, V_0(p, z) \qquad (5.154)$$

we find

$$[L_n, V(p, z)] = -\, z^{-n}(z\frac{d}{dz} + n(p^2 - \frac{1}{2}))\, V(p, z) \qquad (5.155)$$

so that $V(p, z)$ is a generalised projective vector $(S = -1)$

provided that

$$p^2 = -\frac{1}{2} \tag{5.156}$$

This ensures that the commutator is a perfect differential

$$[L_n, V(p, z)] = -z \frac{d}{dz} (z^{-n} V(p, z)) \tag{5.157}$$

as necessary for the generalised projective gauge conditions.

We are ready now to write the amplitude for N pions which is

$$A_N = \int \prod_{i=1}^{N} d_{zi} \, (dV_3)^{-1} \prod_{i=1}^{N} z_i^{-1/2}$$

$$\langle 0| \, V(p_1, z_1) \, V(p_2, z_2), \ldots, V(p_N, z_N) \, |0\rangle \tag{5.158}$$

First we may confirm projective invariance under

$$z_i \rightarrow z_i' = \left(\frac{az_i + b}{cz_i + d}\right) = \Lambda z \tag{5.159}$$

$$\Lambda = e^{i\underline{\xi} \cdot \underline{L}} \tag{5.160}$$

since we find

$$\Lambda \left[\frac{V(p, z)}{\sqrt{z}}\right] \Lambda^{-1} = (a - cz')^2 \left[\frac{V(p, z')}{\sqrt{z'}}\right] \tag{5.161}$$

similarly to the earlier result for $V_0(p, z)$, since in general a field on z with generalised projective spin S picks up a factor $(a - cz')^{-2S}$ under such a transformation.

This observation, together with the fact that

$$dz_i = \frac{dz_i'}{(a - cz_i')^2} \tag{5.162}$$

enables us to confirm projective invariance.

Cyclic symmetry is proved by a straight-forward extension of the earlier case; now we use both

$$V_0(p, z)\, V_0(p', z') = V_0(p', z')\, V_0(p, z)(-1)^{-2p\cdot p'} \tag{5.163}$$

and

$$p\cdot H(z)\, p'\cdot H(z') = p'\cdot H(z')\, p\cdot H(z) - 2\pi\, p\cdot p'\, \delta(\tau - \tau')$$

The delta function term in the second equation gives zero contribution to the integral, and the new phase $(-1)^{N-1} = -1$ from $H_\mu(z)$ anticommutation is compensated by the factor

$$(-1)^{-2p_1 \sum_{i=1}^{N} p_i} = (-1)^{2p_1{}^2} = -1 \tag{5.165}$$

arising in the $V_0(p, z)$ commutation.

To see the factorisation properties explicitly we use

$$z^{-L_0}\, V(p, 1)\, z^{L_0} = V(p, z)\sqrt{z} \tag{5.166}$$

$$\lim_{z_1 \to 0} [\, \langle 0|V(p_1, z_1)] = \langle 0|\, e^{ip_1\cdot Q}\, p_1\cdot b_{1/2} \tag{5.167}$$

$$\lim_{z_N \to \infty} [V(p_N, z_N)|0\rangle] = p_N\cdot b^+_{1/2}\, e^{ip_N\cdot Q}\, |0\rangle \tag{5.168}$$

and make the change of variables

$$U_{1j} = \frac{z_j}{z_{j+1}} \qquad j = 2, 3, \ldots, N-2 \tag{5.169}$$

precisely as in our earlier discussion, to arrive eventually at

$$A_N = \langle 0, p_1|p_1\cdot b_{1/2}\, V(p_2, 1)\, \frac{1}{L_0 - 1}\, V(p_3, 1) \cdots$$

$$\cdots \frac{1}{L_0 - 1}\, V(P_{N-1}, 1)\, p_N\cdot b^+_{1/2}|0, p_N\rangle \tag{5.170}$$

Where we have taken a limit $z_1 \to 0$, $z_{N-1} \to 1$, $z_N \to \infty$.

The generalised projective gauges follow from

$$(L_0 - L_n - 1)\frac{1}{L_0 - 1} = \frac{1}{L_0 + n-1}(L_0 + n-1 - L_n) \quad (5.171)$$

$$(L_0 + n-1-L_n)\, V(p, 1) = V(p, 1)\,(L_0 - L_n - 1) \quad (5.172)$$

together with

$$(L_0 - L_n - 1)\, p\cdot b^{(1/2)+}\, |0\rangle = 0 \quad (5.173)$$

implying

$$L_n\, |\phi\rangle = 0 \quad (5.174)$$

Where $|\phi\rangle$ is a physical state coupling to pions

$$|\phi\rangle = \frac{1}{L_0 - 1} V(p_N, 1) \frac{1}{L_0 - 1} V(p_{N-1}, 1) \cdots \cdot$$

$$\cdot\, V(p_2, 1)\, P_1\, b^{(1/2)^+} |0\rangle \quad (5.175)$$

As a preliminary remark to the study of the spectrum, note that since $V(p, z)$ is linear in $b_\mu^{(n)}$ it follows that A_N vanishes for N odd, and that we may define G-parity by

$$G = (-1)^{\sum_r b^{(r)+}\cdot b^{(r)}} \quad (5.176)$$

so that $G = -1$ for the pion state

$$|\pi\rangle = p\cdot b^{(1/2)+}|0\rangle \quad (5.177)$$

The twisting operator for the model requires a phase dependent on the G-parity as follows

$$\Omega = (-1)^{R_a+R_b}\, e^{-L_-}\, (-1)^{-1/2(1-G)} \quad (5.178)$$

where

$$R_a + R_b = -\sum_{n=1}^{\infty} n a^{(n)+} a^{(n)} - \sum_{r=1/2}^{\infty} n b^{(r)+} b^{(r)} \quad (5.179)$$

Since the eigenvalue of Ω acting on a physical state is the charge conjugation C we have

$$C = (-1)^{M + 1 - 1/4(1-G)} \tag{5.180}$$

When we use a Chan-Paton multiplicative factor for isospin, we deduce that the isospins of the levels in the model are

$$G = +1 \quad M^2 = 0, 2, 4, 6, \ldots T = 1 \tag{5.181}$$

$$M^2 = 1, 3, 5, 7, \ldots T = 0 \tag{5.182}$$

$$G = -1 \quad M^2 = \frac{1}{2}, \frac{5}{2}, \frac{9}{2}, \ldots \quad T = 0 \tag{5.183}$$

$$M^2 = \frac{3}{2}, \frac{7}{2}, \frac{11}{2}, \ldots \quad T = 1 \tag{5.84}$$

A cursory first examination of the spectrum

$$(L_0 - 1)\,|\phi\rangle = (-\sum_{n=1}^{\infty} n a^{(n)+} a^{(n)} - \sum_{r=1/2}^{\infty} n b^{(r)} \cdot b^{(r)} - $$

$$- p^2 - 1)\,|\phi\rangle \tag{5.185}$$

$$= 0 \tag{5.186}$$

reveals the apparent presence of all states ($R_b = 0$) of the unit-intercept Veneziano model plus additional states ($R_b \neq 0$) lying on trajectories integer and half-integer spaced relative to the orbital ($R_b = 0$) ones. Note in particular that acting with $b_\mu^{(1/2)+}$ on the orbital $\alpha(o) = 1$ parent trajectory gives rise to an apparent ancestor $\alpha_A(o) = 3/2$ trajectory. We shall see shortly that both this ancestor trajectory and the $\alpha = 0$ tachyon on the $\alpha_\rho(o) = 1$ the trajectory are spurious.

Before examination of the spectrum, we calculate the four-pion amplitude

$$A_4 = \langle 0| \; p_1 \cdot b^{(1/2)} V(p_2, 1) \frac{1}{L_0 - 1} V(p_3, 1) \cdot$$

$$\cdot \; p_4 \cdot b^{(1/2)+} |0\rangle \tag{5.187}$$

$$= \int_0^1 dx \; x^{-s-2} \; {}_a\langle 0| \; V_0(p_2, 1) \; x^{R_a} \; V_0(p_3, 1) \; |0\rangle_a \cdot$$

$$\cdot \; {}_b\langle 0| \; p_1 \cdot b^{(1/2)} \; p_2 \cdot H(1) x^{R_b} \; p_3 \cdot H(1) \; p_4 \cdot b^{(1/2)+} |0\rangle_b \tag{5.188}$$

$$= \int_0^1 dx \; x^{-2-s} \; (1 - x)^{-1-t} \; \cdot$$

$$\cdot \; [(p_1 \cdot p_2)(p_3 \cdot p_4) + (p_2 \cdot p_3)(p_1 \cdot p_4) \; (\frac{x}{1 - x}) \; -$$

$$- \; (p_1 \cdot p_3)(p_2 \cdot p_4) \; x] \tag{5.189}$$

$$= \frac{1}{4} \; [\alpha_s{}^2 \; B(-\alpha_s, \; 1-\alpha_t) + \alpha_t{}^2 \; B(1-\alpha_s, \; -\alpha_t) \; -$$

$$- \; \alpha_u{}^2 \; B(1-\alpha_s, \; 1-\alpha_t)] \tag{5.190}$$

$$= - \; \frac{1}{4} \; \frac{(1-\alpha_s) \; (1-\alpha_t)}{(1-\alpha_s \; -\alpha_t)} \tag{5.191}$$

where in the final step we used $\alpha_s + \alpha_t + \alpha_u = +1$.

Thus the present model gives a fully-factorised multiparticle extension of the Lovelace-Shapiro four-pion amplitude,[10,11] for the particular intercept values $\alpha_\rho(o) = 1$, $\alpha_\pi(o) = 1/2$.

Going back to the integral representation for A_N we may write

$$A_N = \int \prod_{i=1}^{N} dz_i \; (dV_3)^{-1} \; \prod_{i \neq j} \; (z_i - z_j)^{-p_i \cdot p_j} \; \cdot$$

$$\cdot \prod_{i=1}^{N} \; z_i^{-1/2} \; {}_b\langle 0| \; p_1 \cdot H(z_1) p_2 \cdot H(z_2) \; \ldots \; p_N \cdot H(z_N) |0\rangle_b \tag{5.192}$$

and, evaluating the b vacuum expectation value gives[12,13]

$$A_N = \int \prod_{i=1}^{N} dz_i \, (dV_3)^{-1} \prod_{i \neq j} (z_i - z_j)^{-p_i \cdot p_j} \, \cdot$$

$$\cdot \sum_{\{q_1 \cdot\cdot q_N\}} (-1)^P \frac{p_{q_1} \cdot p_{q_2}}{(z_{q_1} - z_{q_2})} \frac{p_{q_3} \cdot p_{q_4}}{(z_{q_3} - z_{q_4})} \cdots$$

$$\cdots \frac{p_{q_{N-1}} \cdot p_{q_N}}{(z_{q_{N-1}} - z_{q_N})} \qquad (5.193)$$

where the sum is over all permutations of $\{q_1 q_2 \cdot\cdot q_N\}$ = 1, 2, 3, ..., N and P is the parity of the permutation.

Note that, just as for the unit intercept conventional model, the integrand is invariant under arbitrary permutations of the N argument pairs $\{p_i, z_i\}$ i = 1, 2, ..., N; <u>again</u> we are led to associate this symmetric group (S_N) invariance, for all N, with the existence of the L_n algebra.

A slightly different form, treating all z-differences as linearly independent, is

$$A_N = \int \prod_{i=1}^{N} dz_i \, (dV_3)^{-1}$$

$$\sum_{\{q_1 \cdot\cdot q_N\}} (-1)^P \frac{\partial}{\partial(z_{q_1} - z_{q_2})} \frac{\partial}{\partial(z_{q_3} - z_{q_4})} \cdots$$

$$\cdots \frac{\partial}{\partial(z_{q_{N-1}} - z_{q_N})} \, \cdot$$

$$\cdot \left[\prod_{i \neq j} (z_i - z_j)^{-p_i \cdot p_j} \right] \qquad (5.194)$$

5.5 SPECTRUM

To fully analyse the spectrum, and to prove the absence of ghosts, we need to introduce[14] G gauges defined by

$$G_r = \langle z^{-r} : P(z)\cdot H(z) :\rangle \tag{5.195}$$

$$= \sqrt{2}\, p\cdot b^{(r)} - i \sum_{m=1}^{\infty} \sqrt{m}\, (a^{(m)}\cdot b^{(m-r)+} - a^{(m)+}\cdot b^{(m+r)}) \tag{5.196}$$

These operators satisfy

$$[L_m, G_r] = (\frac{m}{2} - r)\, G_{m+r} \tag{5.197}$$

$$\{G_r, G_s\}_+ = 2L_{r+s} + \frac{d}{2}(r^2 - \frac{1}{4})\delta_{r+s,0} \tag{5.198}$$

The proofs are so close to those for the fermion model that we need not elaborate; we remark only that the anomaly term in the anticommutator is from

$$-\sum_{p=1}^{r-1/2} \sum_{q=1}^{s-1/2} \sqrt{pq}\, \{a^{(p)}\cdot b^{(r-p)}, a^{(q)+}\cdot b^{(s-q)+}\}_+ \tag{5.199}$$

which contains a piece

$$d \sum_{p=1}^{r-1/2} p = \frac{d}{2}(r^2 - \frac{1}{4}) \tag{5.200}$$

Consider now the commutator

$$[G_r, V_0(p,1)] = \langle z^{-r} H_\mu(z)\, [P_\mu(z), e^{\sqrt{2}ip\cdot Q(1)}] \tag{5.201}$$

$$= -\sqrt{2}\, p\cdot H(1)\, V_0(p, 1) \tag{5.202}$$

$$= -\sqrt{2}\, V(p, 1) \tag{5.203}$$

It follows that

$$- \sqrt{2}\,\{G_r,\ V(p,1)\}_+ = \{G_r,\ [G_r,\ V_0(p,\ 1)]\}_+ \tag{5.204}$$

$$= [L_{2r},\ V_0(p,1)] \tag{5.205}$$

$$= (L_0 + r - \tfrac{1}{2})\ V_0(p,\ 1) - V_0(p,\ 1)(L_0 - \tfrac{1}{2}) \tag{5.206}$$

where we have used

$$[L_{2r} - L_0,\ V_0(p,\ 1)] = r\ V_0(p,\ 1) \tag{5.207}$$

Now we observe that

$$G_{-\frac{1}{2}}\ |0\rangle = \sqrt{2}\ p\cdot b^{(1/2)+}\ |0\rangle = \sqrt{2}\ |\pi\rangle \tag{5.208}$$

and hence we may write the N-pion amplitude

$$A_N = \langle 0|\ G_{\frac{1}{2}}\ V(p_2,\ 1)\frac{1}{L_0 - 1}\ V(p_3,\ 1)\ \cdots\cdot$$

$$\cdots \frac{1}{L_0 - 1}\ V(p_{N-1},\ 1)\ G_{-\frac{1}{2}}\ |0\rangle \tag{5.209}$$

Now commute $G_{-\frac{1}{2}}$ to the left, making use of

$$V(p,\ 1)G_{-\frac{1}{2}} = -\ G_{-\frac{1}{2}}\ V(p,\ 1) - \frac{1}{\sqrt{2}}\,(L_0 - 1)\ V_0(p,\ 1) +$$

$$+ \frac{1}{\sqrt{2}}\ V_0(p,\ 1)\ (L_0 - \tfrac{1}{2}) \tag{5.210}$$

The second and third of these three terms give vanishing contribution. We need also

$$\frac{1}{L_0 - 1}\ G_{-\frac{1}{2}} = G_{-\frac{1}{2}}\ \frac{1}{L_0 - 1/2} \tag{5.211}$$

which follows from

$$L_0\ G_{-\frac{1}{2}} = G_{-\frac{1}{2}}\ (L_0 + \tfrac{1}{2}) \tag{5.212}$$

$$\therefore\ (L_0 - 1)G_{-\frac{1}{2}} = G_{-\frac{1}{2}}(L_0 - \tfrac{1}{2}) \tag{5.213}$$

Hence we easily arrive at

$$A_N = \langle 0| G_{\frac{1}{2}} G_{-\frac{1}{2}} V(p_2, 1) \frac{1}{L_0 - 1/2} V(p_3, 1) \frac{1}{L_0 - 1/2}$$

$$\cdots \frac{1}{L_0 - 1/2} V(p_{N-1}, 1) \ |0\rangle \qquad (5.214)$$

Now use

$$\langle 0| G_{\frac{1}{2}} G_{-\frac{1}{2}} = \langle 0| \{G_{\frac{1}{2}} G_{-\frac{1}{2}}\}_+ \qquad (5.215)$$

$$= \langle 0| \ 2L_0 = \langle 0| \qquad (5.216)$$

to arrive at the final form

$$A_N = \langle 0| \ V(p_2, 1) \frac{1}{L_0 - 1/2} V(p_3, 1) \cdots V(p_{N-1}, 0)|0\rangle \qquad (5.217)$$

This enables us to abandon the original Fock space (F_1) where the ground state is $|0\rangle$ in favor of a second Fock space (F_2) built on the pion state $|\pi\rangle$. This greatly simplifies the analysis of the spectrum. Firstly, we see that the rho trajectory tachyon and the ancestor $\alpha_A(0) = \frac{3}{2}$ trajectory are <u>outside</u> of F_2.

Next we should outline the no-ghost proof for this case.[15-17] In F_2 we have the mass-shell condition

$$(L_0 - \tfrac{1}{2}) \ |\phi\rangle = 0 \qquad (5.218)$$

and gauges

$$G_r \ |\phi\rangle = 0 \qquad (5.219)$$

$r = \frac{1}{2}, \frac{3}{2}, \frac{5}{2}, \ldots$. The G-gauge conditions imply the L-gauges through the gauge algebra; they follow from the consideration of

$$\langle \psi, L_0 = \tfrac{1}{2} - r| \ G_r \ V(p_1, 1) \frac{1}{L_0 - 1/2} V(p_2, 1) \ldots$$

$$\ldots V(p_{N-1}, 1) |0\rangle = 0 \quad (5.220)$$

The passage from the gauge conditions to a no-ghost theorem closely parallels that of the conventional model. Adopting again the notation S_{ℓ_0} for the spurious subspace at the level $L_0 = \ell_0$ we write a general spurious state as

$$|s, n\rangle = G_{\frac{1}{2}}^{+} |\phi, n-\tfrac{1}{2}\rangle + \tilde{G}_{\frac{3}{2}}^{+} |\phi, n-\tfrac{3}{2}\rangle \ \varepsilon\ S_{\frac{1}{2}} \quad (5.221)$$

with

$$\tilde{G}_{\frac{3}{2}} = \alpha\, L_1 G_{\frac{1}{2}} + \beta\, G_{\frac{3}{2}} \quad (5.222)$$

We can now try to show that

$$G_{\frac{1}{2}} |s,n\rangle \ \varepsilon\ S_0 \quad (5.223)$$

$$G_{\frac{3}{2}} |s,n\rangle \ \varepsilon\ S_{-1} \quad (5.224)$$

When this is done, it is possible to show that any on-shell state satisfying

$$G_{\frac{1}{2}} |\phi\rangle = G_{\frac{3}{2}} |\phi\rangle = 0 \quad (5.225)$$

(and thence all other gauge conditions) must be a positive-norm transverse state, modulo the addition of a possible null spurious state [we refer to the literature for the detailed proof, References 15, 16, 17].

Firstly we find that

$$G_{\frac{1}{2}} |s,n\rangle = G_{\frac{1}{2}} [G_{\frac{1}{2}}^{+} |\phi, n-\tfrac{1}{2}\rangle + \tilde{G}_{\frac{3}{2}}^{+} |\phi, n-\tfrac{3}{2}\rangle] \quad (5.226)$$

$$= - G_{\frac{1}{2}}^{+} G_{\frac{1}{2}} |\phi, n-\tfrac{1}{2}\rangle +$$

$$+ \{ \alpha(L_1^+ L_1 + G_{\frac{1}{2}}^+ G_{\frac{1}{2}}) + \beta(-G_{\frac{3}{2}}^+ G_{\frac{1}{2}} + 2L_1^+)\} |\phi, n-\tfrac{3}{2}\rangle \tag{5.227}$$

$$\epsilon\ S_0 \tag{5.228}$$

for any α, β and space-time dimensionality d. Next we use

$$\{\tilde{G}_{\frac{3}{2}}, G_{\frac{1}{2}}^+\}_+ = \alpha(-L_1 G_{\frac{1}{2}}^+ G_{\frac{1}{2}} + 2L_1 L_0 + G_{\frac{1}{2}}^+ L_1 G_{\frac{1}{2}}) + $$
$$+ 2\beta\, L_1 \tag{5.229}$$

$$\{\tilde{G}_{\frac{3}{2}}, \tilde{G}_{\frac{3}{2}}^+\}_+ = \alpha^2[2L_1 L_0 L_1^+ - L_1 L_1^+ - G_{\frac{1}{2}}^+ L_1 G_{\frac{1}{2}}^+ -$$
$$- 2G_{\frac{1}{2}}^+ L_0 G_{\frac{1}{2}}] +$$
$$+ 4\alpha\beta(L_1 L_1^+ + G_{\frac{1}{2}}^+ G_{\frac{1}{2}}) +$$
$$+ \beta^2(2L_0 + d) \tag{5.230}$$

Acting on $|s,n\rangle$ we find

$$\tilde{G}_{\frac{3}{2}}(G_{\frac{1}{2}}^+ |\phi, n-\tfrac{1}{2}\rangle) = (-\alpha + 2\beta)\, L_1\, |\phi, n-\tfrac{1}{2}\rangle \tag{5.231}$$

so that we must have $\alpha = 2\beta = 1$, say. Next we find

$$G_{\frac{3}{2}}(G_{\frac{3}{2}}^+ |\phi, n-\tfrac{3}{2}\rangle) = [8\, G_{\frac{1}{2}}^+ G_{\frac{1}{2}} + (d - 10)]\, |\phi, n-\tfrac{3}{2}\rangle$$

$$\epsilon\ S_{-1} \tag{5.233}$$

provided that d = 10, the new critical space-time dimensionality. There are no ghosts in the spectrum, provided $d \leq d_c = 10$.

Having established that such a theoretical consistency, we may look at the lowest lying levels in the spectrum. The positive norm physical states are indicated in the Chew-Frautschi plot in Figure 5.2. This should be

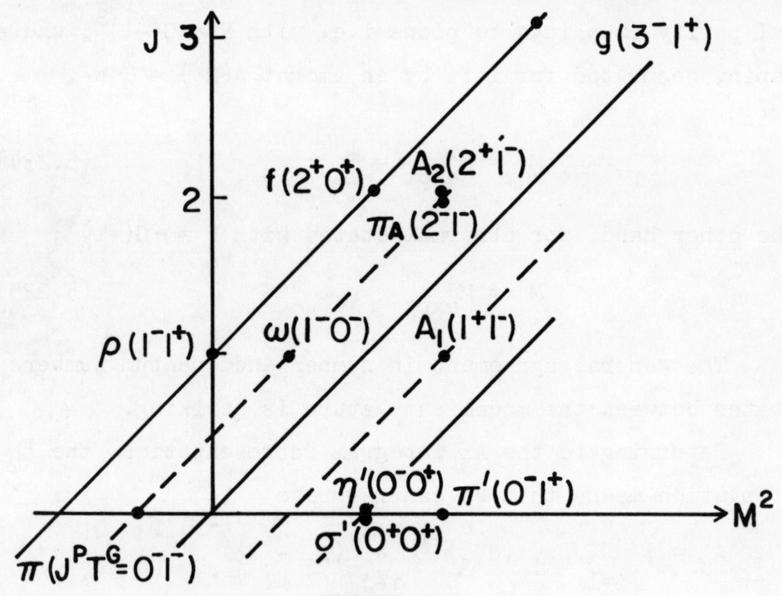

FIGURE 5.2
Chew-Frautschi Plot (Dual Pion Model)

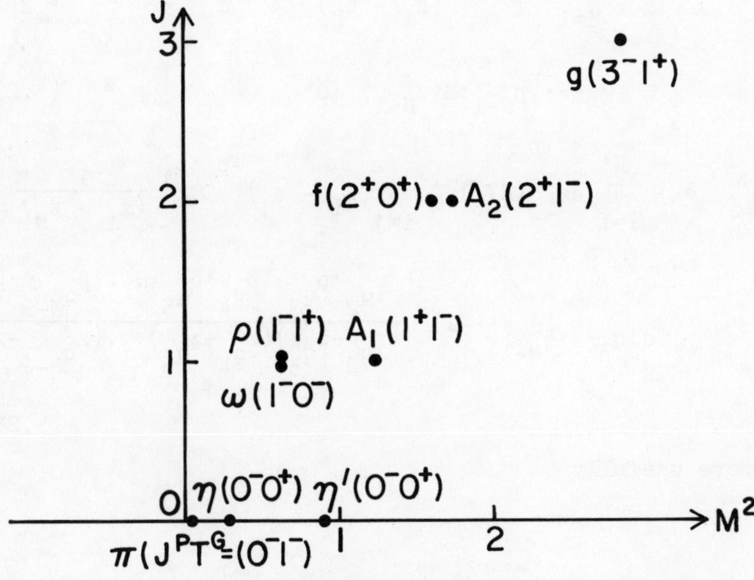

FIGURE 5.3
Chew-Frautschi Plot (Real World)

compared with the spectrum of strangeness-zero mesons observed in Nature, Figure 5.3. We see that the states with normal parity couplings to pions i.e. with $P = G(-1)^S$, where S = spin, occur too far left by an amount $\Delta(M^2) = \frac{1}{2}\alpha'$, that is

$$M^2 = M^2_{expt} - \frac{1}{2} \tag{5.234}$$

On the other hand, for abnormal states with $P = -G(-1)^S$;

$$M^2 = M^2_{expt} \tag{5.235}$$

The general agreement in number and quantum numbers of states between the model and Nature is striking.

Returning to the A_N integral representation, the F_2 reformulation means that we can re-write

$$A_N = \int \prod_{i=1}^{N} dz_i \, (dV_3)^{-1} \prod_{i\neq j} (z_i - z_j)^{-p_i \cdot p_j}$$

$$\prod_{i=1}^{N} z_i^{-1/2} \frac{1}{(z_1 - z_N)} \langle 0|\, p_2\cdot H(z_2)\, p_3\cdot H(z_3) \cdots$$

$$\cdots p_{N-1}\cdot H(z_{N-1})\, |0\rangle \tag{5.236}$$

$$= \int \prod_{i=1}^{N} dz_i \, (dV_3)^{-1} \prod_{i\neq j} (z_i - z_j)^{-p_i \cdot p_j} \frac{1}{(z_1 - z_N)}$$

$$\sum_{\{q_2 q_3 \cdots q_{N-1}\}} (-1)^P \frac{p_{q_2}\cdot p_{q_3}\; p_{q_4}\cdot p_{q_5} \cdots p_{q_{N-2}}\cdot p_{q_{N-1}}}{(z_{q_2} - z_{q_3})(z_{q_4} - z_{q_5})\ldots(z_{q_{N-2}} - z_{q_{N-1}})} \tag{5.237}$$

or, more usefully

$$A_N = \int \prod_{i=1}^{N} dz_i \, (dV_3)^{-1} \frac{1}{(z_1 - z_N)}$$

$$\sum_{\{q_2 \cdots q_{N-1}\}} (-1)^P \frac{\partial}{\partial(z_{q_2} - z_{q_3})} \frac{\partial}{\partial(z_{q_4} - z_{q_5})} \cdots$$

$$\cdots \frac{\partial}{\partial(z_{q_{N-2}} - z_{q_{N-1}})} \left[\prod_{i \neq j} (z_i - z_j)^{-p_i \cdot p_j} \right] \tag{5.238}$$

In these two forms the sum is over all permutations of $\{q_2 q_3 \cdots q_{N-1}\} = 2, 3 \cdots, (N-1)$.

Actually, because of the permutational invariance of the integrand (in its F_1 form) it is clear that the choice of z_1, z_N as neighboring points is unnecessary and we may write finally

$$A_N = \int \prod_{i=1}^{N} dz_i \, (dV_3)^{-1} \frac{1}{(z_a - z_b)}$$

$$\sum_{\{q_1 q_2 \cdot\cdot q_{N-2}\}}' (-1)^P \frac{\partial}{\partial(z_{q_1} - z_{q_2})} \frac{\partial}{\partial(z_{q_3} - z_{q_4})} \cdots$$

$$\cdots \frac{\partial}{\partial(z_{q_{N-3}} - z_{q_{N-2}})} \cdot$$

$$\cdot \left[\prod_{i \neq j} (z_i - z_j)^{-p_i \cdot p_j} \right] \tag{5.239}$$

where a, b are chosen arbitrarily and the sum is over $\{q_1 q_2 \cdot\cdot q_{N-2}\} = 1, 2, 3, \cdots, N$ (excluding a, b). This rewriting is the reflection of the G-gauge algebra in the integral representation, and will be useful in the realistic amplitudes later.

So far we have examined the boson amplitude in isolation, although the similarity of its gauge algebra to that of the free fermion model is evident. Before discussing the fermion-boson couplings, we should mention that a no-ghost theorem holds for the free fermion spectrum satisfying

$$(F_0 - m) \, |\phi\rangle = 0 \tag{5.240}$$

$$(L_0 + m^2) \, |\phi\rangle = 0 \tag{5.241}$$

$$F_n \, |\phi\rangle = 0 \tag{5.242}$$

$$L_n \, |\phi\rangle = 0 \tag{5.243}$$

since it can be shown that the transverse states are complete for d = 10, provided[18,19,20] m = 0. The last condition implies that the leading fermion trajectory has intercept $\alpha_f(o) = \frac{1}{2}$.

Perhaps the simplest way[21] to see the m = 0 condition is to count states at the first excited level. In d space-time dimensions we may make

$$\left.\begin{array}{l} a_\mu^{(1)+} \, |0\rangle \\ b_\mu^{(1)+} \, |0\rangle \end{array}\right\} \qquad \text{2d states} \tag{5.244}$$

Of these, two are spurious

$$F_1^{\,+} \, |0\rangle \tag{5.245}$$

$$L_1^{\,+} \, |0\rangle \tag{5.246}$$

For the transverse states to be complete the two spurious states should be null (i.e. should also be physical) so that their conjugate states also decouple leaving only (d - 4) positive-norm states.

Consider the combinations of the spurious states

$$|\psi_1\rangle = (F_0 + m) \, L_1^{\,+} \, |0\rangle \tag{5.247}$$

$$|\psi_2\rangle = (F_0 + m) \, F_1^{\,+} \, |0\rangle \tag{5.248}$$

so that

$$(L_o + m^2)\,|\psi_i> = 0 \qquad (5.249)$$

$$(F_0 - m)\,|\psi_i> = 0 \qquad (5.250)$$

Now require annihilation by L_1 through

$$L_1|\psi_2> = L_1(F_0 + m)F_1^{+}\,|0> = (\tfrac{5}{2}F_0 - \tfrac{d}{4} + \tfrac{3}{2}mF_0)\,|0> \qquad (5.251)$$

$$= [\tfrac{5}{2}(m^2 + 1) - \tfrac{d}{4} + 3mF_0]\,|0> \qquad (5.252)$$

$$= 0 \qquad (5.253)$$

provided that d = 10 and m = 0. With these conditions satisfied one can confirm that

$$L_1|\psi_1> = F_1|\psi_1> = F_1|\psi_2> = 0 \qquad (5.254)$$

and thus only (2d − 4) states remain, as required. Thus m = 0 is <u>necessary</u> for the completeness of the transverse states.

5.6 BOSON-FERMION COUPLINGS

Consider the diagram, Figure 5.4(a), where a fermion line emits (N-1) ground-state mesons with momenta $p_1, p_2, \cdots, p_{N-1}$. The fermion propagator is

$$\frac{1}{F_0} = -\frac{F_0}{L_0^{\,f}} \qquad (5.255)$$

Now consider the identity[22]

$$[F_0 - F_n + \frac{2}{\sqrt{n}}(L_n^{\,f} - L_0^{\,f})]\,\frac{1}{F_0}$$

$$= \frac{1}{F_0 - \sqrt{n}}\,[F_n - F_0 + \frac{2}{\sqrt{n}}(L_n^{\,f} - L_0^{\,f} - \frac{n}{2})] \qquad (5.256)$$

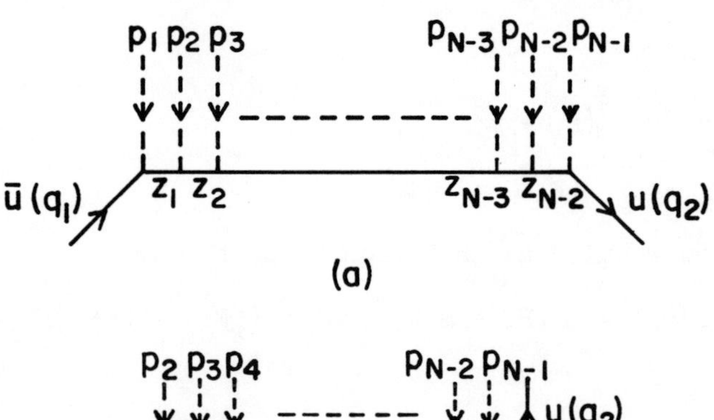

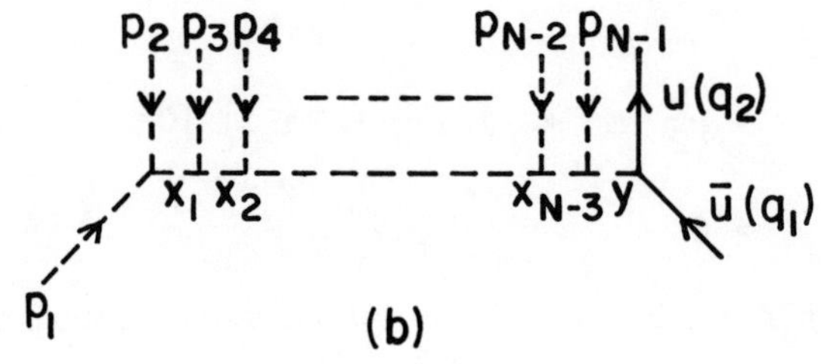

FIGURE 5.4

Boson-Fermion Couplings

which will be useful to construct the emission vertex. This identity is checked by using

$$(F_0 - \sqrt{n})\ [F_0 - F_n + \frac{2}{\sqrt{n}}(L_n{}^f - L_0{}^f)] =$$

$$= - L_0{}^f - F_0F_n + \frac{2}{\sqrt{n}}(F_0L_n{}^f - F_0L_n{}^f) -$$

$$- F_0\sqrt{n} + F_n\sqrt{n} - 2(L_n{}^f - L_0{}^f) \qquad (5.257)$$

$$= - F_0F_n - 2L_n{}^f + L_0{}^f + \frac{2}{\sqrt{n}} \cdot$$

$$\cdot\ (F_0L_n{}^f + \frac{n}{2}F_n - F_0L_0{}^f - F_0\,\frac{n}{2}) \qquad (5.258)$$

$$= [F_n - F_0 + \frac{2}{\sqrt{n}}(L_n{}^f - L_0{}^f - \frac{n}{2})\ F_0] \qquad (5.259)$$

as required.

Now the conventional vertex has the properties

$$[F_0 - F_n, V_0(p, 1)] = 0 \tag{5.260}$$

$$[L_0{}^f - L_n{}^f, V_0(p, 1)] = np^2 V_0(p, 1) \tag{5.261}$$

We need to multiply $V_0(p, 1)$ by an expression that anti-commutes with $(F_0 - F_n)$. The simplest choice (more complicated choices correspond to emission of higher meson states) is[22]

$$\Gamma_5 = \gamma_5 (-1)^{\sum b^{(n)+}\cdot b^{(n)}} \tag{5.262}$$

since

$$F_n = \frac{1}{\sqrt{2}} < z^{-n} : \Gamma\cdot P :> \tag{5.263}$$

$$\Gamma_\mu = \gamma_\mu + i\sqrt{2}\, \gamma_5 \sum_{n=1}^{\infty} (b^{(n)} z^n + b^{(n)+} z^{-n}) \tag{5.264}$$

$$\{\gamma_5, \gamma_\mu\}_+ = 0 \tag{5.265}$$

$$\{(-1)^{\sum_n b^{(n)+}\cdot b^{(n)}}, b_\mu^{(n)+}\}_+ = 0 \tag{5.266}$$

and hence

$$\{\Gamma_5, \Gamma_\mu(z)\}_+ = 0 \tag{5.267}$$

$$\{\Gamma_5, F_n - F_0\}_+ = 0 \tag{5.268}$$

It follows now that

$$[F_0 - F_n + \frac{2}{\sqrt{n}} (L_n{}^f - L_0{}^f)] \frac{1}{F_0} \Gamma_5 V_0(p, 1) =$$

$$= \frac{1}{F_0 - \sqrt{n}} [F_n - F_0 + \frac{2}{\sqrt{n}} (L_n{}^f - L_0{}^f - \frac{n}{2})] \Gamma_5 V_0(p, 1) \tag{5.269}$$

$$= \frac{1}{F_0 - \sqrt{n}} \Gamma_5 V_0(p, 1) [F_0 - F_n + \frac{2}{\sqrt{n}} (L_n{}^f - L_0{}^f - \frac{n}{2} - np^2)] \tag{5.270}$$

so if $p^2 = -\frac{1}{2}$, as is appropriate to the pion state, we have gauge relations.

The amplitude for the diagram of Figure 5.4(a) (see page 276) is therefore

$$A = \bar{u}(q_1) \langle 0| \Gamma_5 V_0(p_1, 1) \frac{1}{F_0} \Gamma_5 V_0(p_2, 1) \cdots$$

$$\cdots \frac{1}{F_0} \Gamma_5 V_0(p_{N-1}, 1) |0\rangle u(q_2) \quad (5.271)$$

$$= \bar{u}(q_1) \langle 0| \Gamma_5 V_0(p_1, 1) \frac{F_0}{L_0^f} \Gamma_5 V_0(p_2, 1) \cdots$$

$$\cdots \frac{F_0}{L_0^f} \Gamma_5 V_0(p_{N-1}, 1) |0\rangle u(q_2) \quad (5.272)$$

$$= (-1)^{\frac{N}{2}+1} \bar{u}(q_1) \gamma_5 \langle 0| V_0(p_1, 1) \frac{F_0}{L_0^f} V_0(p_2, 1) \cdots$$

$$\cdots \frac{F_0}{L_0^f} V_0(p_{N-1}, 1) |0\rangle u(q_2) \quad (5.273)$$

where we have used

$$[\Gamma_5, L_0^f] = 0 \quad (5.274)$$

$$\{\Gamma_5, F_0\}_+ = 0 \quad (5.275)$$

$$\langle 0| \Gamma_5 = \gamma_5 \langle 0| \quad (5.276)$$

and have noted that the Γ_5 anticommutation gives (N-2)/2 sign changes.

Now since

$$[F_0, V_0(p, 1)] = \frac{1}{\sqrt{2}} \langle \Gamma_\mu(z) [P_\mu(z), e^{\sqrt{2}ip\cdot Q(1)}]\rangle \quad (5.277)$$

$$= - p\cdot\Gamma(1) V_0(p, 1) \quad (5.278)$$

we may take the F_0's to the right, dropping terms with

cancelled propagators to obtain

$$A = - \bar{u}(q_1)\, \gamma_5 \, \langle 0 |\, V_0(p_1, 1) \frac{1}{L_0^f} p_2 \cdot \Gamma(1)\, V_0(p_2, 1) \cdot$$

$$\cdot \frac{1}{L_0^f} \cdots \frac{1}{L_0^f} p_{N-1} \cdot \Gamma(1)\, V_0(p_{N-1}, 1)\, | 0 \rangle\, u(q_2) \tag{5.279}$$

We may write now

$$\frac{1}{L_0^f} = \int_0^1 dz\, z^{R_a + R_b - p^2 - 1} \tag{5.280}$$

whereupon the a modes give precisely the generalised conventional integrand. To deal with the b modes, we would like to make a cyclic change of variables to Figure 5.4(b) (see page 276). With the integration variables defined as implied by the Figure 5.4, the appropriate transformation is

$$x_i = \left(\frac{1 - z_1 z_2 \cdots z_i}{1 - z_1 z_2 \cdots z_{i+1}}\right) \; ; \; 1 \leq i \leq (N-3) \tag{5.181}$$

$$y = 1 - \prod_{j=1}^{N-2} z_j \tag{5.182}$$

By going to the pion pole at $(q_1 - q_2)^2 = -\frac{1}{2}$ it is possible to show that

$$A \underset{(q_1 - q_2)^2 \to -\frac{1}{2}}{\sim} \bar{u}(q_1)\gamma_5 u(q_2) \frac{1}{(q_1 - q_2)^2 + 1/2} \cdot$$

$$\cdot A_N^{(NS)}(p_1, p_2, \cdots, p_N) \tag{5.283}$$

with $A_N^{(NS)}$ the N-pion boson amplitude defined by

$$A_N^{(NS)} = \langle 0|\ p_2\cdot H(1)\ V_0(p_2,\ 1)\ \frac{1}{L_0 - 1/2}\ p_3\cdot H(1)\ \cdot$$
$$\cdot\ V_0(p_3,\ 1)\ \cdots\ \frac{1}{L_0 - 1/2}\ p_{N-1}\cdot H(1)\ \cdot$$
$$\cdot V_0(p_{N-1},\ 1)\ |0\rangle \qquad (5.284)$$

This shows that at the ground-state meson pole the RNS theory factorises correctly.

Making the cyclic change of variables carefully, in particular converting from the integer fermion modes to half-integer boson modes, it was shown (by Thorn, Reference 23) that

$$A = \langle 0|\ p_2\cdot H(1)\ V_0(p_2,\ 1)\ \frac{1}{L_0 - 1/2}\ p_3\cdot H(1)\ V_0(p_3,\ 1)\ \cdot$$
$$\cdot\ \cdots\cdot\ \frac{1}{L_0 - 1/2}\ p_{N-1}\cdot H(1)\ V_0(p_{N-1},\ 1)\ \frac{1}{L_0 - 1/2}\ \cdot$$
$$\cdot\ \bar{u}(q_1)\ V_0(q_2,\ 1)\ e^T\ |0\rangle\ \gamma_5\ u(q_2) \qquad (5.285)$$

in which T is given by

$$T = -\frac{i}{\sqrt{2}}\gamma_5\gamma_\mu \sum_{r=1/2}^{\infty} b^{(r)+} \binom{-1/2}{r-1/2} (-1)^{r-1/2} +$$
$$+ \sum_{r,s=1/2}^{\infty} \frac{1}{4}(-1)^{r+s}\ C_{r-\frac{1}{2},s-\frac{1}{2}}\ b^{(r)+}\cdot b^{(s)+} \qquad (5.286)$$

with

$$C_{r-1/2,s-1/2} = \sum_{u=0}^{r-1/2} \binom{1/2}{r+s-u}\binom{-1/2}{u} -$$
$$- \sum_{v=0}^{s-1/2} \binom{1/2}{r+s-v}\binom{-1/2}{v} \qquad (5.287)$$
$$= -C_{s-\frac{1}{2},\ r-\frac{1}{2}} \qquad (5.288)$$

Thus the vertex for emitting a general meson state ($|\lambda_a, \lambda_b>$) from a ground-state fermion-antifermion pair is

$$<\lambda_a|\ V_0(q_2,\ 1)\ |0>_a \cdot <\lambda_b|\ \bar{u}(q_1)\ e^T\ \gamma_5\ u(q_2)|0>_b \tag{5.289}$$

Note that, for the first time, the vertex contains an exponent quadratic in creation operators and this renders the further development within the operator formalism technically cumbersome.

To write a general tree involving several fermion lines (Figure 5.5) we need a vertex for ground-state fermion

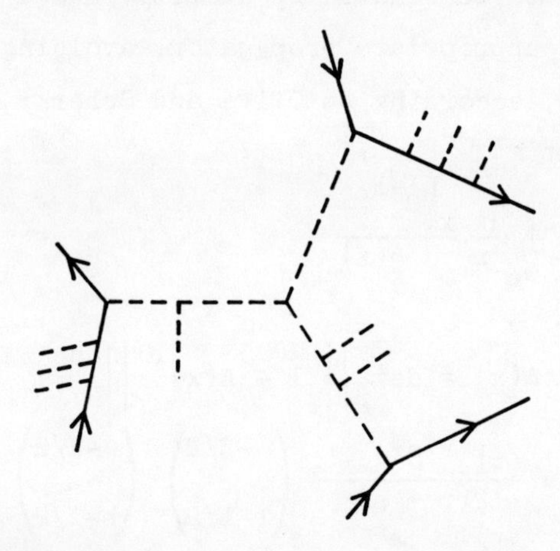

FIGURE 5.5

Multifermion Amplitude

emission from excited meson-excited fermion; such a vertex has been constructed consistent with both the boson and the fermion gauge conditions (References 24, 25, 26).

A very interesting situation is that for four external ground-state fermions, Figures 5.6(a) and 5.6(b),

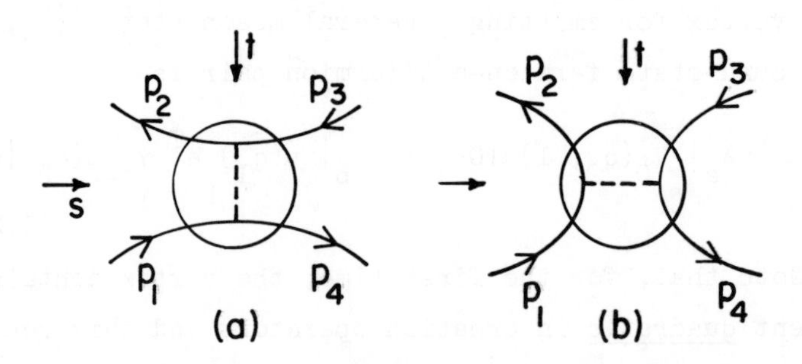

FIGURE 5.6

Two Fermions-Two Antifermions Amplitude

which we expect to be related by duality plus a Fierz transformation. The appropriate propagator, avoiding spurious meson states is, according to Olive and Scherk[27], not $(L_0 - 1/2)^{-1}$ but

$$P(p^2) = \int_0^1 \frac{dx}{x} \frac{x^{L_0 - 1/2}}{\Delta(x)} \tag{5.290}$$

where

$$\Delta(x) = \det \left\| 1 - A(x)^2 \right\| \tag{5.291}$$

$$A(x)_{rs} = \frac{(-1)^{r+s+1} x^r}{2(r+s)} \begin{pmatrix} -3/2 \\ r-1/2 \end{pmatrix} \begin{pmatrix} -1/2 \\ s-1/2 \end{pmatrix} \tag{5.292}$$

with $r, s = \frac{1}{2}, \frac{3}{2}, \frac{5}{2}, \cdots$.

Thus we would expect that (Figure 5.6)

$$A_4 = \langle 0| \bar{u}(p_4)\, \gamma_5\, e^{T+}\, u(q_1)\, V_0(-p_4, 1) \cdot$$

$$\cdot\, P(t)\, V_0(p_3, 1)\, \bar{u}(p_2)\, e^{T}\, \gamma_5\, u(p_3)\, |0\rangle \tag{5.293}$$

$$= \langle 0| \bar{u}(p_4)\, \gamma_5\, e^{T+}\, u(q_3)\, V_0(-p_4, 1) \cdot$$

$$\cdot\, P(s)\, V_0(p_1, 1)\, \bar{u}(p_2)\, e^{T}\, \gamma_5\, u(p_1)\, |0\rangle \tag{5.294}$$

Only after labyrinthine algebra can this relationship be rigorously confirmed in the operator formalism[28]; fortunately, however, a more powerful method is at hand. Using functional integration techniques (starting from the generalised rubber string of Kikkawa and Iwasaki [29]) Mandelstam (References 30-32) has shown that the GGNW amplitudes[33]* for the process of Figure 5.6 are given as

$$\phi_1 = f^s_{++,++} = s\,B(-s-\tfrac{1}{2}, -t) \tag{5.295}$$

$$\phi_2 = f^s_{++,--} = -\tfrac{1}{2}(s-t)\,(s+t+1)^{-1}\,B(-s-\tfrac{1}{2},-t-\tfrac{1}{2}) \tag{5.296}$$

$$\phi_3 = f^s_{+-,+-} = (s+t)\,B(-s, -t) \tag{5.297}$$

$$\phi_4 = f^s_{+-,-+} = t\,B(-s, -t-\tfrac{1}{2}) \tag{5.298}$$

$$\phi_5 = f^s_{++,+-} = 0 \tag{5.299}$$

and that the correct crossing properties are maintained.

Concerning the RNS theory we should add the remarks:

i) Although multifermion Born Amplitudes have not yet been studied carefully, it seems likely that the RNS theory is fully consistent for d = 10 dimensions, leading meson intercepts $\alpha_\rho(o) = 1$, $\alpha_\pi(o) = \frac{1}{2}$ and leading fermion intercept $\alpha_f(o) = \frac{1}{2}$. There are no ghosts, but all fermions except the ground state are parity-doubled. The main importance of the theory is that it shows that the introduction of half-integer spin presents no essential difficulty in dual theories.

---

*A four-dimensional subspace of the full d = 10 space-time has been selected by suitably restricting the excitations.

ii) The crossing properties for the fermion-antifermion amplitude reveal[31] that the odd-G-parity mesons have opposite parity (P) in the s- and t-channels; that is, the odd-G meson sector must be parity doubled in the full RNS theory.

iii) The meson spectrum has some qualitative similarity to Nature, but the fermion spectrum is more obscure. In view of this lack of realism, it is rather academic to discuss whether these fermions are more like nucleons or quarks. We should mention, however, that the fermions certainly have essentially <u>different</u> duality properties (i.e. topology of dual diagrams) from three-quark baryons since non-planar tree diagrams are absent.

iv) At a purely technical level, for the computation of multifermion amplitudes the functional integration method appears to be much superior to the operator formalism approach.

5.7 GENERALISED PROJECTIVE ALGEBRA

For the UIR of SU(1,1) we have previously derived, for the J = -k representations, that

$$D^{(J+)}_{mn}(L_+) = \sqrt{n(n - 1 - 2J)}\ \delta_{m,n-1} \tag{5.300}$$

$$D^{(J+)}_{mn}(L_-) = \sqrt{(n + 1)(n - 2J)}\ \delta_{m,n+1} \tag{5.301}$$

$$D^{(J+)}_{mn}(L_0) = (m - J)\ \delta_{mn} \tag{5.302}$$

Now, our gauge operators for the Neveu-Schwarz

amplitude can be written

$$L_0^{\ b} = - \sum_{m=0}^{\infty} (m + \tfrac{1}{2})\, b^+_{m+1/2} \cdot b_{m+1/2} \tag{5.303}$$

$$L_1^{\ b} = - \sum_{m=0}^{\infty} (m + 1)\, b^+_{m+1/2}\, b_{m+1/2+1} \tag{5.304}$$

$$L_{-1}^{b} = - \sum_{m=0}^{\infty} (m + 1)\, b^+_{m+1/2+1}\, b_{m+1/2} \tag{5.305}$$

corresponding to $J = -\frac{1}{2}$ in the general formulae, as expected.

Consider now the multireggeon vertex in the third-quantised form (discussed earlier)

$$A_N = {}_c\langle 0|\, \langle 0_{j_1} \cdots 0_{j_N}| \;:\; \prod_{i=1}^{N} e^{\sum_{k=0}^{\infty} Q_k^{(i)}\, b_{ik}} \;:\; |0\rangle_c \tag{5.306}$$

where the operators b_{ik} are associated with the leg i (k is the mode number) and satisfy anticommutation rules

$$\{b^{\mu}_{ik},\, b^{\nu+}_{j\ell}\}_+ = - g^{\mu\nu}\, \delta_{ij}\, \delta_{k\ell} \tag{5.307}$$

Let us consider the model[34] where this vertex is multiplied by the conventional (orbital) vertex. Taking the ground state to be

$$e^{ip\cdot Q}\; p\cdot b_0^{\ +}\, |0\rangle \tag{5.308}$$

we find a new contribution from the anticommuting part which is

$${}_c\langle 0|\, \prod_{j=1}^{N} (-p_j \cdot Q_0^{(j)})\, |0\rangle_c \tag{5.309}$$

the negative sign being from the $(-g^{\mu\nu})$ of the definition. Here

$$Q^{(i)}_{0\mu} = i \sum_{n=0}^{\infty} [C^{+}_{n\mu} D^{(-1/2,+)}_{n0} (y_j) + C_{n\mu} D^{(-1/2,+)}_{n0} \cdot$$

$$\cdot (\Gamma y_j)] \tag{5.310}$$

Now recall that for the transformation

$$A = \begin{pmatrix} a & b \\ c & d \end{pmatrix} \tag{5.311}$$

width $ad - bc = 1$ we have

$$D^{(J+)}_{n0} (A) = \frac{N_n^J}{N_0^J} \left(\frac{b}{d}\right)^n d^{2J} \tag{5.312}$$

and, since

$$\Gamma A = \begin{pmatrix} c & d \\ a & b \end{pmatrix} \tag{5.313}$$

we have

$$D^{(J+)}_{n0} (\Gamma A) = \frac{N_n^J}{N_0^J} \left(\frac{d}{b}\right)^n b^{2J} \tag{5.314}$$

Substituting

$$b = \frac{z_i (z_{i-1} - z_{i+1})}{\sqrt{(z_{i-1} - z_{i+1})(z_i - z_{i+1})(z_i - z_{i-1})}} \tag{5.315}$$

$$d = \frac{(z_{i-1} - z_{i+1})}{\sqrt{(z_{i-1} - z_{i+1})(z_i - z_{i+1})(z_i - z_{i-1})}} \tag{5.316}$$

as appropriate to Y_i we find

$$Q^{(i)}_{0\mu} = i \left[\frac{z_i (z_{i-1} - z_{i+1})}{(z_i - z_{i+1})(z_i - z_{i-1})}\right]^J H^{(J)}_{\mu} (z_i) \tag{5.317}$$

with

$$H^{(J)}_{\mu} = \sum_{n=0}^{\infty} \frac{N_n^J}{N_0^J} (C^{+}_{n\mu} z^{n-J} + C_{n\mu} z^{-n+J}) \tag{5.318}$$

$$= \sum_{n=0}^{\infty} \frac{\sqrt{\Gamma(n-2J)}}{\sqrt{\Gamma(n+1)\Gamma(-2J)}} (C^{+}_{n\mu} z^{n-J} + C_{n\mu} z^{-n+J}) \quad (5.319)$$

where we use

$$N_n^J = \frac{\sqrt{\Gamma(n-2J)}}{\sqrt{n!}} \quad (5.320)$$

In particular, for $J = -\frac{1}{2}$ we find

$$H_\mu(z) = \sum_{r=1/2}^{\infty} (b_{r\mu} z^r + b^{+}_{r\mu} z^{-r}) \quad (5.321)$$

in which we define $b^{+}_{r\mu} = c_{r-1/2,\mu}$; this is precisely the Neveu-Schwarz field.

The contribution from the b oscillators to the N point function is therefore (within a phase)

$$\prod_{i=1}^{N} \left[\frac{z_i(z_{i-1} - z_{i+1})}{(z_i - z_{i+1})^2}\right]^{-1/2} {}_c\langle 0| \prod_{i=1}^{N} p_i \cdot H(z_i) |0\rangle_c \quad (5.322)$$

Combining this with the contribution, already calculated, from the a oscillators, we obtain

$$A_N = \int \prod_{i=1}^{N} dz_i \, (dV_3)^{-1} \prod_{i \neq j} (z_i - z_j)^{-p_i \cdot p_j} \prod_{i=1}^{N} z^{-1/2} \langle 0| \, p_1 \cdot H(z_1) \, p_2 \cdot H(z_2) \cdots$$

$$\cdots p_N \cdot H(z_N) \, |0\rangle \quad (5.323)$$

which is the Neveu-Schwarz amplitude.

It turns out that the choice of projective spin is severely restricted by the requirement that the extension to a generalised projective algebra exists. Let us write, rather generally,

$$L_n = \sum_{m=-\infty}^{\infty} \gamma_m^{n,J} \, a^{(m)+} \cdot a^{(m+n)} \tag{5.324}$$

Then the hermiticity

$$L_{-n} = (L_n)^+ \tag{5.325}$$

implies

$$\gamma_p^{-n,J} = \gamma_{p-n}^{n,J} \tag{5.326}$$

The algebra

$$[L_m, L_n] = (m - n) L_{m+n} \tag{5.327}$$

implies (for a operators commuting <u>or</u> anticommuting)

$$\gamma_{p+n}^{m,J} \gamma_p^{n,J} - \gamma_p^{m,J} \gamma_{p+m}^{n,J} = (m - n) \gamma_p^{m+n,J} \tag{5.328}$$

For the SU(1,1) sub-algebra we know that

$$- \gamma_m^{0,J} = (m - J) \tag{5.329}$$

$$- \gamma_m^{1,J} = \sqrt{(m + 1)(m - 2J)} \tag{5.330}$$

Putting m = 2, n = -2 in the coefficient relation, we obtain

$$(\gamma_p^{2,J})^2 - (\gamma_{p-2}^{2,J})^2 = 4(p - J) \tag{5.331}$$

Putting m = 2, n = -1 gives

$$- \gamma_{p-1}^{2,J} \sqrt{p(p-1 - 2J)} + \gamma_p^{2,J} \sqrt{(p + 2)(p+1 - 2J)}$$

$$= - 3 \sqrt{(p + 1)(p - 2J)} \tag{5.332}$$

Putting successively p = 0, 1, 2 we obtain

$$\gamma_0^{2,J} = \frac{-3\sqrt{-2J}}{\sqrt{2(1-2J)}} \tag{5.333}$$

$$\gamma_1^{2,J} = \frac{3(3J-1)}{\sqrt{3(1-J)(1-2J)}} \tag{5.334}$$

$$\gamma_2^{2,J} = \frac{3(6J-4)}{\sqrt{6(3-2J)(1-J)}} \tag{5.335}$$

Substituting these results into the earlier $m = 2$, $n = -2$ relation gives

$$J(J+1)\,(2J+1)\,(8J-7) = 0 \tag{5.336}$$

or

$$J = 0,\ -\frac{1}{2},\ -1,\ +\frac{7}{8}$$

Going to $p = 3$

$$\gamma_3^{2,J} = \frac{30(J-1)}{\sqrt{5(4-2J)(3-2J)}} \tag{5.337}$$

and one finds

$$(\gamma_3^{2,J})^2 - (\gamma_1^{2,J})^2 = 4(3-J) \tag{5.338}$$

holds for $J = 0, -\frac{1}{2}, -1$ but not for $J = \frac{7}{8}$ which is thereby eliminated. For the remaining three solutions we can find explicit solutions

$$\gamma_p^{n,0} = -\sqrt{p(p+n)} \tag{5.339}$$

$$\gamma_p^{n,-1/2} = -\left(p + \frac{1}{2} + \frac{n}{2}\right) \tag{5.340}$$

$$\gamma_p^{n,-1} = -\sqrt{(p+1)(p+n+1)} \tag{5.341}$$

To avoid vanishing L_n it is essential that the a-operators commute for J = 0, -1 and anticommute for $J = -\frac{1}{2}$. For example, with J = 0 we have

$$L_n = -\sum_{p=0}^{\infty} \sqrt{p(p+n)}\, a^{(p)+}\cdot a^{(p+n)} \tag{5.342}$$

$$= -\sum_{P} \sqrt{(P+n/2)(P-n/2)}\, a^{+}_{P-n/2}\cdot a_{P+n/2} \tag{5.343}$$

$$= -\sum_{P} \sqrt{(P+n/2)(P-n/2)}\, a_{P+n/2}\, a^{+}_{P-n/2} \tag{5.344}$$

by substituting $P = (p - \frac{n}{2})$ and $P \to -P$. Thus unless the a's commute there is an inconsistency. Similar arguments relate "spin and statistics" of the oscillators when $J = -\frac{1}{2}, -1$.

Finally we note that the J = -1 solution is simply a re-labelling of the J = 0 one.

In summary, if we assume simple bilinear representations of the L_n operators, we are restricted to the two inequivalent cases[34,35]

$$J = 0 \qquad \text{commuting operators} \tag{5.345}$$

$$J = -\frac{1}{2} \qquad \text{anticommuting operators} \tag{5.346}$$

The simplest additive combination gives the Neveu-Schwarz amplitude, which can thus be derived group-theoretically within the operator formalism.

We remark here that the Neveu-Schwarz spectrum can also be obtained by quantising a modified classical rubber string on which Lorentz vector quantities are continuously distributed; see Iwasaki and Kikkawa[29] and related works.[36,37]

5.8 NON-PLANAR EXTENSION

It was pointed out by Aldrovandi and Neveu[38] and independently by Schwarz[15], that there exists a non-planar extension of the dual pion model analogous to that of Shapiro and Virasoro for the conventional model.

We introduce oscillators

$$[a_\mu^{(m)}, a_\nu^{(n)+}] = - g_{\mu\nu} \delta_{mn} \tag{5.347}$$

$$[\bar{a}_\mu^{(m)}, \bar{a}_\nu^{(n)+}] = - g_{\mu\nu} \delta_{mn} \tag{5.348}$$

$$\{b_\mu^{(r)}, b_\nu^{(s)+}\}_+ = - g_{\mu\nu} \delta_{rs} \tag{5.349}$$

$$\{\bar{b}_\mu^{(r)}, \bar{b}_\nu^{(s)+}\}_+ = - g_{\mu\nu} \delta_{rs} \tag{5.350}$$

and define, in an obvious notation L_n, $\bar{L}_n$. Then the non-planar model has propagator and vertex defined by the replacements

$$(L_0 - 1)^{-1} \rightarrow (L_0 + \bar{L}_0 - 2)^{-1} \tag{5.351}$$

$$p\cdot H(1)\, V_0(p, 1) \rightarrow p\cdot H(1)\, p\cdot \bar{H}(1)\, V_0'(p, 1)\, \bar{V}_0'(p, 1) \tag{5.352}$$

where the barred operators are constructed from barred oscillators, and where $V_0'(p, z) = :\exp(ip\cdot Q(z,1):$ compared to $V_0(p, z) = :\exp(\sqrt{2}\, ip\cdot Q(z, 1):$. This completely defines the model.

We can go into the equivalent of an F_2 representation by redefining the vacuum as

$$|0'\rangle = G_{\frac{1}{2}}^{+}\, \bar{G}_{\frac{1}{2}}^{+}\, |0\rangle \tag{5.353}$$

and using the propagator $(L_0 + \bar{L}_0 - 1)^{-1}$. In this F_2

reformulation it becomes clear that the $\alpha = 0$ tachyon on the leading trajectory of intercept $\alpha(o) = 2$ is missing.

As an example, we calculate the four-point amplitude in the F_2 formulation. It is (putting $z_1 = 0$, $z_3 = 1$, $z_4 = \infty$)

$$A_4 = \int d^2z \; |z|^{-s-3} \langle 0'| \; p_2 \cdot H \; p_2 \cdot \bar{H} \; V_0'(p_2, 1) \; \bar{V}_0'(p_2, 1)$$

$$z^R \bar{z}^{\bar{R}} \; p_3 \cdot H \; p_3 \cdot \bar{H} \; V_0'(p_3, 1) \; V_0'(p_3, 1) \; |0'\rangle \tag{5.354}$$

where we have re-written

$$2\pi(L_0 + \bar{L}_0 - 1)^{-1} = \int d^2z \; z^{L_0 - 3/2} \; \bar{z}^{L_0 - 3/2} \tag{5.355}$$

Noting that

$${}_b\langle 0'| \; p_2 \cdot H(z) \; p_2 \cdot \bar{H}(z) \; p_3 \cdot H(1) \; p_3 \cdot \bar{H}(1) \; |0'\rangle$$

$$= | \; {}_b\langle 0'| \; p_2 \cdot H(z) \; p_3 \cdot H(1) \; |0'\rangle \; |^2 \tag{5.356}$$

$$= (p_2 \cdot p_3)^2 \; \frac{|z|}{|1 - z|^2} \tag{5.357}$$

we arrive at (using $p_i^{\,2} = -1$, $s + 2 = 2p_1 \cdot p_2$, etc.)

$$A_4 = \int d^2z \; |z|^{-2p_1 \cdot p_2} \; |1 - z|^{-2p_2 \cdot p_3 - 2} \; (p_2 \cdot p_3)^2 \tag{5.358}$$

This is obtained from our earlier integral for the Shapiro-Virasoro model through $p_2 \cdot p_3 \to p_2 \cdot p_3 + 1$. Hence in terms of gamma functions it is (using $p_2 \cdot p_3 = \frac{1}{2} \alpha_t$, $\alpha_s + \alpha_t + \alpha_u = 2$)

$$A_4 = \pi(\tfrac{1}{2} \alpha_t)^2 \; \frac{\Gamma(1 - \frac{\alpha_s}{2}) \Gamma(-\frac{\alpha_t}{2}) \; \Gamma(\frac{\alpha_s + \alpha_t}{2})}{\Gamma(1 - \frac{\alpha_s + \alpha_t}{2}) \Gamma(\frac{\alpha_t}{2}) \; \Gamma(1 + \frac{\alpha_t}{2})} \tag{5.359}$$

$$= -\frac{\pi\,\Gamma(1 - \frac{\alpha_s}{2})\,\Gamma(1 - \frac{\alpha_t}{2})\,\Gamma(1 - \frac{\alpha_u}{2})}{\Gamma(1 - \frac{\alpha_s+\alpha_t}{2})\Gamma(1 - \frac{\alpha_t+\alpha_u}{2})\Gamma(1 - \frac{\alpha_u+\alpha_s}{2})} \qquad (5.360)$$

which is related to the four-point Shapiro-Virasoro formula by the addition of one in the arguments, similarly to the way the Lovelace-Shapiro formula is related to the Euler B function.

5.9 SUMMARY

The step from the conventional multiparticle dual model, taken with unit intercept, to the Ramond-Neveu-Schwarz (RNS) model represents an improvement in several properties.

i) In the boson sector, we now have introduced G-parity with different families of trajectories in the even and odd G-parity channels.

ii) The rho trajectory has positive intercept and no tachyon.

iii) Extra degrees of freedom have been added without introducing ghost states.

iv) Most encouragingly, there is some qualitative similarity between the low-mass boson spectrum and experiment.

v) There is now a fermion sector, hitherto absent, and here again extra anticommuting modes lead to no added ghosts.

To be fair one should emphasise the shortcomings (which are reasonably obvious): the leading intercepts

$\alpha_\rho(o) = 1$, $\alpha_\pi(o) = \frac{1}{2}$ and $\alpha_f(o) = \frac{1}{2}$ are all unphysical and all fermions above the ground state are parity doubled.

In spite of these difficulties, the importance of RNS theory is firstly that it shows there is no essential theoretical difficulty in introducing fermions, and secondly that it provides food for thought towards the construction of a realistic theory.

REFERENCES

1. S. Mandelstam, Phys. Rev. 184, 1625 (1969).
2. K. Bardakci and M. B. Halpern, Phys. Rev. 183, 1456 (1969).
3. We follow the conventions of J. D. Bjorken and S. D. Drell, Relativistic Quantum Fields, McGraw-Hill (1965).
4. A. Salam, R. Delbourgo and J. Strathdee, Proc. Roy. Soc. (London) A284, 146 (1965).
5. F. J. Dyson, Symmetry Groups in Nuclear and Particle Physics, Benjamin (1966).
6. A. P. Balachandran, L. N. Chang and P. H. Frampton, Nuovo Cimento 1A, 545 (1971).
7. P. Ramond, Nuovo Cimento 4A, 544 (1971).
8. P. Ramond, Phys. Rev. D3, 2415 (1971).
9. A. Neveu and J. H. Schwarz, Nucl. Phys. B31, 86 (1971).
10. C. Lovelace, Phys. Letters 38B, 264 (1968).
11. J. Shapiro, Phys. Rev. 179, 1345 (1969).
12. D. B. Fairlie, Nucl. Phys. B42, 253 (1972).
13. D. B. Fairlie and D. Martin, Nuov. Cim. 18A, 373 (1973).
14. A. Neveu, J. H. Schwarz and C. B. Thorn, Physics Letters 35B, 529 (1971).
15. J. H. Schwarz, Nucl. Phys. B46, 61 (1972).
16. P. Goddard and C. B. Thorn, Phys. Letters 10B, 235 (1972).
17. R. C. Brower and K. A. Friedman, Phys. Rev. D7, 535 (1973).
18. J. H. Schwarz, Physics Reports 8C, 269 (1973).

19. C. Rebbi, Riv. Nuov. Cim. 3, 194 (193).
20. C. B. Thorn, quoted by Reference 21.
21. E. F. Corrigan and P. Goddard, Nuov. Cim. 18A, 339 (1973).
22. A. Neveu and J. H. Schwarz, Phys. Rev. D4, 1109 (1971).
23. C. B. Thorn, Phys. Rev. D4, 1112 (1971).
24. J. H. Schwarz, Physics Letters 37B, 315 (1971).
25. E. Corrigan and D. I. Olive, Nuovo Cimento 11A, 749 (1972).
26. L. Brink, D. I. Olive, C. Rebbi and J. Scherk, Phys. Letters 45B, 379 (1973).
27. D. I. Olive and J. Scherk, Nucl. Phys. B64, 334 (1973).
28. J. H. Schwarz and C. C. Wu, Phys. Letters 47B, 453 (1973); E. Corrigan, P. Goddard, D. I. Olive and R. A. Smith, Nucl. Phys. B67, 477 (1973).
29. Y. Iwasaki and K. Kikkawa, Phys. Rev. D8, 440 (1973).
30. S. Mandelstam, Nucl. Phys. B64, 205 (1973).
31. S. Mandelstam, Phys. Letters 46B, 447 (1973).
32. S. Mandelstam, Nucl. Phys. B69, 77 (1974).
33. M. L. Goldberger, M. T. Grisaru, S. W. MacDowell and D. Y. Wong, Phys. Rev. 120, 2250 (1960).
34. E. F. Corrigan and C. Montonen, Nucl. Phys. B36, 58 (1972).
35. It is worth remarking that our present derivation differs from and is considerably more general than, that of Reference 32.
36. J. F. Willemsen, Phys. Rev. D9, 507 (1974).
37. J. Wess and B. Zumino, (unpublished).

38. R. Aldrovandi and A. Neveu, (unpublished).

6

SYMMETRIC GROUP

6.1 INTRODUCTION

We now consider the problem of finding a meson Born amplitude which is superior to the Neveu-Schwarz amplitude. The latter satisfies all of our basic axioms including factorisation without ghosts; its principal defect is that the Regge intercepts are misplaced. To be specific it requires that $\alpha_\rho(o) = 1$ and $\alpha_\pi(o) = \frac{1}{2}$ whereas the empirical values are very close to $\alpha_\rho(o) = \frac{1}{2}$ and $\alpha_\pi(o) = 0$. We are thus faced with the (apparently) straightforward task of simply _shifting the Regge intercepts_ downward by one half unit in angular momentum, while preserving the other desirable properties including absence of ghost-states.

We begin by mentioning the different methods that have been used to attack the intercept problem; since only one method has so far yielded an explicit proposal for the S-matrix we thereafter concentrate on it, the symmetric-group method.

The symmetric group is introduced at the four-particle level where a general amplitude is obtained. One important physical property of this amplitude is the absence of odd daughter trajectories, for general intercepts.

The method of extending to a multiparticle production amplitude is given. Here the symmetric group provides us with a framework into which the known integer-intercept dual models fit and, further, suggests how to extrapolate to non-integer intercepts. Using this technique we write an explicit N-point function which achieves our principal objective of accommodating the intercepts $\alpha_\rho(o) = \frac{1}{2}$ and $\alpha_\pi(o) = 0$. This amplitude has nice consistency properties; its known properties are: cyclic symmetry (proved), factorisability (proved) and absence of ghosts (proved for $N = 4$; untested for $N > 4$). The one outstanding test is thus the important no-ghost theorem which may require sharpening our existing techniques.

We then close this study of the construction of improved dual resonance models by offering suggestions for future research.

6.2 METHODS OF ATTACKING THE INTERCEPT PROBLEM

To go in the direction of a realistic dual theory[1] the first objective is to construct a Born amplitude with Regge intercepts and a mass spectrum close to the empirical ones. We first list and briefly discuss four methods which have been considered for accomplishing this goal. They are

i) Operator formalism approach. Here the procedure is to make a realisation of the generalised projective algebra on some Fock space F spanned by certain sets of creation and annihilation operators. One then constructs vertices such that ghost-eliminating gauges are ensured for the model. This approach led to the dual pion model already described, and has been further investigated, particularly by Gervais and Neveu,[2,3] as we shall indicate briefly below.

ii) Strings. The rubber string, as we have seen, provides an attractive picture for the spectrum of free states in the known models. So far this picture has not led to any new explicit model, although some understanding of the interaction of strings has now been achieved.[4-8]

iii) Spontaneous symmetry breaking. Some attempts have been made to introduce spontaneous symmetry breaking into dual theories in order that the massless vector state at $\alpha = 1$ acquire a mass through the Higgs mechanism (see, for example, Bardakci, Reference 9) and hence that the Regge intercept be shifted. So far these potentially important developments are still at a preliminary stage.

iv) Symmetric group. This method (References 10-12) starts from an integral representation for a four-particle amplitude and then makes an extension to any number of external particles. This is similar to the historical development of the conventional generalised Euler B function model. The underlying principle is to impose very high symmetry properties on the integrand, similar to those obtained in the models known to be ghost-free.

As already mentioned, we here concentrate on method (iv). In 6.3 to 6.7 we discuss the symmetric group for general Regge intercepts in the four-point function and in 6.8 - 6.12 the discussion is extended to an N-point function; the reader primarily interested in the general method could study only 6.3, 6.4, 6.8 at first reading.

6.3 FOUR-MESON AMPLITUDE; SYMMETRIC GROUP[13]

In the simplest dual theory, namely the generalised Euler B function model, there is an absence of ghosts only if the leading intercept satisfies $\alpha(o) = 1$. To proceed towards realistic intercepts we begin by carefully studying the unit-intercept four-particle amplitude of this theory for scalar mesons (with momenta $p_1 + p_2 \rightarrow (-p_3) + (-p_4)$) in the form

$$T = B_{st} + B_{tu} + B_{us} \tag{6.1}$$

with

$$B_{st} = \int_0^1 dx\, x^{-\alpha_t - 1} (1 - x)^{-\alpha_s - 1} \tag{6.2}$$

where $\alpha_s = \alpha(o) + \alpha' s$, etc. and $s = (p_1 + p_2)^2$, $t = (p_2 + p_3)^2$, and $u = (p_1 + p_3)^2$.

For the special case $\alpha(o) = 1$ we should recall the following four properties of the Euler B function.

i) The absence of odd daughter trajectories i.e. the Regge trajectories are spaced by two units in angular momentum. To see this we may look at the asymptotic behaviour or, more simply, at the residues in

$$B_{st} = \sum_{n=0}^{\infty} \frac{R_n(\alpha_t)}{n - \alpha_s} \tag{6.3}$$

with

$$R_n(\alpha_t) = \frac{1}{n!} (\alpha_t + 1)(\alpha_t + 2) \cdots (\alpha_t + n) \tag{6.4}$$

and hence, since $\alpha_t = -1 - n - \alpha_u$ on mass shell,

$$R_n(\alpha_t) = (-1)^n R_n(\alpha_u) \tag{6.5}$$

This implies that only alternate powers of

$$\cos\theta_s = \frac{\alpha_t - \alpha_u}{2 - (\alpha_t + \alpha_u)} \tag{6.6}$$

are present.

ii) The summability condition that we may add the three terms

$$T = B_{st} + B_{tu} + B_{us} \tag{6.7}$$

$$= \int_{-\infty}^{\infty} dx\ |x|^{-\alpha_t - 1}\ |1 - x|^{-\alpha_s - 1} \tag{6.8}$$

as can be easily seen by making the transformations $x \to (1 - x)^{-1} \to (1 - \frac{1}{x})$.

iii) The supplementary condition on the trajectory functions

$$\alpha_s + \alpha_t + \alpha_u + 1 = \gamma = 0 \tag{6.9}$$

when we identify $\alpha' p_i{}^2 = -\alpha(o) = -1$.

iv) The phase relations (Plahte[14]) of the form

$$B_{st} - e^{\pm i\pi\alpha_t} B_{tu} - e^{\mp i\pi\alpha_s} B_{us} = 0 \tag{6.10}$$

and cyclic permutations thereof.

To obtain these identities we close the contour in the summed form for $(B_{st} + B_{tu} + B_{us})$, and exploit the analyticity of the integrand away from the real axis, taking care of phases due to the branch points at $x = 0$ and $x = 1$.

These four properties are special to the case $\alpha(o) = 1$; in an approach to realistic intercepts we decide to keep the first two properties and reject the third and fourth.

The reason for keeping the summability condition is that, as we have already seen, the earlier ghost-free meson Born amplitudes possessed invariance under the N! permutations of the variables $\{p_i, z_i\}$ in the integrand. In fact for $\alpha(o) \neq 1$ in the Euler B function model we have seen that the factor

$$\prod_{i=1}^{N} (z_i - z_{i+1})^{\alpha(o)-1} \tag{6.11}$$

is precisely what destroys generalised projective gauge invariance. As we shall see shortly, this invariance (symmetric group invariance) is what is the immediate extension of the summability condition (ii) for N = 4. Once the summability condition is satisfied, the absence of odd daughters (i) will follow. Hence we keep both (i) and (ii).

On the other hand physical trajectories do not, in general, satisfy the supplementary condition (iii): for example, in $\pi\pi \to \pi\pi$ one has $\gamma = \alpha_s + \alpha_t + \alpha_u + 1 = 3\alpha_\rho(o) - 4\alpha_\pi(o) + 1 \simeq \frac{5}{2}$. Therefore this must be relaxed. Concerning the phase identities note that from the linear combination

$$B_{st} - e^{+i\pi\alpha_t} B_{tu} - e^{-i\pi\alpha_s} +$$

$$+ e^{i\pi\alpha_t} (B_{tu} - e^{i\pi\alpha_u} B_{us} - e^{-i\pi\alpha_t} B_{st}) = 0 \tag{6.12}$$

we deduce that

$$e^{i\pi(\alpha_s+\alpha_t+\alpha_u)} = -1 \tag{6.13}$$

or

$$\gamma = \alpha_s + \alpha_t + \alpha_u + 1 \tag{6.14}$$

$$= 0, \text{ modulo } 2 \tag{6.15}$$

We may refer to this as a generalised supplementary condition. Its deduction implies that we must abandon also the phase identities (iv).

Since the phase relations were previously derived from summability (which we have decided to keep) plus analyticity, it is clear that the analytic properties of the integrand have to be altered in an essential way by the introduction of complex singularities in the Euler x-plane. In the following, we shall be able to specify rather precisely the nature of these new singularities.

We now modify the integrand by multiplication with an arbitrary function which we choose to write in the form

$$A_4 = \int_0^1 dx\, x^{-\alpha_t - 1} (1 - x)^{-\alpha_s - 1} (x - e^{i\pi/3})^{\frac{\gamma}{2}} \cdot$$

$$\cdot (x - e^{-i\pi/3})^{\frac{\gamma}{2}} \phi_4(\alpha_s, \alpha_t, \alpha_u; x) \quad (6.16)$$

The factor $(1 - x + x^2)^{\gamma/2}$ has been separated out so that the analytic properties of ϕ_4 are simplified, as we shall show later.

It turns out that the necessary and sufficient conditions for property (i) (no odd daughters) and for property (ii) (summability) coincide. For example, we may consider the asymptotic behaviour of A_4 by converting to a Laplace transform through the change of variables

$$x = 1 - e^{-W} \quad (6.17)$$

$$\nu = \frac{1}{2}(\alpha_s - \alpha_u) \quad (6.18)$$

to arrive at

$$A_4 = \int_0^\infty dW\, W^{-\alpha_t - 1} e^{\nu W} \left[\sum_{n=0}^{\infty} \frac{1}{(2n+1)!} \left(\frac{W}{2}\right)^{2n} \right]^{-\alpha_t - 1} \cdot$$

$$\cdot \left[1 + 2 \sum_{n=1}^{\infty} \frac{W^{2n}}{(2n)!}\right]^{\frac{\gamma}{2}} \phi_4(\alpha_s, \alpha_t, \alpha_u; 1 - e^{-W}) \tag{6.19}$$

Using

$$\int_0^\infty dW\, W^{-\alpha_t - 1 - r} e^{\nu W} = (-a\nu)^{\alpha_t - r} \Gamma(-\alpha_{t+r}) \tag{6.20}$$

we see that odd daughter terms (odd r) are absent provided that ϕ_4 is even in W. This implies that

$$\phi_4(\alpha_s, \alpha_t, \alpha_u; x) = \phi_4(\alpha_u, \alpha_t, \alpha_s; \frac{-x}{1 - x}) \tag{6.21}$$

Combining this with the crossing symmetry requirement that

$$\phi_4(\alpha_s, \alpha_t, \alpha_u; x) = \phi_4(\alpha_t, \alpha_s, \alpha_u; 1 - x) \tag{6.22}$$

we find that ϕ_4 satisfies invariance under a six element S_3 symmetric group.

This is most clearly expressed by introducing a function $\beta_s(x)$ which is invariant under inversion (we will specify an explicit function in the following section)

$$\beta_s(x) = \beta_s\left(\frac{1}{x}\right) \tag{6.23}$$

Now define the entities

$$\beta_t(x) = \beta_s(1 - x) \tag{6.24}$$

$$\beta_u(x) = \beta_s\left(\frac{-x}{1 - x}\right) \tag{6.25}$$

With these definitions, the S_3 invariance of ϕ_4 may be expressed by the requirement that

$$\phi_4(\alpha_s, \alpha_t, \alpha_u; x) = \Phi_4\left(\frac{\alpha_s}{\beta_s(x)}; \frac{\alpha_t}{\beta_t(x)}; \frac{\alpha_u}{\beta_u(x)}\right) \tag{6.26}$$

be invariant under arbitrary permutations of the three argument pairs $\{\alpha_i, \beta_i(x)\}$ with i = s, t, u.

It is a straightforward exercise to check that we have also ensured the summability condition

$$A_4(-\alpha_s, -\alpha_t) + A_4(-\alpha_t, -\alpha_u) + A_4(-\alpha_u, -\alpha_s) =$$
$$= \int_{-\infty}^{\infty} dx\ |x|^{-\alpha_t - 1}\ |1 - x|^{-\alpha_s - 1}\ (1 - x + x^2)^{\frac{\gamma}{2}} \cdot$$
$$\cdot\ \phi_4(\alpha_s, \alpha_t, \alpha_u; x) \tag{6.27}$$

by imposing this S_3-invariance requirement on ϕ_4. This A_4 is therefore the most general form consistent with our requirements (i) and (ii). To proceed further, we impose the requirement of Regge behaviour on this general amplitude.

6.4 REGGE BEHAVIOUR AND ANALYTIC PROPERTIES

It is necessary to ensure that the four-particle amplitude satisfy Regge asymptotic behaviour as $|\nu| \to \infty$ (at fixed α_t) in any complex direction of the ν-plane. By using Watson's lemma,[15] applied to the rotation of the contour in the Laplace transform given above, we deduce that ϕ_4 must be analytic in the domain Re W > 0, or equivalently $|1 - x| < 1$. For fixed α_s Regge behaviour, we similarly require analyticity in $|x| < 1$. The combination gives an analyticity domain

$$\mathcal{D}(x) = (|x| < 1) \cup (|1 - x| < 1) \tag{6.28}$$

in the x-plane, as indicated in Figure 6.1. The function ϕ_4 must not contain singularities in x within this "opera-glass" shape $\mathcal{D}(x)$.

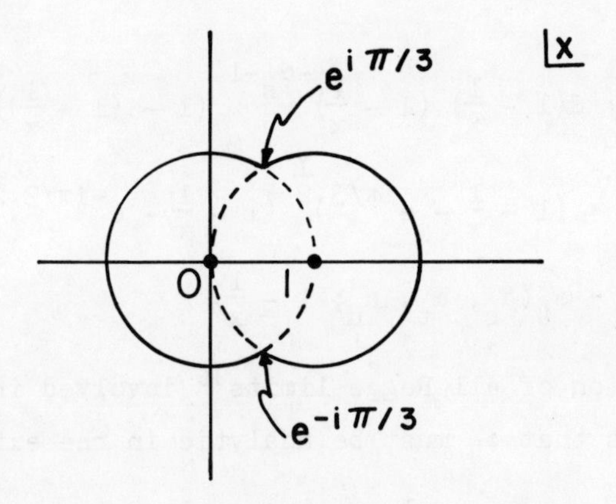

FIGURE 6.1

Fixed-point Singularities

This required domain of analyticity can be greatly extended by considering simultaneously the three permutations in the summed form

$$T = A_4(-\alpha_s, -\alpha_t) + A_4(-\alpha_t, -\alpha_u) + A_4(-\alpha_u, -\alpha_s) \tag{6.29}$$

$$= \int_{-\infty}^{\infty} dx\, |x|^{-\alpha_t - 1} |1 - x|^{-\alpha_s - 1} (1 - x + x^2)^{\frac{\gamma}{2}} \cdot \phi_4(\alpha_s, \alpha_t, \alpha_u; x) \tag{6.30}$$

$$= \int_0^1 dx\, x^{-\alpha_t - 1} (1 - x)^{-\alpha_s - 1} (x - e^{i\pi/3})^{\frac{\gamma}{2}} (x - e^{-i\pi/3})^{\frac{\gamma}{2}} \cdot$$

$$\cdot\, \phi_4(\alpha_s, \alpha_t, \alpha_u; x) +$$

$$+ \int_0^1 d\left(\frac{1}{1 - x}\right) \left(\frac{1}{1 - x}\right)^{-\alpha_u - 1} \left(1 - \frac{1}{1 - x}\right)^{-\alpha_t - 1} \cdot$$

$$\cdot \left(\frac{1}{1 - x} - e^{i\pi/3}\right)^{\frac{\gamma}{2}} \left(\frac{1}{1 - x} - e^{-i\pi/3}\right)^{\frac{\gamma}{2}} +$$

$$+ \int_0^1 d(1 - \frac{1}{x}) \, (1 - \frac{1}{x})^{-\alpha_s - 1} \, (1 - (1 - \frac{1}{x}))^{-\alpha_u - 1} \, \cdot$$

$$\cdot \, (1 - \frac{1}{x} - e^{i\pi/3})^{\frac{\gamma}{2}} \, (1 - \frac{1}{x} - e^{-i\pi/3})^{\frac{\gamma}{2}} \, \cdot$$

$$\cdot \, \phi_4(\alpha_s, \alpha_t, \alpha_u; 1 - \frac{1}{x}) \tag{6.31}$$

Consideration of all Regge limits[16] involved in the three terms shows that ϕ_4 must be analytic in the extended domain

$$\mathcal{D}(x) \cup \mathcal{D}(\frac{1}{1 - x}) \cup \mathcal{D}(1 - \frac{1}{x}) \tag{6.32}$$

which, remarkably, comprises the whole x-plane, including the point at infinity, except for two points $x = e^{\pm i\pi/3}$ (see Figure 6.1). That is,

$$\mathcal{D}(x) \cup \mathcal{D}(\frac{1}{1 - x}) \cup \mathcal{D}(1 - \frac{1}{x}) =$$

$$= \{x | \; x \neq e^{\pm i\pi/3}\} \tag{6.33}$$

Since we have already shown that the integrand <u>must</u> contain complex singularities, we now deduce that they are required to be precisely at the two conjugate points $x = e^{\pm i\pi/3}$.

By considerations of the S_3 group alone we can already see that these two points play a special role, as follows. In general, starting from any point x and making the S_3 transformations

$$x \to 1 - x \to \frac{1}{x} \to 1 - \frac{1}{x} \to \frac{1}{1 - x} \to \frac{x}{x - 1} \tag{6.34}$$

we arrive in general at six different points (a six-point orbit). For special cases, only three points are involved, namely

$$x = 0,\ 1,\ \infty \tag{6.35}$$

$$x = -1,\ \frac{1}{2},\ 2 \tag{6.36}$$

Finally for only one case there is a two-point orbit, namely

$$x = e^{\pm i\pi/3} \tag{6.37}$$

Thus these special values correspond to the most degenerate orbit of the S_3 group.

6.5 EXPLICIT CONSTRUCTION

We have deduced that ϕ_4 must satisfy the two fundamental properties

i) $\phi_4 \begin{pmatrix} \alpha_s & ; & \alpha_t & ; & \alpha_u \\ \beta_s(x) & & \beta_t(x) & & \beta_u(x) \end{pmatrix}$ is S_3 invariant under permutation of the 3 arguments $\{\alpha_i, \beta_i(x)\}$, i = s, t, u.

ii) ϕ_4 must be analytic in the entire x-plane, except possibly for poles at $x = e^{\pm i\pi/3}$.

The analyticity properties suggest that we introduce the definition

$$\beta_s(x) = x(x - e^{i\pi/3})^{-1}\ (x - e^{-i\pi/3})^{-1} \tag{6.38}$$

$$= x(1 - x + x^2)^{-1} = \beta_s(\frac{1}{x}) \tag{6.39}$$

with simple poles at the point $x = e^{\pm i\pi/3}$. It follows that

$$\beta_t(x) = \frac{1 - x}{(1 - x + x^2)} \tag{6.40}$$

$$\beta_u(x) = \frac{-x(1 - x)}{(1 - x + x^2)} \tag{6.41}$$

The analyticity requirement (ii) on ϕ_4 may now be

re-expressed by stating that ϕ_4 is analytic in the $\beta_i(x)$; it must also be analytic in the α_i, or equivalently in s, t, u from the usual analyticity requirements of $A_4(-\alpha_s, -\alpha_t)$ as a scattering amplitude.

In order to distinguish between an infinite number of possibilities, we must adopt some simplicity criterion, and we shall assume that the poles in ϕ_4 at $x = e^{\pm i\pi/3}$ are of low order*. Up to double poles we may write, with full generality,

$$\phi_4 = \sum_{i=0}^{4} \lambda_i \phi_4^{(i)} \tag{6.42}$$

with

$$\phi_4^{(0)} = 1 \tag{6.43}$$

$$\phi_4^{(1)} = \alpha_s\beta_s + \alpha_t\beta_t + \alpha_u\beta_u \tag{6.44}$$

$$\phi_4^{(2)} = \alpha_s\beta_t\beta_u + \alpha_t\beta_u\beta_s + \alpha_u\beta_s\beta_t \tag{6.45}$$

$$\phi_4^{(3)} = \alpha_s\alpha_t\beta_s\beta_t + \alpha_t\alpha_u\beta_t\beta_u + \alpha_u\alpha_s\beta_u\beta_s \tag{6.46}$$

$$\phi_4^{(4)} = \alpha_s^2\beta_s + \alpha_t^2\beta_t + \alpha_u^2\beta_u \tag{6.47}$$

If we are considering, in particular, pion-pion elastic scattering we must impose $\lambda_0 = 0$ to avoid a tachyon on the positive intercept rho trajectory.

A more general formula for ϕ_4 (see Balachandran and Rupertsberger, Reference 17) is the following

*This simplicity criterion is an important assumption and it is expected that it may eventually be replaced by, or be shown to be equivalent to, the criterion of minimum degeneracy in the factorisation of the corresponding N-point amplitude.

$$\phi_4 = S_1 + [\tfrac{1}{3}(2\alpha_s - \alpha_t - \alpha_u)(2\beta_s - \beta_t - \beta_u) +$$

$$+ (\alpha_t - \alpha_u)(\beta_t - \beta_u)]S_2 +$$

$$+ [\tfrac{1}{3}(2\alpha_s{}^2 - \alpha_t{}^2 - \alpha_u{}^2)(2\beta_s - \beta_t - \beta_u) +$$

$$+ (\alpha_t{}^2 - \alpha_u{}^2)(\beta_t - \beta_u)]S_3 +$$

$$+ [\tfrac{1}{3}(2\alpha_s - \alpha_t - \alpha_u)(2\beta_s{}^2 - \beta_t{}^2 - \beta_u{}^2) +$$

$$+ (\alpha_t - \alpha_u)(\beta_t{}^2 - \beta_u{}^2)]S_4$$

$$+ [\tfrac{1}{3}(2\alpha_s{}^2 - \alpha_t{}^2 - \alpha_u{}^2)(2\beta_s{}^2 - \beta_t{}^2 - \beta_u{}^2) +$$

$$+ (\alpha_t{}^2 - \alpha_u{}^2)(\beta_t{}^2 - \beta_u{}^2)]S_5$$

$$+ (\alpha_s - \alpha_t)(\alpha_t - \alpha_u)(\alpha_u - \alpha_s) \cdot$$

$$\cdot (\beta_s - \beta_t)(\beta_t - \beta_u)(\beta_u - \beta_s)S_6 \qquad (6.48)$$

where the six functions S_i are power series in the three symmetric quantities

$$\bar{\alpha} = \alpha_s\alpha_t + \alpha_t\alpha_u + \alpha_u\alpha_s \qquad (6.49)$$

$$\underline{\alpha} = \alpha_s\alpha_t\alpha_u \qquad (6.50)$$

$$z = \beta_s\beta_t\beta_u \qquad (6.51)$$

The variable z (which was discussed already in a paper by Dixon[18] in 1904) is fully invariant under the S_3 group.

To derive[17] this general expression for ϕ_4 we note that the S_3 invariance implies that we may write ϕ_4 as a superposition of functions of the form ϕ where

$$\phi(\alpha_s, \alpha_t, \alpha_u; x) = \psi^{(s)} (\alpha_s, \alpha_t, \alpha_u) \chi^{(s)}(x) + \\ + \psi^{(A)}(\alpha_s, \alpha_t, \alpha_u) \chi^{(A)}(x) + \\ + \sum_{\rho,\sigma=1}^{2} \psi_\rho^{(M)} (\alpha_s, \alpha_t, \alpha_u) g_{\rho\sigma} \chi_\sigma^{(M)}(x) \qquad (6.52)$$

Here $\psi^{(s)}$ is fully symmetric in α_s, α_t, α_u and $\chi^{(s)}$ is invariant under $x \to (1-x)^{-1} \to (x-1)x^{-1} \to x^{-1} \to (1 - x) \to x(x -1)^{-1}$. Similarly $\psi^{(A)}$ and $\chi^{(A)}$ carry the antisymmetric representation of S_3, while $\psi_\rho^{(M)}$ and $\chi_\rho^{(M)}$ are the basis for the two-dimensional representation with metric g.

The method of constructing the $\psi^{(i)}$ is already known from work on two-variable expansions of scattering amplitudes. Hence $\psi^{(s)}$ and $\psi^{(A)}$ are of the form

$$\psi^{(s)} (\alpha_s, \alpha_t, \alpha_u) = \sum_{p,q>0} E_{pq}^{(s)} \bar{\alpha}^p \underline{\alpha}^q \qquad (6.53)$$

$$\psi^{(A)} = (\alpha_s, \alpha_t, \alpha_u) = (\alpha_s - \alpha_t) (\alpha_t - \alpha_u) (\alpha_u - \alpha_s) \cdot \\ \cdot \sum_{p,q>0} E_{pq}^{(A)} \bar{\alpha}^p \underline{\alpha}^q \qquad (6.54)$$

while the two-dimensional representations are carried by functions either of the form

$$(\psi_1^{(M)}(\alpha_s, \alpha_t, \alpha_u), \psi_2^{(M)}(\alpha_s, \alpha_t, \alpha_u)) = \\ = (2\alpha_s - \alpha_t - \alpha_u, \alpha_t - \alpha_u) \sum_{p,q>0} \cdot \\ \cdot G_{pq}^{(1)} \bar{\alpha}^p \underline{\alpha}^q \qquad (6.55)$$

or of the form

$$(\psi_1{}^{(M)}(\alpha_s, \alpha_t, \alpha_u), \psi_2{}^{(M)}(\alpha_s, \alpha_t, \alpha_u)) =$$

$$= (2\alpha_s{}^2 - \alpha_t{}^2 - \alpha_u{}^2, \alpha_t{}^2 - \alpha_u{}^2) \sum_{p,q>0} \cdot$$

$$\cdot\ G_{pq}^{(2)} \bar{\alpha}^p \underline{\alpha}^q \tag{6.56}$$

From these formulae the metric is checked to be

$$g = \begin{bmatrix} 1/3 & 0 \\ 0 & 1 \end{bmatrix} \tag{6.57}$$

To find $\chi^{(s)}$, we can first show that $\chi^{(s)}(x) - \chi^{(s)}(0)$ must have zeros of order at least two at $x = 0, 1, \infty$ by considering

$$\chi^{(s)}(x) - \chi^{(s)}(0) = \chi^{(s)}\left(\frac{x}{x-1}\right) - \chi^{(s)}(0) \tag{6.58}$$

$$= \chi^{(s)}\left(\frac{1}{1-x}\right) - \chi^{(s)}(0) \tag{6.59}$$

$$= \chi^{(s)}\left(\frac{1}{x}\right) - \chi^{(s)}(0) \tag{6.60}$$

and the first and second derivatives thereof.

By considering, in addition,

$$[\chi^{(s)}(x) - \chi(0)]^{-1} = [\chi^{(s)}\left(\frac{1}{1-x}\right) - \chi^{(s)}(0)]^{-1} \tag{6.61}$$

and its derivatives, we find that $[\chi^{(s)}(x) - \chi^{(s)}(0)]$ has poles of order at least three at $x = e^{\pm i\pi/3}$. Hence we may write

$$\chi^{(s)}(x) = \chi^{(s)}(0) + z\ \chi_1{}^{(s)}(x) \tag{6.62}$$

where

$$z = \beta_s(x)\ \beta_t(x)\ \beta_u(x) = -\frac{x^2(1-x)^2}{(1-x+x^2)^3} \qquad (6.63)$$

By repeating the process we deduce that $\chi^{(s)}(x)$ is a polynomial

$$\chi^{(s)}(x) = \sum_{p=0}^{\infty} c_p\ z^p \qquad (6.64)$$

in the variable z.

Concerning $\chi^{(A)}(x)$, antisymmetry dictates that it change sign under $x \to \frac{1}{x}$, $x \to (1-x)$, $x \to x(x-1)^{-1}$ and therefore it must have zeros at $x = -1, 0, \frac{1}{2}, 1, 2$ and ∞. By considering $(\chi^{(A)}(x))^{-1}$ we may show that $\chi^{(A)}(x)$, like $\chi^{(s)}(x)$, has poles of order at least three at $x = e^{\pm i\pi/3}$. Therefore

$$\chi^{(A)} = -2(x+1)\ x(x-\tfrac{1}{2})\ (x-1)\ (x-2)\ \overline{\chi}^{(s)}(x) \qquad (6.65)$$

$$= (\beta_s - \beta_t)\ (\beta_t - \beta_u)\ (\beta_u - \beta_s)\ \overline{\chi}^{(s)}(x) \qquad (6.66)$$

where $\overline{\chi}^{(s)}(x)$ is fully symmetric, and hence a polynomial in z.

At this stage we may proceed further deductively but by now it is an obvious guess that the correct mixed symmetry $\chi^{(M)}(x)$ are either of the form (by analogy with $\psi^{(M)}$)

$$(2\beta_s - \beta_t - \beta_u,\ \beta_t - \beta_u)\ \hat{\chi}^{(s)}(x) \qquad (6.67)$$

or of the form

$$(2\beta_s^{\ 2} - \beta_t^{\ 2} - \beta_u^{\ 2},\ \beta_t^{\ 2} - \beta_u^{\ 2})\ \hat{\hat{\chi}}^{(s)}(x) \qquad (6.68)$$

where $\hat{\chi}^{(s)}$, $\hat{\hat{\chi}}^{(s)}$ are polynomials in z. Indeed, it is possible to show that this exhausts all possible forms for $\chi^{(M)}$.

Combining the results for $\psi^{(s)}$, $\psi^{(A)}$, $\psi^{(M)}$ with those for $\chi^{(s)}$, $\chi^{(A)}$, $\chi^{(M)}$ then demonstrates that the formula for ϕ_4, given earlier, is the most general representation consistent with the S_3 invariance and the given analytic properties.

6.6 FOUR-PION AMPLITUDE

To select between the above candidates for the four-pion amplitude, we must bear in mind that the ultimate goal is an N-pion amplitude which factorises on a ghost-free spectrum. A necessary condition is therefore that the four-pion amplitude should have on-shell residues such that the partial-wave projections give rise to a positive-definite imaginary part.

This question has been examined in Reference 10. The result is that any linear combination

$$\phi_4 = \lambda_2 \phi_4^{(2)} + \lambda_3 \phi_4^{(3)} \tag{6.69}$$

of the double-pole models contains negative-squared coupling constants and is thereby eliminated. Similarly the model corresponding to $\phi_4^{(4)}$ has ghosts for physical intercepts. This leaves only the (simplest) model $\phi_4^{(1)}$.

For this model, corresponding to $\phi_4^{(1)}$, the amplitude is

$$A_4 = \int_0^1 dx\, x^{-\alpha_t - 1} (1 - x)^{-\alpha_s - 1} (1 - x + x^2)^{\frac{\gamma}{2}} \cdot$$

$$\cdot (\alpha_s \beta_s + \alpha_t \beta_t + \alpha_u \beta_u) \tag{6.70}$$

Setting $\alpha_\rho(o) - \alpha_\pi(o) = \frac{1}{2}$ exactly, this amplitude becomes

$$A_4 = \int_0^1 dx\, x^{-\alpha_t - 1} (1 - x)^{-\alpha_s - 1} (1 - x + x^2)^{\frac{1 - \alpha_\rho(o)}{2}} \cdot$$

$$\cdot (\alpha_s x + \alpha_t (1 - x) - \alpha_u x(1 - x)) \qquad (6.71)$$

For a wide range of values of $\alpha_\rho(o)$ (at least $\frac{1}{3} \leq \alpha_\rho(o) \leq 1$) it can be shown numerically that the partial waves are all positive for the lowest-lying mass levels.

We should note that in the unphysical limit $\alpha_\rho(o) \to 1$, the amplitude becomes

$$A_4 \underset{\alpha_\rho(o)\to 1}{\to} - 3 \frac{\Gamma(1 - \alpha_s)\Gamma(1 - \alpha_t)}{\Gamma(1 - \alpha_s - \alpha_t)} \qquad (6.72)$$

which is precisely the four-pion amplitude of the Neveu-Schwarz model, known to be ghost-free.

For other values of $\alpha_\rho(o)$, we can give an analytic proof[19] of the Gribov-Pomeranchuk identities[20], namely

$$\frac{\partial^n}{\partial s^n} \operatorname{Im} A_4(s, t)\Big|_{s=0} \geq 0 \qquad \text{for all } n \qquad (6.73)$$

This condition follows from unitarity, and is a necessary condition for the absence of ghost states.

To prove these inequalities for $\frac{1}{2} \leq \alpha_\rho(o) \leq 1$ we re-write

$$A_4 = \int_0^1 dx\, x^{-\alpha_t - 1} (1 - x)^{-s} \left[\frac{1 + x^3}{(1 - x)(1 - x^2)}\right]^{\frac{1 - \alpha_\rho(o)}{2}}$$

$$\cdot (1 - x)^{1 - 2\alpha_\rho(o)} \left[\alpha_s \frac{x}{1 - x} + \alpha_t - \alpha_u x\right]$$

$$(6.74)$$

[Here $\alpha' = 1$].

By examining in turn the factors, we deduce that

$$(1 - x)^{-s} \qquad (6.75)$$

has a positive-definite power series in x for all s. Similarly

$$(1 - x)^{1 - 2\alpha_\rho(o)} \tag{6.76}$$

is positive-definite for $\alpha_\rho(o) \geq \frac{1}{2}$. We can re-write

$$\alpha_s \frac{x}{1 - x} + \alpha_t - \alpha_u x =$$

$$= (\alpha_\rho(o) + t) + (4\alpha_\rho(o) + 2s + t - 2) + \\ + (\alpha_\rho(o) + s) \sum_{n=2}^{\infty} x^n \tag{6.77}$$

which is positive for all s, provided $\alpha_\rho(o) \geq \frac{1}{2}$. Finally we note that

$$\left[\frac{1 + x^3}{(1 - x)(1 - x^2)}\right]^{\frac{1-\alpha_\rho(o)}{2}} =$$

$$= \exp\{(\frac{1 - \alpha_\rho(o)}{2}) \{\ell n(1 + x^3) - \ell n(1 - x) - \\ - \ell n(1 - x^2)\}\} \tag{6.78}$$

$$= \exp\{(\frac{1 - \alpha_\rho(o)}{2}) [- \sum_{n=1}^{\infty} \frac{(-x^3)^n + x^n + (x^2)^n}{n}]\} \tag{6.79}$$

which is positive-definite for $\alpha_\rho(o) \leq 1$. This completes the proof of the inequalities for $\frac{1}{2} < \alpha_\rho(o) \leq 1$ since we have shown that

$$A_4 = \int_0^1 dx\, x^{-\alpha_t - 1} \sum_{n=1}^{\infty} R_n(\alpha_s)\, x^n \tag{6.80}$$

and

$$\left.\frac{\partial^r}{\partial s^r} R_n(\alpha_s)\right|_{s=0} \geq 0 \tag{6.81}$$

as required.

To extend this proof to $\frac{1}{3} \leq \alpha_\rho(o) \leq \frac{1}{2}$ we write A_4 in the alternative form

$$A_4 = \int_0^1 dx\, x^{-\alpha_t - 1} (1 - x)^{-s} \left(\frac{1 + x^3}{1 - x^2}\right)^{\frac{1-\alpha_\rho(o)}{2}} \cdot$$
$$\cdot\, (1 - x)^{\frac{1-3\alpha_\rho(o)}{2}} \left[\alpha_s \frac{x}{1 - x} + \alpha_t - \alpha_u x\right] \tag{6.82}$$

Here the three factors

$$(1 - x)^{-s} \tag{6.83}$$

$$(1 - x)^{\frac{1-3\alpha_\rho(o)}{2}} \tag{6.84}$$

$$(\alpha_s \frac{x}{1 - x} + \alpha_t - \alpha_u x) \tag{6.85}$$

have positive powers series in x for all s, provided $\alpha_\rho(o) \geq \frac{1}{3}$.

Consider the function

$$F(x) = \left(\frac{1 + x^3}{1 - x^2}\right)^{\frac{1-\alpha_\rho(o)}{2}} = \sum_{n=0}^{\infty} C_n\, x^n \tag{6.86}$$

Then

$$(1 + x^3)(1 - x^2)\, F'(x) = \left(\frac{1 - \alpha_\rho(o)}{2}\right) F(x)\, x(2 - x) \tag{6.87}$$

Hence (putting $A = \frac{1 - \alpha_\rho(o)}{2}$)

$$(n + 1)\, C_{n+1} = 2nC_n + 2(A + 1 - n)C_{n-1} + $$
$$+ (n - 2 - A)C_{n-2} \tag{6.88}$$

and from this we deduce that

$$(n + 4)\, C_{n+4} = 2A\, C_{n+2} + 2A\, C_n + 3A\, C_{n-1} + (n - 2\,A) \cdot$$
$$\cdot\, C_{n-2} \tag{6.89}$$

with asymptotically ($n \to \infty$) positive coefficients and that the first few Taylor coefficients are

$$C_0 = 1 \tag{6.90}$$

$$C_1 = 0 \tag{6.91}$$

$$C_2 = A \tag{6.92}$$

$$C_3 = A \tag{6.93}$$

$$C_4 = \frac{1}{2} A(A + 1) \tag{6.94}$$

$$C_5 = A^2 \tag{6.95}$$

$$C_6 = \frac{1}{6} A(A^2 + 6A - 1) \tag{6.96}$$

From this we deduce that

$$C_6 \geq 0 \tag{6.97}$$

requires $A \geq \sqrt{10} - 3$, or

$$\alpha(o) \leq 7 - 2\sqrt{10} = 0.68 \tag{6.98}$$

If this is satisfied, then it follows that $C_n \geq 0$ for all n by using the recursion formula given above. Thus we have positivity for $(7 - 2\sqrt{10}) \geq \alpha_\rho(o) \geq \frac{1}{3}$.

Collecting these results, we have demonstrated that A_4 satisfies the Gribov-Pomeranchuk inequalities for at least the range $\frac{1}{3} \leq \alpha_\rho(o) \leq 1$ and $\alpha_\rho(o) - \alpha_\pi(o) = \frac{1}{2}$.

Next we consider the Adler consistency condition which states that for the four-pion amplitude

$$\lim_{p_1^\mu \to 0} [T(s, t, u)] = 0 \tag{6.99}$$

We may re-write A_4 as

$$A_4(-\alpha_s, -\alpha_t) = -\frac{3\Gamma(1-\alpha_s)\Gamma(1-\alpha_t)}{\Gamma(1-\alpha_s-\alpha_t)} +$$

$$+ (\alpha_\rho(o) - 1)\, B(1-\alpha_s, 1-\alpha_t) -$$

$$- \sum_{R=0}^{\infty} \binom{\frac{1-\alpha_\rho(o)}{2}}{R+1} (-1)^R \begin{bmatrix} \alpha_s\, B(R+1-\alpha_s, R+2-\alpha_t) \\ +\alpha_t\, B(R+2-\alpha_s, R+1-\alpha_t) \\ -\alpha_u\, B(R+2-\alpha_s, R+2-\alpha_t) \end{bmatrix} \tag{6.100}$$

The leading trajectory contribution is now entirely contained in the first term of the right hand side, and this leading term vanishes at the Adler point $s = t = u = \mu^2$ provided that

$$\alpha_\rho(\mu^2) = \alpha_\rho(o) - \alpha_\pi(o) = \frac{1}{2} \tag{6.101}$$

Thus at the Adler point, the only contributions are then from the daughter levels which are more distant singularities. This particular intercept constraint is well satisfied physically, and, as we shall see later, it leads also to a simplification of the N-pion generalisation of this N = 4 amplitude.

Finally, we may briefly mention some phenomenological implications of this amplitude. For the elastic decay widths, into two pions, of the parent resonances the predictions coincide with those of the Lovelace-Shapiro formula[22], which contains only the leading term. At the daughter levels, however, the predictions are quite different (see Figure 6.2). The s-wave (ε) daughter of the ρ is absent; this is consistent with recent phenomenological analysis by, for example, Protopopescu et al.[23] The p-wave (ρ') daughter of the f is absent, in agreement with experiment, but there is an s-wave (ε') at this mass with $\Gamma_{\pi\pi}$ ~ 260 MeV in agreement with $\pi\pi$ phase shift analysis[24]. The p-wave (ρ'') daughter of the g has $\Gamma_{\pi\pi}$ ~ 140 MeV, consistent with experiment[25] (Figure 6.2(a)).

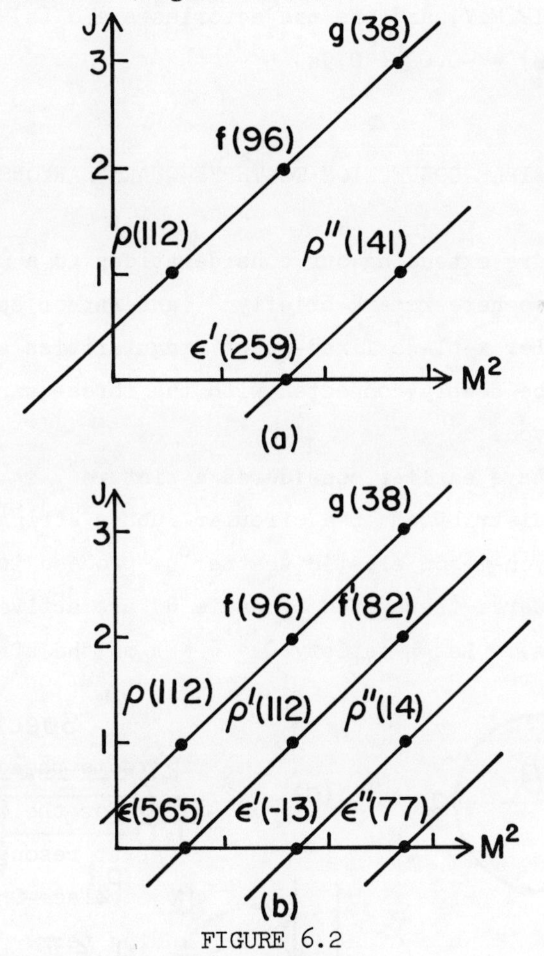

FIGURE 6.2

Elastic Widths

By contrast, the Lovelace-Shapiro formula predicts a very broad ε resonance, a strongly coupled ρ' and a slightly ghost-like ε' resonance (Figure 6.2(b)). We conclude, therefore, that the present model is phenomenologically preferable, at the daughter levels, to the Lovelace-Shapiro model.

[Note: In Figure 6.2, the widths are normalised to $\Gamma_{\rho \to 2\pi} = 112$ MeV, and the trajectories are $\alpha_\rho(s) = 0.48 + 0.9s$, $\alpha_\pi(s) = -0.02 + 0.9s$]

6.7 POSSIBLE CONNECTION TO THREE-QUARK BARYONS

Before extending our considerations to a many-meson amplitude, we here remark briefly[26] (and rather speculatively) that the Euler x-plane fixed-point singularities at $x = e^{\pm i\pi/3}$ may be deeply connected with the three-quark structure of baryons.

We have earlier considered a picture[28] in which quarks are distributed on a circular rubber string. Consider a baryon-meson elastic scattering process in which two different quarks (separated by angle θ) are active, as in Figure 6.3(a). Let θ satisfy $0 \leq \theta \leq \pi$ on the string which

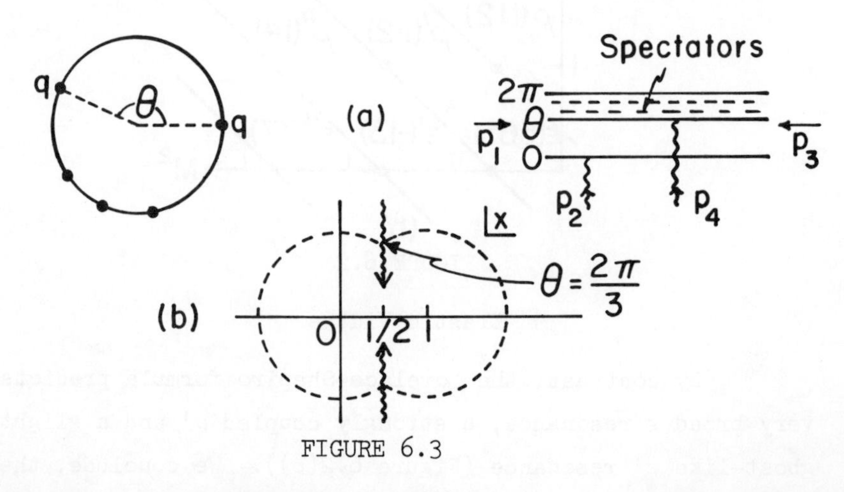

FIGURE 6.3

Three-Quark Baryons

is of total period 2π, and suppose there is in addition some arbitrary number of spectator quarks.

The amplitude is given by

$$A_4^{(BM)}(-\alpha_u, -\alpha_s) = \int_0^1 dy\, y^{-\alpha_s - 1} \cdot$$

$$\cdot \langle 0|\, \exp(\sqrt{2}\, ip_2 \cdot \sum_{n=1}^{\infty} \frac{a^{(n)}}{\sqrt{n}} \cos n\theta)\, y^R \cdot$$

$$\cdot \exp(\sqrt{2}\, ip_4 \cdot \sum_{n=1}^{\infty} \frac{a^{(n)+}}{\sqrt{n}})\, |0\rangle + (p_2 \leftrightarrow p_4) \qquad (6.102)$$

$$= \int_0^1 dy\, y^{-\alpha_s - 1} \exp[p_2 \cdot p_4 \sum_{n=1}^{\infty} \frac{y^n}{n} (e^{in\theta} + e^{-in\theta})] +$$

$$+ (p_2 \leftrightarrow p_4) \qquad (6.103)$$

$$= \int_0^1 dy\, y^{-\alpha_s - 1} (y^2 - 2y\cos\theta + 1)^{-\frac{1}{2}(1+\alpha_t)} +$$

$$+ (\alpha_s \leftrightarrow \alpha_u) \qquad (6.104)$$

To combine the two terms in $A_4(-\alpha_u, -\alpha_s)$ one puts

$$u = y(y^2 - 2y\cos\theta + 1)^{-1/2} \qquad (6.105)$$

$$v = \frac{u}{y} \qquad (6.106)$$

whereupon

$$A_4^{(BM)}(-\alpha_u, -\alpha_s) = \int_0^{\frac{1}{2\sin\theta/2}} \frac{du}{v - u\cos\theta}\, u^{-\alpha_s - 1}\, v^{-\alpha_u - 1} +$$

$$+ \int_{\frac{1}{2\sin /2}}^{1} \frac{dv}{u - v\cos\theta}\, v^{-\alpha_s - 1}\, u^{-\alpha_u - 1} \qquad (6.107)$$

$$= 2 \int_0^1 du \int_0^1 dv\, u^{-\alpha_s - 1}\, v^{-\alpha_u - 1}\, \delta(u^2 + v^2 - 2uv\cos\theta - 1) \qquad (6.108)$$

Now put

$$u = x[x^2 + (1 - x)^2 - 2x(1 - x)\cos\theta]^{1/2} \qquad (6.109)$$

to obtain

$$A_4^{(BM)}(-\alpha_u, -\alpha_s) = \int_0^1 dx\, x^{-\alpha_s - 1} (1 - x)^{-\alpha_u - 1} \cdot$$

$$\cdot [1 - 4x(1 - x)\cos^2 \frac{\theta}{2}]^{\frac{\alpha_s + \alpha_u + 2}{2}} \qquad (6.110)$$

This is remarkably similar to the forms we have considered for meson-meson scattering. In particular, the integrand of $A_4^{(BM)}$ has branch points situated symmetrically about the real axis at

$$x_0 = \frac{1}{2} \pm \frac{i}{2} (\sec^2\frac{\theta}{2} - 1)^{1/2} \qquad (6.111)$$

Consider now what happens as we change continuously the angular separation, θ, of the active quarks. For $\theta = \pi$, the branch points are absent (i.e. are at infinity); for decreasing θ the branch points move to the finite x-plane and eventually reach the special points $x = e^{\pm i\pi/3}$ when $\theta = 2\pi/3$. (See Figure 6.3(b), page 322).

Now assume that our earlier result, that the only singularities are at $x = 0, 1, \infty$ and $e^{\pm i\pi/3}$, extends to this baryonic case. It follows that the angular separation must be $\theta = 2\pi/3$ and hence that, within such dual schemes, the number of quarks in the (symmetric) baryon should be a precise multiple of three.

To summarise, we may conjecture that the singularities at $x = e^{\pm i\pi/3}$ that we have introduced to change the intercept of the <u>meson</u> amplitude, could have a much deeper significance: it may ultimately be shown that such singularities play a central role in a consistent bootstrap solution for mesons and baryons together.

6.8 MULTIPARTICLE EXTENSION

Now we return to the question of how to extend our four-particle amplitude to any number of particles. We have previously written (section 6.3) the four-meson amplitude in the general form

$$A_4(-\alpha_s, -\alpha_t) = \int_0^1 dx\, x^{-\alpha_t - 1} (1 - x)^{-\alpha_s - 1} \cdot$$
$$\cdot (x - e^{i\pi/3})^{\frac{\gamma}{2}} (x - e^{-i\pi/3})^{\frac{\gamma}{2}} \phi_4 \qquad (6.112)$$

where $\phi_4(\alpha_s, \alpha_t, \alpha_u; x)$ is invariant under an S_3 group acting on the objects $\{\alpha_i, \beta_i(x)\}$, $i = s, t, u$.

We would like to extend this formula to any (even) number of external pions. In particular, we wish to maintain the symmetric group properties of the integrand in the multiparticle extension; as we shall see, this greatly resolves the ambiguity inherent in the generalisation from the four-particle amplitude.

For A_4, we introduce the variables[29,30] z_i ($i = 1, 2, 3, 4$) lying on a circle in the complex z-plane, and make the identification

$$x = (z_2, z_1; z_3, z_4) = \frac{(23)}{(13)} \qquad (6.113)$$

$$1 - x = (z_1, z_4; z_2, z_3) = \frac{(12)}{(13)} \qquad (6.114)$$

where the notation for an anharmonic ratio is

$$(a, b; c, d) = \frac{(a - c)(b - d)}{(a - d)(b - c)} \qquad (6.115)$$

and we have introduced the convenient shorthand, to make the later development more manageable

$$(12) = (z_1 - z_2)\,(z_3 - z_4) \qquad (6.116)$$

$$(23) = (z_2 - z_3)\,(z_1 - z_4) \tag{6.117}$$

$$(12) = (z_1 - z_3)\,(z_2 - z_4) \tag{6.118}$$

First we note that the Euler B function may be written (for $\alpha_\rho(o) \neq \alpha_\pi(o)$) in the form

$$B(-\alpha_s, -\alpha_t) = \int \prod_{i=1}^{4} dz_i \left[\frac{dz_a\, dz_b\, dz_c}{(z_a - z_b)(z_b - z_c)(z_c - z_a)}\right]^{-1}$$

$$\prod_{i<j} (z_i - z_j)^{-2p_i \cdot p_j} \prod_{i=1}^{4} (z_i - z_{i+1})^{\alpha_\rho(o)-1} \cdot$$

$$\cdot \left[\prod_{i<j} (z_i - z_j)^{(-1)^{i-j}}\right]^{2(\alpha_\rho(o)-\alpha_\pi(o))} \tag{6.119}$$

where z_a, z_b, z_c are fixed values for three out of the four z_i (arbitrarily chosen) and the integration domain is the boundary of the circle with the restriction that the cyclic ordering of the points z_1, z_2, z_3, z_4 is preserved.

It is useful to note the identity

$$\frac{1}{2}\gamma = -\frac{1}{2}(\alpha_\rho(o) - 1) + 2[\alpha_\rho(o) - \alpha_\pi(o)] \tag{6.120}$$

This identity motivates us to establish two fundamental identities which enable us to extend the factor $(1 - x + x^2)^{\frac{\gamma}{2}}$ containing the branch point singularities at $x = e^{\pm i\pi/3}$. First we note that

$$\prod_{i=1}^{4} (z_i - z_{i+1})^2 = (12)^2\,(23)^2 \tag{6.121}$$

and then consider the symmetric expression

$$\sum_{\{q_1 q_2 q_3 q_4\}} \prod_{i=1}^{4} (z_{q_i} - z_{q_{i+1}})^{-2} \tag{6.122}$$

where the sum is over the 4! permutations of $\{q_1 q_2 q_3 q_4\} =$

{1, 2, 3, 4}.

This is equal to

$$\sum_{\{q_1 q_2 q_3 q_4\}} \prod_{i=1}^{4} (z_{q_i} - z_{q_{i+1}})^{-2} =$$

$$= 8\left[\frac{1}{(12)^2(23)^2} + \frac{1}{(12)^2(13)^2} + \frac{1}{(23)^2(13)^2}\right] \quad (6.123)$$

$$= 16 \prod_{i=1}^{4} (z_i - z_{i+1})^{-2} (x - e^{i\pi/3})(x - e^{-i\pi/3}) \quad (6.124)$$

In other words, there is the identity

$$(x - e^{i\pi/3})(x - e^{-i\pi/3}) =$$

$$= \frac{1}{16} \Pi (z_i - z_{i+1})^2 \sum_{\{q_1 q_2 q_3 q_4\}} \prod_{i=1}^{4} (z_{q_i} - z_{q_{i+1}})^{-2} \quad (6.125)$$

The second identity concerns the expression

$$\prod_{i<j} (z_i - z_j)^{(-1)^{i-j}} = \frac{(13)}{(12)(23)} \quad (6.126)$$

The corresponding symmetric form is

$$\sum_{\{q_1 q_2 q_3 q_4\}} (-1)^P \prod_{k<\ell} (z_{q_k} - z_{q_\ell})^{(-1)^{k-\ell}} =$$

$$= 8\left[\frac{(13)}{(12)(23)} + \frac{(23)}{(12)(13)} + \frac{(12)}{(23)(13)}\right] \quad (6.127)$$

which leads to the identity

$$(x - e^{i\pi/3})(x - e^{-i\pi/3}) = \frac{1}{16} \prod_{i<j} (z_i - z_j)^{-(-1)^{i-j}}$$

$$\sum_{\{q_1 q_2 q_3 q_4\}} (-1)^P \prod_{k<\ell} (z_{q_k} - z_q)^{(-1)^{k-\ell}} \quad (6.128)$$

In these expressions, P is the parity of the permutation.

Collecting these results together leads to

$$(x - e^{i\pi/3})^{\frac{\gamma}{2}} (x - e^{-i\pi/3})^{\frac{\gamma}{2}} =$$

$$= 4^{-\gamma} \prod_{i=1}^{4} (z_i - z_{i+1})^{1-\alpha_\rho(o)} \cdot$$

$$\cdot \Big[\prod_{i<j} (z_i - z_j)^{(-1)^{i-j}} \Big]^{-2(\alpha_\rho(o)-\alpha_\pi(o))} \cdot$$

$$\cdot \Big[\sum_{\{q_1 q_2 q_3 q_4\}} \prod_{k=1}^{4} (z_{q_k} - z_{q_{k+1}})^{-2} \Big]^{\frac{1-\alpha_\rho(o)}{2}} \cdot$$

$$\cdot \Big[\sum_{\{q_1 q_2 q_3 q_4\}} (-1)^P \cdot$$

$$\cdot \prod_{k\neq\ell} (z_{q_k} - z_{q_\ell})^{(-1)^{k-\ell}} \Big]^{2(\alpha_\rho(o)-\alpha_\pi(0))} \qquad (6.129)$$

We may insert such an expression directly into A_4. We see that all asymmetric factors are thereby cancelled, and replaced by symmetric ones.

The remaining question is to deal with ϕ_4. It is not difficult to see that the S_3 invariance of

$$\phi_4\left({\alpha_s \atop \beta_s(x)} ; {\alpha_t \atop \beta_t(x)} ; {\alpha_u \atop \beta_u(x)} \right) \qquad (6.130)$$

becomes, when re-written in terms of Koba-Nielsen variables, the S_4 invariance of

$$\Phi_4\left({p_1 \atop z_1} ; {p_2 \atop z_2} ; {p_3 \atop z_3} ; {p_4 \atop z_4} \right) \qquad (6.131)$$

under the 4! permutations of the argument pairs $\{p_i, z_i\}$, $i = 1, 2, 3, 4$.

Collecting results, we arrive at an expression for A_4 in a form immediately extendable to A_N, $N > 4$. The result is

$$A_N = \int \prod_{i=1}^{N} dz_i \, (dV_3)^{-1} \prod_{i<j} (z_i - z_j)^{-2p_i \cdot p_j} \cdot$$

$$\cdot \, (S_N^{(1)}(z))^{\frac{1-\alpha_\rho(o)}{2}} \cdot$$

$$\cdot \, (S_N^{(2)}(z))^{2(\alpha_\rho(o)-\alpha_\pi(o))}$$

$$\cdot \, \Phi_N\left(\frac{p_1}{z_1}; \frac{p_2}{z_2}; \ldots \frac{p_N}{z_N}\right) \tag{6.132}$$

where the whole integrand, including the final factor Φ_N, is invariant under an S_N symmetric group acting on the N argument pairs $\{p_i, z_i\}$, i = 1, 2, 3, ..., N. We have introduced the notations

$$S_N^{(1)}(z) = \sum_{\{q_1 q_2 \cdots q_N\}} \prod_{k=1}^{N} (z_{q_k} - z_{q_{k+1}})^{-2} \tag{6.133}$$

$$S_N^{(2)}(z) = \sum_{\{q_1 q_2 \cdots q_N\}} (-1)^P \prod_{k<\ell} (z_{q_k} - z_{q_\ell})^{(-1)^{k-\ell}} \tag{6.134}$$

The crucial point is that there is little freedom in making the multiparticle extension, when we insist on the symmetric group invariance. This is a very strong result.

6.9 CLASSIFICATION OF DUAL RESONANCE MODELS

The advantages of the symmetric group become clear, when we realise that the two classical dual models, together with new ones proposed by Gervais and Neveu[2,3] from an operator formalism approach, can be readily accommodated by the general formula, Equation (6.132).

This fact will then give more confidence in using the method to propose a multipion amplitude for realistic intercepts.

Let us deal with the earlier models in turn:

i) The conventional (generalised Euler B function) model corresponds to putting $\alpha_\rho(o) = \alpha_\pi(o) = 1$ and $\phi_N = 1$ in our general formula. This proposal was important historically as the first closed-form Born amplitude. Indeed the mechanism for producing the resonance poles is unaltered, in general, by the new singularities of the present integrand.

ii) The Neveu-Schwarz dual pion model, which has $\alpha_\rho(o) = 1$ and $\alpha_\pi(o) = \frac{1}{2}$ may be written

$$A_N = \int \prod_{i=1}^{N} dz_i \, (dV_3)^{-1} \prod_{i<j} (z_i - z_j)^{-2p_i \cdot p_j} \cdot \sum_{\{q_1 q_2 \cdots q_N\}} (-1)^P \frac{(p_{q_1} \cdot p_{q_2})}{(z_{q_1} - z_{q_2})} \frac{(p_{q_3} \cdot p_{q_4})}{(z_{q_3} - z_{q_4})} \cdot \cdots \frac{(p_{q_{N-1}} \cdot p_{q_N})}{(z_{q_{N-1}} - z_{q_N})} \tag{6.135}$$

and therefore corresponds to the specific choice

$$\phi_N = (S_N^{(2)})^{-1} \sum_{\{q_1 q_2 \cdots q_N\}} (-1)^P \cdot \frac{(p_{q_1} \cdot p_{q_2})}{(z_{q_1} - z_{q_2})} \frac{(p_{q_3} \cdot p_{q_4})}{(z_{q_3} - z_{q_4})} \cdots \frac{(p_{q_{N-1}} \cdot p_{q_N})}{(z_{q_{N-1}} - z_{q_N})} \tag{6.136}$$

An alternative form for A_N is (treating all z-differences as independent)

$$A_N = \int \prod_{i=1}^{N} dz_i \, (dV_3)^{-1}$$

$$\sum_{\{q_1 q_2 \cdots q_N\}} (-1)^P \frac{\partial}{\partial(z_{q_1} - z_{q_2})} \cdots$$

$$\cdots \frac{\partial}{\partial(z_{q_{N-1}} - z_{q_N})} [\prod_{i<j} (z_i - z_j)^{-2p_i \cdot p_j}] \tag{6.137}$$

and corresponds to a generalised logarithmic derivative choice for ϕ_N

$$\phi_N = (S_N^{(2)}(z))^{-1} (\prod_{i<j} (z_i - z_j)^{-2p_i \cdot p_j})^{-1} \cdot$$

$$\sum_{\{q_1 q_2 \cdots q_N\}} (-1)^P \frac{\partial}{\partial(z_{q_1} - z_{q_2})} \cdots \frac{\partial}{\partial(z_{q_{N-1}} - z_{q_N})} \cdot$$

$$\cdot [\prod_{i<j} (z_i - z_j)^{-2p_i \cdot p_j}] \tag{6.138}$$

We have seen earlier how the second-Fock space F_2 reformulation of the Neveu-Schwarz model corresponds to re-writing A_N in the forms

$$A_N = \int \prod_{i=1}^{N} dz_i \, (dV_3)^{-1} \prod_{i<j} (z_i - z_j)^{-2p_i \cdot p_j} \cdot$$

$$\cdot \frac{1}{(z_a - z_b)} \sum_{\{q_1 q_2 \cdots q_{N-2}\}} (-1)^P \cdot$$

$$\cdot \frac{\partial}{\partial(z_{q_1} - z_{q_2})} \cdots \frac{\partial}{\partial(z_{q_{N-3}} - z_{q_{N-2}})} \cdot$$

$$\cdot [\prod_{i<j} (z_i - z_j)^{-2p_i \cdot p_j}] \tag{6.139}$$

or, equivalently,

$$A_N = \int \prod_{i=1}^{N} dz_i \, (dV_3)^{-1} \prod_{i<j} (z_i - z_j)^{-2p_i \cdot p_j} \cdot$$

$$\cdot \frac{1}{(z_a - z_b)} \sum_{\{q_1 q_2 \cdots q_{N-2}\}}' (-1)^P \cdot$$

$$\cdot \frac{(p_{q_1} \cdot p_{q_2})}{(z_{q_1} - z_{q_2})} \cdots \frac{p_{q_{N-3}} \cdot p_{q_{N-2}}}{(z_{q_{N-3}} - z_{q_{N-2}})} \qquad (6.140)$$

These correspond to

$$\phi_N = (S_N^{(2)})^{-1} \left(\prod_{i<j} (z_i - z_j)^{-2p_i \cdot p_j} \right)^{-1} \frac{1}{(z_a - z_b)} \cdot$$

$$\cdot \sum_{\{q_1 \cdots q_{N-2}\}}' (-1)^P \frac{\partial}{\partial (z_{q_1} - z_{q_2})} \cdots$$

$$\cdots \frac{\partial}{\partial (z_{q_{N-3}} - z_{q_{N-2}})} \left[\prod_{i<j} (z_i - z_j)^{-2p_i \cdot p_j} \right] \qquad (6.141)$$

and

$$\phi_N = (S_N^{(2)})^{-1} \frac{1}{(z_a - z_b)} \sum_{\{q_1 \cdots q_{N-2}\}}' (-1)^P \cdot$$

$$\cdot \frac{p_{q_1} \cdot p_{q_2}}{(z_{q_1} - z_{q_2})} \cdots \cdot \frac{p_{q_{N-3}} \cdot p_{q_{N-2}}}{(z_{q_{N-3}} - z_{q_{N-2}})} \qquad (6.142)$$

respectively.

Here the primed summations are over all permutations of $\{q_1 \cdots q_{N-2}\} = 1, 2, 3, \ldots, N$ (excluding a, b which are chosen arbitrarily).

At the level $N = 4$, the two forms are easily identifiable to be (original formulation)

$$\phi_4^{(4)} = \alpha_s{}^2 \beta_s + \alpha_t{}^2 \beta_t + \alpha_u{}^2 \beta_u \qquad (6.143)$$

and (F_2 reformulation)

$$\phi_4^{(1)} = \alpha_s\beta_s + \alpha_t\beta_t + \alpha_u\beta_u \qquad (6.144)$$

respectively.

Thus, from the symmetric group standpoint, we can understand both the Neveu-Schwarz model and its F_2 reformulation in an illuminating way.

iii) Gervais and Neveu[2] have obtained an operatorial factorisation of the amplitude

$$A_N = \int \prod_{i=1}^{N} dz_i \, (dV_3)^{-1} \prod_{i<j} (z_i - z_j)^{-2p_i \cdot p_j} \cdot (S_N^{(1)}(z))^{\frac{1 - \alpha_\rho(o)}{2}} \qquad (6.145)$$

which corresponds to $\alpha_\rho(o) = \alpha_\pi(o)$ and $\phi_N = 1$ in our general formula. For N = 4 this reduces to

$$A_4 = \int_0^1 dx\, x^{-\alpha_t - 1} (1 - x)^{-\alpha_s - 1} (1 - x + x^2)^{\frac{\gamma}{2}} \qquad (6.146)$$

as expected, and as was proposed by Mandelstam[13] some time ago.

Actually the factorisation was obtained for $\alpha_\rho(o) = -1$, where there was a generalised projective gauge algebra. Unfortunately, for this low intercept the model contained ghosts[31].

Full factorisation for $\alpha_\rho(o)$ in the range near $\alpha_\rho(o) \simeq \frac{1}{2}$ has not yet been exhibited, although the importance of doing so for this amplitude will become evident in the later development.

iv) In a subsequent work, Gervais and Neveu[3] have investigated the possibility of constructing dual vertices, from the oscillators $a_\mu^{(n)}$, $a_\mu^{(n)+}$ of the conventional model, that will lead to ghost-free amplitudes for general $\alpha(o) < 1$. They do this by multiplying the conventional vertex by a function of the gauge operators L_n.

The solution is found, only for special cases, in closed form. In particular a model with leading intercept $\alpha(o) = 0$ and critical dimension $d_c = 25$ was constructed. For this model the four-particle amplitude is

$$A_4 = \int_0^1 dx\, x^{-\alpha_t - 1} (1 - x)^{-\alpha_s - 1} (1 - x + x^2)^{\frac{1}{2}(1-\alpha(o))} \cdot$$

$$\cdot \left[2\, {}_2F_1\left(\tfrac{1}{6}, -\tfrac{1}{6}; \tfrac{3}{2}; y\right) - {}_2F_1\left(\tfrac{1}{6}, -\tfrac{1}{6}; \tfrac{1}{2}; y\right)\right] \tag{6.147}$$

with

$$y = \frac{27}{4} \frac{x^2(1 - x)^2}{(1 - x + x^2)^3} \tag{6.148}$$

This is a special case of our general formula for A_4, corresponding to a particular non-polynomial choice for the function S_1.

6.10 MULTIPION AMPLITUDE

We have previously selected as an explicit new four-pion amplitude the formula (section 6.6)

$$A_4 = \int_0^1 dx\, x^{-\alpha_t - 1} (1 - x)^{-\alpha_s - 1} (1 - x + x^2)^{\frac{\gamma}{2}} \cdot$$

$$\cdot \left(\alpha_s \beta_s + \alpha_t \beta_t + \alpha_u \beta_u\right) \tag{6.149}$$

If we put $\alpha_\rho(o) - \alpha_\pi(o) = \frac{1}{2}$, as suggested by soft-pion consistency at leading order, this amplitude becomes

$$A_4 = \int_0^1 dx\, x^{-\alpha_t - 1} (1 - x)^{-\alpha_s - 1} (1 - x + x^2)^{\frac{1-\alpha_\rho(o)}{2}} \cdot$$

$$\cdot \left[\alpha_s x + \alpha_t (1 - x) - \alpha_u x(1 - x)\right] \qquad (6.150)$$

Now we may re-write

$$A_4 = \int_0^1 dx\, x^{-\alpha_t - 1} (1 - x)^{-\alpha_s - 1} (1 - x + x^2)^{\frac{\gamma}{2}} \cdot$$

$$\cdot (\hat{\alpha}_s \beta_s + \hat{\alpha}_t \beta_t + \hat{\alpha}_u \beta_u) \qquad (6.151)$$

where (putting $a = 1 - \alpha_\rho(o)$, $\alpha_\rho(o) - \alpha_\pi(o) = \frac{1}{2}$)

$$\hat{\alpha}_s = \alpha_s + \frac{a\, \beta_t \beta_u}{2\beta_s} \qquad (6.152)$$

$$\hat{\alpha}_t = \alpha_t + \frac{a\, \beta_u \beta_s}{2\beta_t} \qquad (6.153)$$

$$\hat{\alpha}_u = \alpha_u + \frac{a\, \beta_s \beta_t}{2\beta_u} \qquad (6.154)$$

since we have added to the original amplitude a quantity which vanishes identically, namely

$$\frac{a}{2}(\beta_s \beta_t + \beta_t \beta_u + \beta_u \beta_s) = 0 \qquad (6.155)$$

More generally it is useful to define, for an N-point function,

$$\hat{\alpha}_{ij} = \hat{\alpha}_{ij} + \frac{a\, S_N^{(1)}\{ij\}}{S_N^{(1)}} \qquad (6.156)$$

in which $S_N^{(1)}\{ij\}$ contains only those terms of the sum over permutations $\{q_1 q_2 \cdots q_N\} = 1, 2, 3, \ldots, N$ in which i, j are adjacent. These quantities α_{ij} possess many intercept-independent properties, and are easily checked to coincide with the definitions of $\hat{\alpha}_s$, $\hat{\alpha}_t$, $\hat{\alpha}_u$ given above for N = 4. We also may check, for example, that

$$\hat{\gamma} = \hat{\alpha}_s + \hat{\alpha}_t + \hat{\alpha}_u + 1 \tag{6.157}$$

$$= 0 \tag{6.158}$$

for any value of a.

We may also write*

$$A_4 = 3 \int_0^1 dx\, x^{-\alpha_t - 1} (1 - x)^{-\alpha_s - 1} (1 - x + x^2)^{\frac{\gamma}{2}} \cdot$$

$$\cdot \left[\hat{\alpha}_s{}^2 \beta_s + \hat{\alpha}_t{}^2 \beta_t + \hat{\alpha}_u{}^2 \beta_u - \frac{3a}{2} \beta_s \beta_t \beta_u\right] \tag{6.159}$$

Notice that in the unphysical limit $\gamma \to 0$, $a \to 0$, $\hat{\alpha}_{ij} \to \alpha_{ij}$ this formula becomes that corresponding to $\phi_4^{(4)}$ given earlier. The reason for rewriting the $a \neq 0$ case in this way is that it makes it considerably easier to write the multiparticle amplitude, to which we now turn.

In keeping with the general method of extension we write the multiparticle amplitude as $(\alpha_\rho(o) - \alpha_\pi(o) = \frac{1}{2})$

$$A_N = \int \prod_{i=1}^{N} dz_i \, (dV_3)^{-1} \cdot$$

$$\sum_{\{q_1 q_2 \cdots q_N\}} (-1)^P \frac{\partial}{\partial(z_{q_1} - z_{q_2})} \cdots$$

$$\cdots \frac{\partial}{\partial(z_{q_{N-1}} - z_{q_N})} \cdot$$

$$\cdot \left[\prod_{i<j} (z_i - z_j)^{-2p_i \cdot p_j} (S_N^{(1)}(z))^{a/2}\right] \tag{6.160}$$

Here the integrand is S_N-invariant as required.

We should first note that for $a \to 0$ this becomes the Neveu-Schwarz dual pion model, for which a no-ghost

---

*For a proof of Equation (6.159), see P. H. Frampton and K. A. Geer, Phys. Rev. D10, 1284 (1974).

theorem is well established.

Let us check that for N = 4, this coincides with our earlier A_4 formula. We find

$$A_4 = \int \prod_{i=1}^{4} dz_i \, (dV_3)^{-1} \prod_{i<j} (z_i - z_j)^{-2p_i \cdot p_j} \cdot$$

$$\cdot \, (S_4{}^{(1)})^{\frac{a}{2}} \sum_{\{q_1 q_2 q_3 q_4\}} (-1)^P \frac{\theta_{q_1 q_2 q_3 q_4}}{(z_{q_1} - z_{q_2})(z_{q_3} - z_{q_4})} \tag{6.161}$$

where $\theta_{q_1 q_2 q_3 q_4}$ is given by

$$\begin{aligned}\theta_{q_1 q_2 q_3 q_4} &= (-\,2p_{q_1} \cdot p_{q_2})\,(-\,2p_{q_3} \cdot p_{q_4}) \\ &+ (-2p_{q_1} \cdot p_{q_2})\,(S_4{}^{(1)})^{-\frac{a}{2}} \frac{\partial}{\partial(z_{q_3} - z_{q_4})} (S_4{}^{(1)})^{\frac{a}{2}} \\ &+ (-\,2\,p_{q_3} \cdot p_{q_4})\,(S_4{}^{(1)})^{-\frac{a}{2}} \frac{\partial}{\partial(z_{q_1} - z_{q_2})} (S_4{}^{(1)})^{\frac{a}{2}} \\ &+ (S_4{}^{(1)})^{-\frac{a}{2}} \frac{\partial}{\partial(z_{q_1} - z_{q_2})} \frac{\partial}{\partial(z_{q_3} - z_{q_4})} (S_4{}^{(1)})^{\frac{a}{2}}\end{aligned} \tag{6.162}$$

Now

$$(S_N{}^{(1)})^{-\frac{a}{2}} \frac{\partial}{\partial(z_{q_1} - z_{q_2})} (S_N{}^{(1)})^{\frac{a}{2}} = -\,\frac{a\, S_N{}^{(1)}\{q_1 q_2\}}{S_N{}^{(1)}} \tag{6.163}$$

where $S_N{}^{(1)}\{q_1 q_2\}$ contains only those permutations with $q_1 q_2$ adjacent and

$$(S_N{}^{(1)})^{-\frac{a}{2}} \frac{\partial}{\partial(z_{q_1} - z_{q_2})} \frac{\partial}{\partial(z_{q_3} - z_{q_4})} (S_N{}^{(1)})^{\frac{a}{2}} =$$

$$= \frac{2a\, S_N^{(1)}\{q_1q_2q_3q_4\}}{S_N^{(1)}} + a(a-2)\, \frac{S_N^{(1)}\{q_1q_2\} S_N^{(1)}\{q_3q_4\}}{(S_N^{(1)})^2} \tag{6.164}$$

where $S_N^{(1)}\{q_1q_2q_3q_4\}$ contains only those terms with (q_1, q_2) <u>and</u> (q_3, q_4) as adjacent pairs.

We can now write

$$\theta_{q_1q_2q_3q_4} = \hat{\alpha}_{q_1q_2}\hat{\alpha}_{q_3q_4} + \frac{2a}{S_4^{(1)}} \cdot \left[S_4^{(1)}\{q_1q_2q_3q_4\} - \frac{S_4^{(1)}\{q_1q_2\}S_4^{(1)}\{q_3q_4\}}{S_4^{(1)}}\right] \tag{6.165}$$

Bearing in mind that (in our earlier shorthand)

$$S_4^{(1)} = 8\left[\frac{1}{(12)^2(23)^2} + \frac{1}{(12)^2(13)^2} + \frac{1}{(23)^2(13)^2}\right] \tag{6.166}$$

$$= \frac{8}{(12)^2(23)^2}\,[1 + x^2 + (1 - x)^2] \tag{6.167}$$

we find that A_4, as derived from the general A_N, corresponds to a choice

$$\phi_4 = \hat{\alpha}_s{}^2\beta_s + \hat{\alpha}_t{}^2\beta_t + \hat{\alpha}_u{}^2\beta_u + 2a\left[(1 - \frac{\beta_t\beta_u}{2\beta_s})\,\frac{\beta_t\beta_u}{2} + (1 - \frac{\beta_u\beta_s}{2\beta_t})\,\frac{\beta_u\beta_t}{2} + (1 - \frac{\beta_s\beta_t}{2\beta_u})\,\frac{\beta_s\beta_t}{2}\right] \tag{6.168}$$

$$= \hat{\alpha}_s{}^2\beta_s + \hat{\alpha}_t{}^2\beta_t + \hat{\alpha}_u{}^2\beta_u - \frac{3a}{2}\,\beta_s\beta_t\beta_u \tag{6.169}$$

where we used

$$\beta_s\beta_t + \beta_t\beta_u + \beta_u\beta_s = 0 \tag{6.170}$$

together with

$$\frac{\beta_t^{\,2}\beta_u^{\,2}}{\beta_s} + \frac{\beta_u^{\,2}\beta_s^{\,2}}{\beta_t} + \frac{\beta_s^{\,2}\beta_t^{\,2}}{\beta_u} = 3\beta_s\beta_t\beta_u \tag{6.171}$$

This completes the proof that

$$A_4 = \int_0^1 dx\, x^{-\alpha_t - 1} (1-x)^{-\alpha_s - 1} (1 - x + x^2)^{1+a/2} \,\cdot$$

$$\cdot\, (\hat{\alpha}_s^{\,2}\beta_s + \hat{\alpha}_t^{\,2}\beta_t + \hat{\alpha}_u^{\,2}\beta_u - \frac{3a}{2}\beta_s\beta_t\beta_u) \tag{6.172}$$

$$= \frac{1}{3}\int_0^1 dx\, x^{-\alpha_t - 1} (1-x)^{-\alpha_s - 1} (1 - x + x^2)^{1+a/2} \,\cdot$$

$$\cdot\, (\alpha_s\beta_s + \alpha_t\beta_t + \alpha_u\beta_u) \tag{6.173}$$

is a special case of our general multipion amplitude, as required.

6.11 SPIN-LOWERING SYMMETRY

The basic conjecture underlying the present approach is that the insistence on the symmetric group invariance will ensure for some $\alpha_\rho(o)$, other than the known case $\alpha_\rho(o) = 1$, that there will be a generalised projective gauge group, to allow ghost elimination. Actually, since there is extra momentum-dependence in the integrand, as is essential to decouple tachyons, we expect to need an enlarged gauge group, analogous to the G-gauges of the Neveu-Schwarz dual pion model.

In the integral representation, we know that the existence of these extra gauges should be reflected by a higher symmetry, which we shall denote as spin-lowering symmetry[32,33] (for reasons to become clear later). To be explicit, we expect to be able to rewrite the amplitude as

$$A_N = \int \prod_{i=1}^{N} dz_i \, (dV_3)^{-1} \frac{1}{(z_a - z_b)} \cdot$$

$$\cdot \sum_{\{q_1 q_2 \cdots q_{N-2}\}}' (-1)^P \frac{\partial}{\partial(z_{q_1} - z_{q_2})} \cdots$$

$$\cdots \frac{\partial}{\partial(z_{q_{N-3}} - z_{q_{N-2}})} \left[\prod_{i \neq j} (z_i - z_j)^{-p_i \cdot p_j} \cdot \right.$$

$$\left. \cdot (S_N^{(1)})^{a/2} \right] \qquad (6.174)$$

where a, b are chosen arbitrarily and the primed summation is over all permutations of $\{q_1 q_2 \cdot\cdot q_{N-2}\} = 1, 2, 3, \ldots, N$ (excluding a, b).

While there is no rigorous proof of this re-writing for general N we shall convince the reader that it is valid by considering specific cases. The validity appears to hold for general a, and therefore the new A_N possesses <u>all</u> of the symmetry properties of the integrand (symmetric group invariance plus spin-lowering symmetry) of the Neveu-Schwarz amplitude.

Firstly we consider N = 4. There are three inequivalent choices of a, b which we may take as (a, b) = (1, 2), (1, 3) and (1, 4). For these three choices we obtain respectively the three formulae

$$A_4 = \int_0^1 dx \, x^{-\alpha_t - 1} (1 - x)^{-\alpha_s - 1} (1 - x + x^2)^{\frac{a}{2} + 1} \hat{\alpha}_s \beta_s \qquad (6.175)$$

$$A_4 = \int_0^1 dx \, x^{-\alpha_t - 1} (1 - x)^{-\alpha_s - 1} (1 - x + x^2)^{\frac{a}{2} + 1} \hat{\alpha}_t \beta_t \qquad (6.176)$$

$$A_4 = \int_0^1 dx \, x^{-\alpha_t - 1} (1 - x)^{-\alpha_s - 1} (1 - x + x^2)^{\frac{a}{2} + 1} \hat{\alpha}_u \beta_u \qquad (6.177)$$

Now consider the perfect differential

$$0 = \int_0^1 dx \frac{d}{dx} [x^{-\alpha_t} (1 - x)^{-\alpha_s} (1 - x + x^2)^{a/2}] \quad (6.178)$$

$$= \int_0^1 dx\, x^{-\alpha_t - 1} (1 - x)^{-\alpha_s - 1} (1 - x + x^2)^{\frac{a}{2}+1} \cdot$$

$$\cdot [\alpha_s \beta_s - \alpha_t \beta_t + \frac{a(2x - 1)x(1 - x)}{2(1 - x + x^2)^2}] \quad (6.179)$$

$$= \int_0^1 dx\, x^{-\alpha_t - 1} (1 - x)^{-\alpha_s - 1} (1 - x + x^2)^{\frac{a}{2}+1} \cdot$$

$$\cdot (\hat{\alpha}_s \beta_s - \hat{\alpha}_t \beta_t) \quad (6.180)$$

to see that the first two forms for A_4 are equal. The equality of the third form, involving $\hat{\alpha}_u \beta_u$, follows from the symmetric group property. Finally note that, using the symbol $\equiv$ to denote equality after integration with

$$\int_0^1 dx\, x^{-\alpha_t - 1} (1 - x)^{-\alpha_s - 1} (1 - x + x^2)^{\frac{a}{2}+1} \cdots$$

we have the equivalences

$$\hat{\alpha}_s \beta_s \equiv \hat{\alpha}_t \beta_t \equiv \hat{\alpha}_u \beta_u \equiv \frac{1}{3}(\hat{\alpha}_s \beta_s + \hat{\alpha}_t \beta_t + \hat{\alpha}_u \beta_u) \quad (6.182)$$

$$= \frac{1}{3}(\alpha_s \beta_s + \alpha_t \beta_t + \alpha_u \beta_u) + \frac{a}{6}(\beta_s \beta_t + \beta_t \beta_u + \beta_u \beta_s) \quad (6.183)$$

$$= \frac{1}{3}(\alpha_s \beta_s + \alpha_t \beta_t + \alpha_u \beta_u) \quad (6.184)$$

to see that all three choices (a, b) = (1, 2), (1, 3), (1, 4) lead to the original amplitude A_4. This is consistent with the proposed spin-lowering symmetry.

When we discuss the bootstrap condition at N = 6, in the following sub-section, we shall show how spin-lowering symmetry decouples unwanted ancestor and tachyon states, similar to the way they are decoupled by G-gauges in the (a = 0) Neveu-Schwarz model.

6.12 FACTORISATION

We now consider factorisation at the internal pion pole of our A_6 amplitude. That is, we consider

$$\lim_{\alpha_{123} \to 0} (\alpha_{123} A_6) \tag{6.185}$$

Later we shall exhibit bootstrap consistency for general N, but it is instructive to study N = 6 in detail since the important role of spin-lowering symmetry becomes evident.

We can re-write A_6 in the form

$$\begin{aligned} A_6 = \int \prod_{i=1}^{6} dz_i \, (dV_3)^{-1} \cdot \\ \cdot \Big[\prod_{i \neq j} (z_i - z_j)^{-p_i \cdot p_j} \prod_{i=1}^{6} (z_i - z_{i+1})^{-a} \cdot \\ \cdot \prod_{i<j} (z_i - z_j)^{(-1)^{i-j}} \Big] \cdot \\ \cdot \Big[\prod_{i=1}^{6} (z_i - z_{i+1}) \; (S_6^{(1)})^{a/2} \Big] \cdot \\ \cdot \Big[\prod_{i<j} (z_i - z_j)^{-(-1)^{i-j}} \sum_{\{q_1 \cdot\cdot q_6\}} (-1)^P \cdot \\ \cdot \frac{\theta_{q_1 q_2, q_3 q_4, q_5 q_6}}{(z_{q_1} - z_{q_2})(z_{q_3} - z_{q_4})(z_{q_5} - z_{q_6})} \Big] \end{aligned} \tag{6.186}$$

which is convenient for the discussion of factorisation.

The first of the three square brackets is simply the integrand for the generalised Euler B function and has, therefore, well-known factorisation properties.

To study the factorisation properties of the second and third square brackets in A_6 we introduce integration variables

$$x = (z_1, z_6; z_2, z_3) \tag{6.187}$$

$$w = (z_1, z_6; z_3, z_4) \tag{6.188}$$

$$y = (z_1, z_6; z_4, z_5) \tag{6.189}$$

satisfying $0 \leq x, w, y \leq 1$ (see Figure 6.4), and take the limit

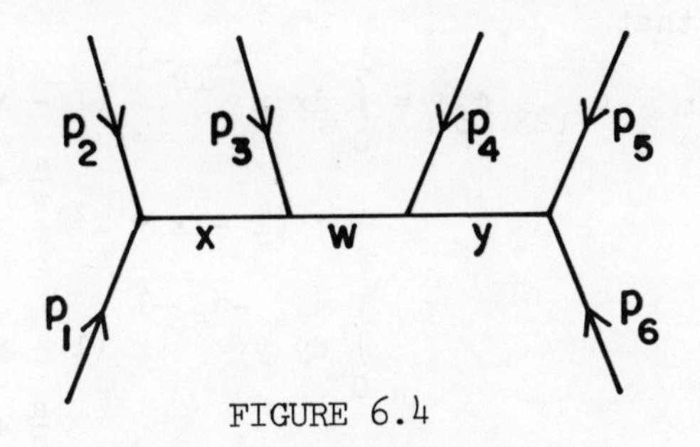

FIGURE 6.4

Six-Point Function Factorisation

$w \to 0$. For the second square bracket we find

$$\lim_{w \to 0} \left[\prod_{i=1}^{6} (z_i - z_{i+1})^2 \, S_6^{(1)}(z) \right] = 96(1 - x + x^2) \cdot$$

$$\cdot (1 - y + y^2) \tag{6.190}$$

consistent with factorisation, when raised to an arbitrary (non-integer) power $(\frac{a}{2})$.

The factorisation of the third and final bracket is more subtle. Let us define, in analogy with the N = 4 notations,

$$\beta_{12}(x) = (1 - x)\,(1 - x + x^2)^{-1} \tag{6.191}$$

$$\beta_{23}(x) = x(1 - x + x^2)^{-1} \tag{6.192}$$

$$\beta_{13}(x) = -x(1 - x)\,(1 - x + x^2)^{-1} \tag{6.193}$$

and

$$\beta_{56}(y) = (1 - y)(1 - y + y^2)^{-1} \tag{6.194}$$

$$\beta_{45}(y) = y(1 - y + y^2) \tag{6.195}$$

$$\beta_{46}(y) = -\,y(1 - y)(1 - y + y^2)^{-1} \tag{6.196}$$

Then we find that

$$\lim_{\alpha_{123} \to 0} (\alpha_{123}\, A_6) = \int_0^1 dx\, x^{-\alpha_{12}-1}\, (1 - x)^{-\alpha_{23}-1} \cdot \cdot\, (1 - x + x^2)^{\frac{a}{2}+1} \int_0^1 dy\, y^{-\alpha_{56}-1}\, (1 - y)^{-\alpha_{45}-1} \cdot \cdot\, (1 - y + y^2)^{\frac{a}{2}+1}\, \rho(x, y) \tag{6.197}$$

where

$$\rho(x, y) = \beta_{12}(x)\beta_{56}(y)\,\bar{\theta}_{12,34,56} + \beta_{12}(x)\beta_{45}(y) \cdot \cdot\bar{\theta}_{12,36,45} + + \beta_{12}(x)\beta_{46}(y)\bar{\theta}_{12,35,46} + \beta_{23}(x)\beta_{56}(y) \cdot \cdot\bar{\theta}_{14,23,56} + + \beta_{23}(x)\beta_{45}(y)\bar{\theta}_{16,23,45} + \beta_{23}(x)\beta_{46}(y) \cdot$$

$$\cdot \bar{\theta}_{15,23,46} +$$

$$+ \beta_{13}(x)\beta_{56}(y)\bar{\theta}_{13,24,56} + \beta_{13}(x)\beta_{45}(y) \cdot$$

$$\cdot \bar{\theta}_{13,26,45} +$$

$$+ \beta_{13}(x)\beta_{46}(y)\bar{\theta}_{13,25,46} \quad (6.198)$$

in which

$$\bar{\theta}_{q_1q_2,q_3q_4,q_5q_6} = \lim_{w \to 0} [\theta_{q_1q_2,q_3q_4,q_5q_6}] \quad (6.199)$$

It is convenient now to define

$$f_{12}(x) = 1 + \frac{\beta_{23}\beta_{13}}{2\beta_{12}} \quad (6.200)$$

$$f_{23}(x) = 1 + \frac{\beta_{12}\beta_{13}}{2\beta_{23}} \quad (6.201)$$

$$f_{13}(x) = 1 + \frac{\beta_{12}\beta_{23}}{2\beta_{13}} \quad (6.202)$$

and similarly

$$f_{56}(y) = 1 + \frac{\beta_{45}\beta_{46}}{2\beta_{56}} \quad (6.203)$$

$$f_{45}(y) = 1 + \frac{\beta_{46}\beta_{56}}{2\beta_{45}} \quad (6.204)$$

$$f_{46}(y) = 1 + \frac{\beta_{45}\beta_{56}}{2\beta_{46}} \quad (6.205)$$

We now find, after some algebra, that

$$\bar{\theta}_{12,34,56} = \hat{\alpha}_{12}\hat{\alpha}_{34}\hat{\alpha}_{56} + 2p_1 \cdot p_2 \, af_{12}f_{56}(1 - f_{56}) + + 2p_5 \cdot p_6 \, af_{12}f_{56}(1 - f_{12}) + 2af_{12}f_{56} + + a(a - 2)f_{12}f_{56}(f_{12} + f_{56}) + + 2a(1 - a)f_{12}{}^2 f_{56}{}^2 \tag{6.206}$$

with

$$\hat{\alpha}_{34} = 2p_3 \cdot p_4 + \frac{a}{2} f_{12} \, f_{56} \tag{6.207}$$

Here we notice that $\rho(x, y)$ will contain terms linear in the momentum transfer, corresponding to a spin-one ancestor, unless there is some delicate decoupling mechanism at work. Such a decoupling does occur, due to spin-lowering symmetry in the form of the equivalences

$$\hat{\alpha}_{12} \, \beta_{12} \equiv \hat{\alpha}_{23} \, \beta_{23} \equiv \hat{\alpha}_{13} \, \beta_{13} \tag{6.208}$$

and

$$\hat{\alpha}_{45} \, \beta_{45} \equiv \hat{\alpha}_{56} \, \beta_{56} \equiv \hat{\alpha}_{46} \, \beta_{46} \tag{6.209}$$

This enables us to recombine the ancestor terms in $\rho(x, y)$ according to

$$\hat{\alpha}_{12}\beta_{12} \, (2p_3 \cdot p_4) \, \hat{\alpha}_{56}\beta_{56} + \hat{\alpha}_{12}\beta_{12} \, (2p_3 \cdot p_6) \, \hat{\alpha}_{45}\beta_{45} + + \hat{\alpha}_{12}\beta_{12} \, (2p_3 \cdot p_5) \, \hat{\alpha}_{46}\beta_{46} + \hat{\alpha}_{23}\beta_{23} \, (2p_1 \cdot p_4) \, \hat{\alpha}_{56}\beta_{56} + + \hat{\alpha}_{23}\beta_{23} \, (2p_1 \cdot p_6) \, \hat{\alpha}_{45}\beta_{45} + \hat{\alpha}_{23}\beta_{23} \, (2p_1 \cdot p_5) \, \hat{\alpha}_{46}\beta_{46} + + \hat{\alpha}_{13}\beta_{13} \, (2p_2 \cdot p_4) \, \hat{\alpha}_{56}\beta_{56} + \hat{\alpha}_{13}\beta_{13} \, (2p_2 \cdot p_6) \, \hat{\alpha}_{45}\beta_{45} + + \hat{\alpha}_{13}\beta_{13} \, (2p_2 \cdot p_5)\hat{\alpha}_{46}\beta_{46} \equiv \equiv (1 - 2a) \, \hat{\alpha}_{12}\beta_{12} \, \hat{\alpha}_{56}\beta_{56} \tag{6.210}$$

By using this relationship we find that

$$\rho(x,y) = \hat{\alpha}_{12}\beta_{12}\,\hat{\alpha}_{56}\beta_{56}\left(1 + \frac{5a}{2}\right) + \frac{3a}{4}\left(\hat{\alpha}_{12}\beta_{12}\,\hat{\phi}_4^{(2)}(456) + \hat{\phi}_4^{(2)}(123)\,\hat{\alpha}_{56}\beta_{56}\right) - \frac{9a}{4}\left(\hat{\alpha}_{12}\beta_{12}\beta_{45}\beta_{56}\beta_{46} + \beta_{12}\beta_{23}\beta_{13}\,\hat{\alpha}_{56}\beta_{56}\right) + \frac{9}{8}\hat{\phi}_4^{(2)}(123)\,\hat{\phi}_4^{(2)}(456) + \frac{9a}{8}\beta_{12}\beta_{23}\beta_{13}\cdot\beta_{45}\beta_{56}\beta_{45} - \frac{3a}{8}\left(\hat{\phi}_4^{(2)}(123)\,\beta_{45}\beta_{56}\beta_{46} + \beta_{12}\beta_{23}\beta_{13}\cdot\hat{\phi}_4^{(2)}(456)\right) \quad (6.211)$$

where we defined

$$\hat{\phi}_4^{(2)}(123) = \hat{\alpha}_{12}\beta_{23}\beta_{13} + \hat{\alpha}_{23}\beta_{12}\beta_{13} + \hat{\alpha}_{13}\beta_{12}\beta_{23} \quad (6.212)$$

$$\hat{\phi}_4^{(2)}(456) = \hat{\alpha}_{56}\beta_{45}\beta_{56} + \hat{\alpha}_{45}\beta_{56}\beta_{46} + \hat{\alpha}_{46}\beta_{45}\beta_{56} \quad (6.213)$$

Now we may use the perfect differentials that give vanishing contribution under the x and y integrals respectively, namely

$$\hat{\alpha}_{12}\beta_{12} + \frac{1}{2}\hat{\phi}_4^{(2)}(123) - \frac{3}{2}\beta_{12}\beta_{23}\beta_{13} \equiv 0 \quad (6.214)$$

and

$$\hat{\alpha}_{56}\beta_{56} + \frac{1}{2}\hat{\phi}_4^{(2)}(456) - \frac{3}{2}\beta_{45}\beta_{56}\beta_{46} \equiv 0 \quad (6.215)$$

and thereby deduce that

$$\rho(x,y) \equiv \left((1+a)\,\hat{\alpha}_{12}\beta_{12} + \frac{a}{2}\hat{\phi}_4^{(2)}(123) - \frac{3a}{2}\cdot\beta_{12}\beta_{23}\beta_{13}\right)\cdot\left((1+a)\,\hat{\alpha}_{56}\beta_{56} + \frac{a}{2}\hat{\phi}_4^{(2)}(456) - \frac{3a}{2}\cdot\beta_{45}\beta_{56}\beta_{46}\right) \quad (6.216)$$

$$\equiv (\hat{\alpha}_{12}\beta_{12})(\hat{\alpha}_{56}\beta_{56}) \tag{6.217}$$

as required for bootstrap consistency.

This result holds for any value of $a = 1 - \alpha_\rho(o)$. It is intriguing to notice, however, that, if we do not exploit the vanishing perfect differentials, bootstrap consistency gives non-linear constraints on a, namely the six equations

$$(1 + a)^2 = 1 + \frac{5a}{2} \tag{6.218}$$

$$\frac{a}{2}(1 + a) = \frac{3a}{4} \tag{6.219}$$

$$\frac{a^2}{4} = \frac{a}{8} \tag{6.220}$$

$$\frac{9a^2}{4} = \frac{9a}{8} \tag{6.221}$$

$$-\frac{3a^2}{4} = -\frac{3a}{8} \tag{6.222}$$

$$-\frac{3a}{2}(1 + a) = -\frac{9a}{4} \tag{6.223}$$

These equations have two solutions: (I) $a = 0$ corresponding to $\alpha_\rho(o) = \alpha_\pi(o) + \frac{1}{2} = 1$ and, more interesting, (II) $a = \frac{1}{2}$ corresponding to $\alpha_\rho(o) = \alpha_\pi(o) + \frac{1}{2} = \frac{1}{2}$. This may indicate to us that the intercept $\alpha_\rho(o) = \frac{1}{2}$ which is, of course, physically acceptable, may play a special role in the full factorisation.

We can now examine further how the spin-lowering symmetry works at $N = 6$ when we write the alternative form

$$A_6' = \int \prod_{i=1}^{6} dz_i \, (dV_3)^{-1} \frac{1}{(z_a - z_b)} \sum_{\{q_1 q_2 \cdot\cdot q_4\}}{}' (-1)^P \cdot$$

$$\cdot \frac{\partial}{\partial(z_{q_1} - z_{q_2})} \frac{\partial}{\partial(z_{q_3} - z_{q_4})} \cdot$$

$$[\prod_{i\neq j} (z_i - z_j)^{-p_i \cdot p_j} (S_6^{(1)})^{\frac{a}{2}}] \quad (6.224)$$

where the sum is over $\{q_1q_2q_3q_4\} = 1, 2, 3, 4, 5, 6$ (excluding a, b). For the factorisation at $\alpha_{123} = 0$ there are six inequivalent ways of choosing a, b which we may take as (a, b) = (1, 4); (1, 5); (1, 6); (2, 5); (5, 6); (4, 6).

Defining $\rho'(x,y)$ by

$$\lim_{\alpha_{123}\to 0} (\alpha_{123} A_6') = \int_0^1 dx\, x^{-\alpha_{12}-1} (1 - x)^{-\alpha_{23}-1} \cdot \cdot (1 - x + x^2)^{\frac{a}{2}+1} \int_0^1 dy\, y^{-\alpha_{56}-1} (1 - y)^{-\alpha_{45}-1} \cdot \cdot (1 - y + y^2)^{\frac{a}{2}+1} \rho'(x,y) \quad (6.225)$$

we find for the first four cases, respectively

$$\rho'(x,y) = \hat{\alpha}_{23}\beta_{23}(x)\, \hat{\alpha}_{56}\beta_{56}(y) \quad \text{for a, b} = 1, 4 \quad (6.226)$$

$$\rho'(x,y) = \hat{\alpha}_{23}\beta_{23}(x)\, \hat{\alpha}_{46}\beta_{46}(y) \quad \text{for a, b} = 1, 5 \quad (6.227)$$

$$\rho'(x,y) = \hat{\alpha}_{23}\beta_{23}(x)\, \hat{\alpha}_{45}\beta_{45}(y) \quad \text{for a, b} = 1, 6 \quad (6.228)$$

$$\rho'(x,y) = \hat{\alpha}_{13}\beta_{13}(x)\, \hat{\alpha}_{46}\beta_{46}(y) \quad \text{for a, b} = 2, 5 \quad (6.229)$$

and bootstrap consistency follows at once.

Taking a, b = 5, 6 one finds

$$\rho'(x,y) = \beta_{12}(x)\beta_{56}(y)[\hat{\alpha}_{12}\hat{\alpha}_{34} + \frac{2a}{S_6^{(1)}} \{S_6^{(1)}\{12,34\} - - \frac{S_6^{(1)}\{12\}S_6^{(1)}\{34\}}{S_6^{(1)}}\}] +$$

$$+ \beta_{23}(x)\beta_{56}(y)[\hat{\alpha}_{23}\hat{\alpha}_{56} + \frac{2a}{S_6^{(1)}}\{S_6^{(1)}\{23,14\} - \frac{S_6^{(1)}\{23\}S_6^{(1)}\{14\}}{S_6^{(1)}}\}] +$$

$$+ \beta_{13}(x)\beta_{56}(y)[\hat{\alpha}_{13}\hat{\alpha}_{56} + \frac{2a}{S_6^{(1)}}\{S_6^{(1)}\{13,24\} - \frac{S_6^{(1)}\{13\}S_6^{(1)}\{24\}}{S_6^{(1)}}\}] \tag{6.230}$$

$$= \beta_{56}(y)\ \hat{\alpha}_{12}\beta_{12}(x)\ 2p_5\cdot p_6 + \alpha\beta_{56}(y)\ f_{56}(y)\ \cdot$$
$$\cdot[\beta_{12}(x)\ (\tfrac{1}{2}\ \hat{\alpha}_{12}f_{12} + f_{12}(1 - f_{12}))$$
$$+ \beta_{23}(x)\ (\tfrac{1}{2}\ \hat{\alpha}_{23}f_{23} + f_{23}(1 - f_{23}))$$
$$+ \beta_{13}(x)\ (\tfrac{1}{2}\ \hat{\alpha}_{13}f_{13} + f_{13}(1 - f_{13}))] \tag{6.231}$$

$$= \hat{\alpha}_{12}\beta_{12}(x)\ \hat{\alpha}_{56}\beta_{56}(y) \tag{6.232}$$

as required. By a similar calculation, one finds the same result $\rho'(x, y)$ corresponding to the choice $(a, b) = (4, 6)$.

Thus the residue at the internal pion pole of A_6' satisfies bootstrap consistency and further it is independent of the choice of a, b as expected; these results are consistent with the spin-lowering symmetry defined by $A_6' = A_6$.

Note that the higher symmetry is clearly associated with the tachyon-killing mechanism for the rho trajectory, since the latter is the primary function of ϕ_N, which the higher symmetry describes. We have also seen how this symmetry decouples unwanted ancestor states. Both properties support the suggestion that spin-lowering symmetry provides an extension, of the G-gauges for the $\alpha_\rho(o) = 1$ dual pion model, to other intercepts $\alpha_\rho(o) \neq 1$.

We have studied N = 6 in some detail to learn about spin-lowering symmetry. Before proceeding to arbitrary N, it is worth noting some intercept-independent properties of the $\hat{\alpha}_{ij}$ introduced earlier. For N = 6, in the $w \to 0$ limit we see that

$$\hat{\alpha}_{34} + \hat{\alpha}_{35} + \hat{\alpha}_{36} = \hat{\alpha}_{12} \tag{6.233}$$

$$\hat{\alpha}_{24} + \hat{\alpha}_{25} + \hat{\alpha}_{26} = \hat{\alpha}_{13} \tag{6.234}$$

$$\hat{\alpha}_{14} + \hat{\alpha}_{15} + \hat{\alpha}_{16} = \hat{\alpha}_{12} \tag{6.235}$$

and similarly

$$\hat{\alpha}_{14} + \hat{\alpha}_{24} + \hat{\alpha}_{34} = \hat{\alpha}_{56} \tag{6.236}$$

$$\hat{\alpha}_{15} + \hat{\alpha}_{25} + \hat{\alpha}_{35} = \hat{\alpha}_{46} \tag{6.237}$$

$$\hat{\alpha}_{16} + \hat{\alpha}_{26} + \hat{\alpha}_{36} = \hat{\alpha}_{45} \tag{6.238}$$

for all intercepts $\alpha_\rho(o)$.

More generally, for any N, we may check, for example, that

$$\sum_{\substack{j \\ j\neq i}} \hat{\alpha}_{ij} = \sum_{\substack{j \\ j\neq i}} \left(2p_i \cdot p_j + \frac{a\, S_N^{(1)}\{ij\}}{S_N^{(1)}}\right) \tag{6.239}$$

$$= -2p_i^2 + 2a \tag{6.240}$$

$$= 1 \tag{6.241}$$

for any intercept $\alpha_\rho(o)$.

Finally, to provide the promised proof[33] of bootstrap consistency for general N, it is simplest to work directly in Koba-Nielsen variables, as follows.

Consider the pion pole in $(p_1 + p_2 + \cdots + p_\ell)^2$ of A_N and define

$$\alpha_{1\ell} = \alpha_\pi((p_1 + p_2 + \cdots + p_\ell)^2) = \alpha_\pi(s_{1\ell}) \qquad (6.242)$$

where ℓ is odd and N is even. We wish to consider the limit

$$\lim_{\alpha_{1\ell}\to 0} (\alpha_{1\ell} A_N) \qquad (6.243)$$

First note that the channel variable $u_{1\ell}$ is given by

$$u_{1\ell} = (z_1, z_N; z_\ell, z_{\ell+1}) \qquad (6.244)$$

$$= \frac{(z_\ell - z_1)(z_{\ell+1} - z_N)}{(z_\ell - z_N)(z_{\ell+1} - z_1)} \qquad (6.245)$$

Consequently if we take as the three fixed values

$$z_1 = 0, \quad z_{\ell+1} = 1, \quad z_N = \infty \qquad (6.246)$$

and define $z_\ell = z$ then we have the simple relationship

$$u_{1\ell} = z \qquad (6.247)$$

Now we define Koba-Nielsen variables for the two sub-systems by writing

$$z_i = x_i z \qquad (1 \le i \le \ell) \qquad (6.248)$$

$$z_i = \frac{1}{y_i} \qquad (\ell+1 \le i \le N) \qquad (6.249)$$

Introducing new variables x_I, y_I corresponding to the intermediate state, the sub-systems have fixed points

$$x_1 = 0, \; x_\ell = 1, \; x_I = \infty \qquad (6.250)$$

$$y_N = 0, \; y_{\ell+1} = 1, \; y_I = \infty \qquad (6.251)$$

respectively.

Now consider the full integral

$$A_N = \int \prod_{i=1}^{N} dz_i \, (dV_3)^{-1} \cdot$$

$$\cdot \sum_{\{q_1 q_2 \cdots q_N\}} (-1)^P \frac{\partial}{\partial(z_{q_1} - z_{q_2})} \frac{\partial}{\partial(z_{q_3} - z_{q_4})} \cdot$$

$$\cdot \ \cdots \frac{\partial}{\partial(z_{q_{N-1}} - z_{q_N})} \cdot$$

$$[\prod_{i \neq j} (z_i - z_j)^{-p_i \, p_j} (S_N^{(1)})^{\frac{a}{2}}] \qquad (6.252)$$

under the change of variables to the x_i and y_i defined above.

The Jacobian is

$$\prod_{i=1}^{N} dz_i \, (dV_3)^{-1} = \prod_{i=1}^{\ell+1} dx_i \, (dV_3(x))^{-1} \prod_{j=\ell}^{N} dy_j \cdot$$

$$\cdot \, (dV_3(y))^{-1} \prod_{i=\ell+1}^{N} y_i^{-2} \, dz \, z^{\ell-2} \qquad (6.253)$$

where we have identified $x_{\ell+1} = x_I$ and $y_\ell = y_I$.

Note also that, under this change of variables

$$\prod_{i \neq j} (z_i - z_j)^{-p_i \cdot p_j} = \prod_{1 \leq i < j \leq \ell+1} (x_i - x_j)^{-2p_i \cdot p_j} \cdot$$

$$\cdot \prod_{\ell \leq i < j \leq N} (y_i - y_j)^{-2p_i \cdot p_j} \cdot$$

$$\cdot \prod_{\substack{1 \leq m \leq \ell \\ \ell+1 \leq n \leq N}} (1 - x_m \, z \, y_n)^{-p_m \cdot p_n} \cdot$$

$$\cdot \prod_{i=\ell+1}^{N} y_i^{-2\mu^2} \, z^{-s_{1\ell} + \ell\mu^2} \qquad (6.254)$$

where $\mu^2 = p_i^{\,2}$.

Next we calculate the leading term (in z) of the factor $(S_N^{(1)}(z))^{a/2}$. After some simple algebra we find that

$$(S_N{}^{(1)}(z))^{\frac{a}{2}} = \Big(\sum_{\{q_1 \cdot\cdot q_{\ell+1}\}} \prod_{k=1}^{\ell+1} (x_{q_k} - x_{q_{k+1}})^{-2}\Big)^{\frac{a}{2}} \cdot$$

$$\cdot \Big(\sum_{\{q_\ell \cdots q_N\}} \prod_{k=\ell}^{N} (y_{q_k} - y_{q_{k+1}})^{-2}\Big)^{\frac{a}{2}} \cdot$$

$$\cdot \prod_{i=\ell+1}^{N} y_i{}^{2a} \; z^{-a(\ell-1)} \; (1 + O(z)) \quad (6.255)$$

as required for factorisation.

Finally we must deal with the partial differential operators. The leading term in z occurs when we take $\frac{1}{2}(\ell-1)$ differentiations with respect to x-differences $(x_i - x_j)$. To see this, we re-write

$$\frac{\partial}{\partial(z_i - z_j)} \equiv \frac{1}{z}\frac{\partial}{\partial(x_i - x_j)} \quad \text{for } 1 \leq i,j \leq \ell \quad (6.256)$$

and

$$\frac{\partial}{\partial(z_i - z_j)} \equiv y_i y_j \frac{\partial}{\partial(y_i - y_j)} \quad \text{for } (\ell+1) \leq i,j \leq N \quad (6.257)$$

Collecting results, we find that after these differentiations the leading term in z for the complete expression is given by

$$A_N = \int \prod_{i=1}^{\ell+1} dx_i \; (dV_3(x))^{-1} \sum_{\{q_1 \cdots q_{\ell+1}\}} (-1)^P \cdot$$

$$\cdot \frac{\partial}{\partial(x_{q_1} - x_{q_2})} \cdots \frac{\partial}{\partial(x_{q_\ell} - x_{q_{\ell+1}})} \cdot$$

$$\cdot \Big[\prod_{i \neq j} (x_i - x_j)^{-p_i \cdot p_j} (S_{\ell+1}^{(1)}(x))^{\frac{a}{2}}\Big] \cdot$$

$$\int \prod_{j=\ell}^{N} dy_j \; (dV_3(y))^{-1} \sum_{\{q_\ell \cdots q_N\}} (-1)^P \cdot$$

$$\cdot \frac{\partial}{\partial(y_{q_\ell} - y_{q_{\ell+1}})} \cdots \frac{\partial}{\partial(y_{q_{N-1}} - y_{q_N})} \cdot$$

$$\cdot \left[\prod_{i \neq j} (y_i - y_j)^{-p_i \cdot p_j} (S^{(1)}_{N-\ell+1}(y))^{a/2} \right] \cdot$$

$$\cdot \prod_{i=\ell+1}^{N} y_i^{2a-2\mu^2-1}$$

$$\cdot \int_0^1 dz\, z^{-1+\mu^2-s_{1\ell}}\, z^{(\ell-1)(\frac{1}{2}-a+\mu^2)} (1 + O(z)) \tag{6.258}$$

Now the amplitude was constructed under the constraint

$$\alpha_\rho(o) - \alpha_\pi(0) - \frac{1}{2} = \frac{1}{2} - a + \mu^2 = 0 \tag{6.259}$$

and hence we have

$$\prod_{i=1}^{\ell+1} y_i^{2a-2\mu^2-1} = 1 \tag{6.260}$$

and

$$z^{(\ell-1)(\frac{1}{2}-a+\mu^2)} = 1 \tag{6.261}$$

This leaves a simple z integration, which gives

$$A_N \underset{\alpha_{1\ell}\to 0}{\sim} A_{\ell+1} \frac{1}{\alpha_{1\ell}} A_{N-\ell+1} \tag{6.262}$$

as required for bootstrap consistency. Note that the leading intercept $\alpha_\rho(o) = \alpha_\pi(o) + \frac{1}{2}$ is not constrained by this requirement.

Finally, we should add the remarks

i) It can be shown [33] that A_N defined by Equation (6.160) factorises, at a general level, with finite degeneracy.

ii) It is straightforward to construct [32] a non-planar extension of the symmetric group amplitude, analogous to the Shapiro-Virasoro formula (subsection 2.) and to the non-planar Neveu-Schwarz model (subsection 5.); one makes two modifications to the planar amplitude of Equation (6.160): firstly, multiply the integrand by its complex conjugate then, secondly, extend the integration region to the full complex z-plane. The resultant non-planar amplitude is projective invariant provided the leading intercept satisfies $\alpha(o) = 2\alpha_\rho(o)$.

6.13 SUMMARY

We have seen how the symmetric group enables us to give explicit proposals on how to write dual resonance models with realistic Regge intercepts.

This ends our formal development of the dual Born amplitudes and it is appropriate to mention some outstanding problems on which further research is needed:

i) The symmetric-group model satisfies most basic axioms; still open, however, is the difficult question of negative probabilities. The establishment of a no-ghost theorem for the multipion A_N taken with leading intercept $\alpha(o) = \frac{1}{2}$ seems to require improved techniques. Of course, if the amplitude does contain ghosts, it must be further modified.

ii) One difficulty which is more than a mere technicality is to introduce abnormal-parity couplings, for example an ω-meson in the 3π channel, for trajectories with acceptable intercept values.

iii) Going beyond the non-strange mesons, we would like

to introduce K-mesons. Just as the intercept quantisation $\alpha_\rho(o) - \alpha_\pi(o) = \frac{1}{2}$ occurs naturally in the dual models, we would like to understand $\alpha_{K*(890)}(o) - \alpha_K(o) = \frac{1}{2}$. (Note that here, empirically, $\alpha_{K*}(o) \simeq \frac{1}{4}$ compared to $\alpha_\rho(o) = -\frac{1}{2}$)

iv) Very important, phenomenologically, is of course to introduce baryons. Here there are two fundamental requirements: firstly the baryons should possess some three-quark structure (compare section 6.7). Secondly, the baryons are fermions with half-integer spins; here it is quite reasonable to expect that the similarity of the symmetric group models to the dual pion model implies that a fermionic sector, analogous to that of Ramond[34], must exist.

REFERENCES

1. In our discussion of the symmetric group we shall often quote properties, given earlier, of the generalised Veneziano amplitude and of the Neveu-Schwarz amplitude; rather than repeat the many references here we refer the reader back to Parts Two through Five where the original literature has been extensively cited.
2. J. L. Gervais and A. Neveu, Nucl. Phys. B47, 422 (1972).
3. J. L. Gervais and A. Neveu, Nucl. Phys. B63, 127 (1973).
4. J. L. Gervais and B. Sakita, Phys. Rev. Letters 30, 716 (1973).
5. S. Mandelstam, Nucl. Phys. B64, 205 (1973).
6. S. Mandelstam, Phys. Letters 46B, 447 (1973).
7. S. Mandelstam, Nucl. Phys. B69, 77 (1974).
8. R. Ademollo, A. d'Adda, R. D'Auria, E. Napolitano, S. Sciuto, P. DiVecchia, F. Gliozzi, R. Musto, and F. Nicodemi, Nuov. Cim. 21A, 77 (1974).
9. K. Bardakci, Nucl. Phys. B68, 331 (1974); ibid B70, 39 (1974). See also E. Cremmer and J. Scherk, Nucl. Phys B72, 117 (1974).
10. P. H. Frampton, Phys. Rev. D7, 3077 (1973).
11. P. H. Frampton, Nuov. Cim. Lett. 8, 525 (1973).
12. For a review, see P. H. Frampton, in Laws of Hadronic Matter, Ed. A. Zichichi, Academic Press (1975) pages 112-136.

13. In subsections 6.3 to 6.6 and 6.8, we follow closely Reference 10. Note that for the simplest special case ($\phi_4 = 1$) the four-meson amplitude collapses to that given by S. Mandelstam, Phys. Rev. Letters 21, 1724 (1968). The amplitude chosen by Mandelstam cannot accommodate realistic intercepts without tachyons. The present symmetric group considerations and the resultant N = 4 amplitudes are, however, much more general than those given by Mandelstam and the difficulty with tachyons is thereby eliminated.

14. E. Plahte, Nuovo Cimento 66A, 713 (1970).
15. G. N. Watson, Bessel Functions, Cambridge University Press (1945).

16. The fixed-angle behaviour of this amplitude has been carefully studied by C. W. Gardiner, Phys. Rev. D9, 2340 (1974) and private communications.

17. A. P. Balachandran and H. Rupertsberger, Phys. Rev. D8, 4524 (1973).
18. A. C. Dixon, Proc. Lond. Math. Soc. 2, 8 (1904).
19. A. P. Balachandran and H. Rupertsberger, Phys. Rev. D9, 4528 (1973).
20. V. N. Gribov and I. Ya. Pomeranchuk, Soviet Phys. JETP 43, 208 (1962) [Translation: 16, 220 (1963)].
21. S. L. Adler, Phys. Rev. 137, B1022 (1965).
22. C. Lovelace, Phys. Letters 28B, 264 (1968).
J. Shapiro, Phys. Rev. 179, 1345 (1969).
23. S. D. Protopopescu, M. Alson-Garnjost, A. Barbaro-Galtieri, S. M. Flatté, J. H. Friedman, T. A. Lasinski, G. R. Lynch, M. S. Robin, and F. T. Solmitz, Phys. Rev. D7, 1279 (1973).

24. P. Estabrooks, A. D. Martin, G. Grayer, B. Hyams, C. Jones, P. Weilhammer, W. Blum, H. Dietl, W. Koch, E. Lorenz, G. Lutjens, W. Manner, J. Meissburger and U. Stierlin, AIP Conf. Proc. 13, 37 (1973).

25. See, for example, the review by L. Montonet, in Laws of Hadronic Matter, Ed. A. Zichichi, Academic Press (1975).

26. The result of the present subsection 6.7 provides a possible answer to a question posed by M. Gell-Mann (Reference 27) who has attached great importance to finding a reason for the number three, in the sense that three quarks make a baryon, within the framework of dual resonance models.

27. E. Gotsman (editor), Proceedings of the International Conference on Duality and Symmetry in Hadron Physics, Tel Aviv, Weizmann Institute Press (1971).

28. G. Frye, C. W. Lee and L. Susskind, Nuovo Cimento 69A, 497 (1970); S. Mandelstam, Phys. Rev. D1, 1720 (1970); P. H. Frampton and P. G. O. Freund, Nucl. Phys. B24, 453 (1970); S. Ellis, P. H. Frampton, P. G. O. Freund and D. Gordon, Nucl. Phys. B24, 465 (1970).

29. Z. Koba and H. B. Nielsen, Nucl. Phys. B12, 633 (1969).

30. The following mathematical references study the problem of representing the symmetric group S_N on N complex variables (≡ Koba-Nielsen variables):
E. H. Moore, American J. Math. 22, 279 (1900);
W. Burnside, Messenger of Mathematics, 30, 148 (1901).

31. P. H. Frampton, unpublished.

32. P. H. Frampton, Lettere al Nuovo Cimento 8, 525 (1973); Phys. Rev. D9, 487 (1974).

33. P. H. Frampton, Phys. Rev. D9, 487 (1974).

34. P. Ramond, Phys. Rev. D3, 2415 (1971).

7

PHENOMENOLOGICAL APPLICATIONS

7.1 INTRODUCTION

Here we outline some fits that have been made to experimental data with an Euler B function model. Because of the shortcomings of the model, some *ad hoc* modifications and approximations must be made to it, as follows:

i) Unitarity. The infinities arising from the narrow-resonance poles must be smeared out by adding an imaginary part to the trajectory functions. This is logically inconsistent because ancestor states are introduced but we overlook this fact because the ancestor couplings are, in general, small.

ii) Ghosts. The model is ghost-free only for unit intercept. For phenomenological use we simply take the model with physical intercepts and overlook the concomitant presence of ghost states.

iii) Diffraction. The pomeron singularity cannot be naturally incorporated into the dual Born amplitude. Therefore either we must select reactions where the pomeron contribution is absent or small, or we must add to the dual amplitude a separate model for the pomeron contribution.

iv) Spin. The model does not accommodate half-integer spins so that the spin of baryons must either be ignored or introduced by some overall kinematic factor.

Despite these four problems, impressive fits have nevertheless been obtained to experimental data in several types of hadronic process; the fits are imprecise in detail but certain gross features are well described, sometimes for the first time.

It is appropriate to add some purely theoretical remarks. So far we have adopted a strictly formal approach insisting that all fundamental axioms be satisfied at each step. This has led to three surviving meson amplitudes: firstly the generalised Euler B function, secondly the Neveu-Schwarz dual pion model, and thirdly the symmetric-group model. Only mutilations of the first (in any case, most imperfect) model have been subjected to extensive phenomenological test. The mathematically-oriented theorist may easily regard as misguided any attempt to compare such an incomplete theory to experiment; he may ask: what do we learn from such work that is relevant to building a better theory? He anticipates the answer which is: very little. But what must be added to that answer is the fact that the results discussed in the following have proved very important and significant in their own right as phenomenological fits which have provided a stimulus to the experimentalists.

As already mentioned, what follows is only an outline or over-view but it should, at least, provide the reader with a useful guide to the original literature.

We first discuss exclusive data fitting [i.e. where all particles are detected in the final state]. We deal in turn with baryon-antibaryon annihilation where certain features of the three-pion Dalitz plot were explained for the first time; with meson-baryon scattering where the use of Euler B functions presents a viable alternative to Regge-pole fits; and with B_5 phenomenology in which an impressive body of experimental data is beautifully correlated.

Then we turn to inclusive reactions, for both single- and many-particle spectra; here there is the remarkable discovery that perhaps the most striking feature of very high energy hadron collisions - the restriction in transverse momenta - is automatically incorporated in the dual resonance model.

7.2 BARYON-ANTIBARYON ANNIHILATION

Experiments on antiproton capture at rest in deuterium have revealed[1,2] a richly-structured three-pion Dalitz plot for the reaction $\bar{p}n \to \pi^+\pi^-\pi^-$. There is a strong enhancement in the low $\pi^-\pi^-$ mass region, and an absence of events near the centre of the plot where the two $(\pi^+\pi^-)$ invariant squared masses are both approximately equal to 1.08 GeV^2. These two features were difficult to interpret by superimposing resonances, even when a strong exotic $\pi^-\pi^-$ contribution was included [Figure 7.1 shows the $\bar{p}n \to \pi^+\pi^-\pi^-$ (at rest) Dalitz plot taken from Reference 1].

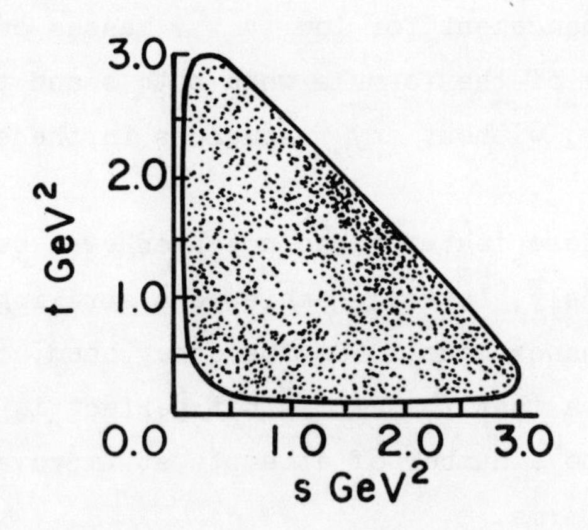

FIGURE 7.1

Dalitz Plot for $\bar{p}n \to \pi^+\pi^-\pi^-$ (Reference 1)

A very early application of the Veneziano formula was made to this process by Lovelace[3,4]. He assumed that the annihilation takes place in a $^1S_0$ ($J^P = 0^-$) state so that it resembles the decay of a heavy pion. Since very little ρ production is observed, he proposed to eliminate completely the ρ-trajectory, and to write the amplitude

$$A(s,t) = \frac{\Gamma(1 - \alpha(s))\Gamma(1 - \alpha(t))}{\Gamma(2 - \alpha(s) - \alpha(t))} \tag{7.1}$$

where

$$s = (p_{\pi^+} + p_{\pi^-(1)})^2 \tag{7.2}$$

$$t = (p_{\pi^+} + p_{\pi^-(2)})^2 \tag{7.3}$$

$$\alpha(x) = 0.483 + 0.885\, x + i\, 0.28 \sqrt{x - 4m_\pi^2} \tag{7.4}$$

$$m_\pi = \text{pion mass.} \tag{7.5}$$

The formula has lines of zeros at constant u, in particular at $\alpha_\rho(s) + \alpha_\rho(t) = 3$, giving rise to a substantial hole in the centre of the Dalitz plot. At the same time, the enhancement for low $(\pi^-\pi^-)$ masses occurs as a natural property of the formula when both s and t are large and positive, without any resonances in the exotic channel.

Because these features had not been even qualitatively understood previously, Lovelace achieved a striking success for the dual resonance model. On the other hand, closer examination reveals that the fit is not perfect in detail, and this has led to a number of attempts at improvement by adding satellite terms.

Firstly Altarelli and Rubinstein[5] suggested the most general form having the important line of zeros at $\alpha_\rho(s) + \alpha_\rho(t) = 3$, namely

$$\begin{aligned} A(s,t) = {} & C_{10}\,\frac{\Gamma(1-\alpha(s))\Gamma(1-\alpha(t))}{\Gamma(1-\alpha(s)-\alpha(t))} + \\ & + C_{11}\,\frac{\Gamma(1-\alpha(s))\Gamma(1-\alpha(t))}{\Gamma(2-\alpha(s)-\alpha(t))} + \\ & + C_{20}\,\frac{\Gamma(2-\alpha(s))\Gamma(2-\alpha(t))}{\Gamma(2-\alpha(s)-\alpha(t))} + \\ & + C_{21}\,\frac{\Gamma(2-\alpha(s))\Gamma(2-\alpha(t))}{\Gamma(3-\alpha(s)-\alpha(t))} + \\ & + C_{30}\,\frac{\Gamma(3-\alpha(s))\Gamma(3-\alpha(t))}{\Gamma(3-\alpha(s)-\alpha(t))} \end{aligned} \tag{7.6}$$

and adjusted the coefficients to improve the fit. The coefficients were later derived[6] by considering factorisation of the five-point function, B_5.

In Reference 7, the imaginary part of the trajectory functions was added in a different way to that of Lovelace, and qualitative success obtained in comparison to experiment.

An important point in fitting $\bar{p}n \to \pi^+\pi^-\pi^-$ is to realise that the fit is <u>essentially</u> two-dimensional and one should not look only at one-dimensional projections. This point was most clearly made by Gopal, Migneron and Rothery[8] who have obtained an excellent fit over the whole Dalitz plot for the first time, by a more careful treatment of the satellite coefficients mentioned earlier under Altarelli and Rubinstein. The best fit of Reference 8 is exhibited in Figure 7.2(b) compared to the data[9] in

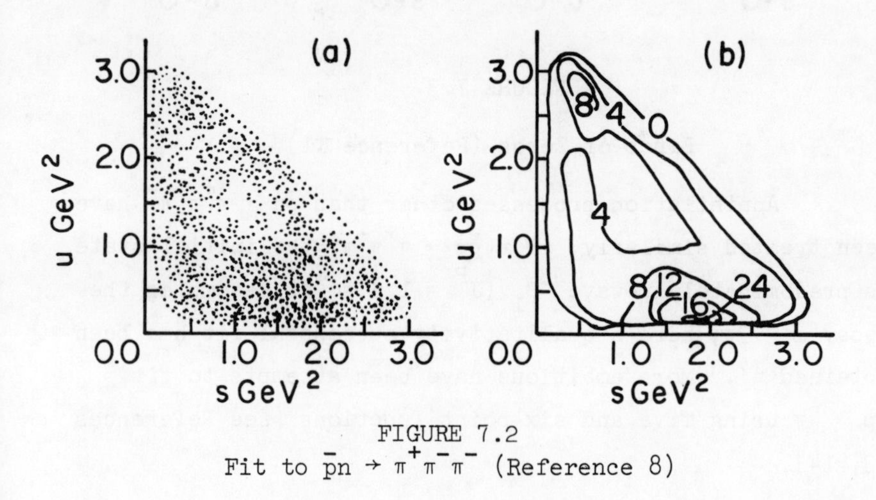

FIGURE 7.2
Fit to $\bar{p}n \to \pi^+\pi^-\pi^-$ (Reference 8)

Figure 7.2(a). [For yet another treatment, see Reference 10].

It is worth mentioning that a somewhat different interpretation of the lines of zeros in the Dalitz plot has been offered by Odorico,[11] who predicts some lines of zeros at constant (s + t), but others at fixed (s - t). Odorico considers a formula

$$A(s,t) = \frac{\Gamma(1 - \alpha(s))\Gamma(1 - \alpha(t))}{\Gamma(2 - \alpha(s) - \alpha(t))} \cdot \frac{\sin \frac{1}{2}\pi(\alpha(s) - \alpha(t))}{\sin \frac{1}{2}\pi(\alpha(s) + \alpha(t))} \qquad (7.7)$$

giving rise to the lines of zeros indicated in Figure 7.3(b), to be compared to the Lovelace interpretation in Figure 7.3(a).

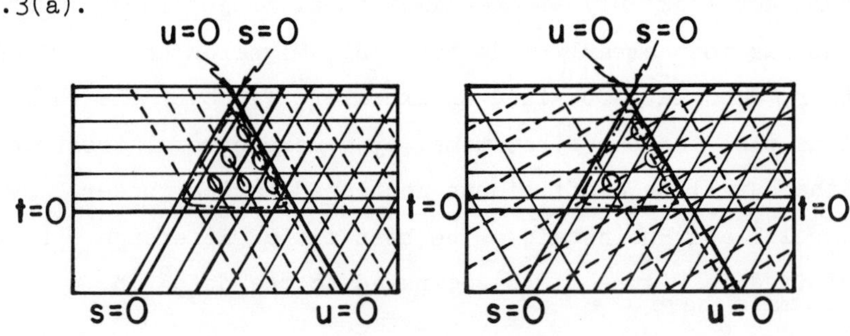

FIGURE 7.3

Lines of Zeros (Reference 11)

Annihilation processes other than $\bar{p}n \to \pi^+\pi^-\pi^-$ have been treated similarly. For $\bar{p}p \to \pi^+\pi^-\pi^o$ the initial state is predominately p-wave $^3S_1$ ($J^P = 1^-$) so it resembles the decay $\omega \to 3\pi$; here a qualitatively successful fit has been obtained [12]. More ambitious have been attempts to fit $\bar{p}p \to 4\pi$ using five and six-point functions [see References 13, 14].

7.3 MESON-NUCLEON SCATTERING

One of the best measured hadronic processes which is also perhaps the most pertinent to an understanding of nuclear binding is meson-nucleon scattering; it is natural therefore that many attempts have been made by theorists to fit this reaction by Euler B functions.

For meson-baryon scattering the parametrisations have universally been made for the invariant amplitudes $A^{\pm}$, $B^{\pm}$ defined for $M(p_1) + N_\alpha(p_2) \to M(-p_3) + N_\beta(-p_4)$ by

$$T = \bar{u}(-p_4)[A_{\beta\alpha}(s,t,u) + \tfrac{1}{2}(\not{p}_1 - \not{p}_3)B_{\beta\alpha}(s,t,u)]u(p_2) \tag{7.8}$$

$$A_{\beta\alpha} = \delta_{\beta\alpha}A^{(+)} + \tfrac{1}{2}[\tau_\beta, \tau_\alpha]\, A^{(-)} \tag{7.9}$$

$$B_{\beta\alpha} = \delta_{\beta\alpha}B^{(+)} + \tfrac{1}{2}[\tau_\beta, \tau_\alpha]\, B^{(-)} \tag{7.10}$$

Here α, β = 1, 2 are isospin indices for the nucleon. These amplitudes have the properties under crossing

$$A^{\pm}(s,t,u) = \pm\, A^{\pm}(u,t,s) \tag{7.11}$$

$$B^{\pm}(s,t,u) = \mp\, B^{\pm}(u,t,s) \tag{7.12}$$

as follows from the change of sign of $(p_1 - p_3)_\mu$ under $s \leftrightarrow u$ crossing. The Regge asymptotic behaviours are, for $|s| \to \infty$ at fixed t [omitting, for the moment, the pomeron for the plus amplitudes]

$$A \sim s^{\alpha_M(t)} \tag{7.13}$$

$$B \sim s^{\alpha_M(t) - 1} \tag{7.14}$$

where $\alpha_M(t)$ is the leading meson trajectory exchanged in the t-channel and, for $|s| \to \infty$ at fixed u

$$A \sim s^{\alpha_F(u)} \tag{7.15}$$

$$B \sim s^{\alpha_F(u)} \tag{7.16}$$

where $\alpha_F(u)$ is the leading fermionic trajectory exchanged in the u-channel.

The choice of A and B is arbitrary when we are using a dual resonance model which does not naturally accommodate fermions. The principal advantages of A and B are the abse ce of kinematic singularities, and simple crossing properties and Regge asymptotic behaviour. For example, we might consider the t-channel non-flip amplitude

$$A' = A + \frac{s - u}{4M(1 - t/4M^2)} B \tag{7.17}$$

[where M = nucleon mass] which is directly related at $t = 0$ to the total cross section but which has a kinematic singularity at $t = 4M^2$. To avoid this singularity we may modify A' further to the form

$$A'' = A + \frac{s - u}{4M} B \tag{7.18}$$

but this has unattractive fixed-u Regge behaviour. Thus A and B do seem to be the simplest choice.

One can now parametrise A and B as sums over Euler B functions. In general, the model will yield parity-doubled fermion trajectories and we must eliminate the low-lying unwanted parity-partners by suitably constraining the relative coefficients of terms.

This formalism is applicable to $\pi N \to \pi N$ and to $KN \to KN$. As we shall discuss, the fits to $\pi N \to \pi N$ are rather unsuccessful; fits to $KN \to KN$ fare better and will be treated later in this sub-section.

For $\pi N \to \pi N$, the trajectories N_α, N_γ, Δ_δ contribute in the s- and u-channels [the subscripts are a convention for signature and parity: $\alpha = \frac{1}{2}^+, \frac{5}{2}^+, \ldots$; $\beta = \frac{1}{2}^-, \frac{5}{2}^-, \ldots$; $\gamma = \frac{3}{2}^-, \frac{7}{2}^-, \ldots$; $\delta = \frac{3}{2}^+, \frac{7}{2}^+, \ldots$; the symbols N and Δ denote isospin $T = \frac{1}{2}$ and $\frac{3}{2}$ respectively]; the trajectories $\rho - f$ (or P') are in the t-channel, in addition of course to the pomeron.

We would like to fit simultaneously backward scattering, the elastic resonance widths, charge-exchange forward scattering and finally forward elastic scattering (although in the last the pomeron must somehow be added).

The difficulty which arises in πN scattering, and because of which we shall not describe the details here, is the following. For backward $\pi^- p$ elastic scattering the sole contributor is the Δ_δ trajectory in the u-channel. At the same time, the model should allow us to continue from the backward region ($u < 0$) to the elastic widths of the Δ_δ resonances($u > 0$) since the Regge residue function and scale factor are prescribed. It seems to be impossible by keeping any reasonable number of Euler B function terms to fit both the (small) backward $\pi^- p$ cross-section and the (large) $\Delta_\delta(1236)$ elastic width. This is a universal feature of all the fits; presumably it reflects the fact that the model is too naive and that, for example, some absorptive Regge cuts must be added for a successful parametrisation. We therefore refer the reader to the literature cited for details [References 15 - 20. Early attempts to treat the πN process were by Igi[15] and Virasoro[16]. Hara[17] considered using the Virasoro formula. Amann[18] studied the baryon elastic widths. Fenster and Wali[19] gave a careful treatment, while Berger and Fox[20] have contributed a good comprehensive review and analysis of the subject].

Next we turn to $\bar{K}N$ scattering. This is much simpler than $\pi N \to \pi N$ because the u-channel (KN) is exotic, and therefore assumed to contain no resonance poles. In the t-channel we include ρ-f-ω-A_2 degenerate trajectories; in the s-channel we include Λ_α-Λ_γ and Σ_β-Σ_δ exchange-degenerate pairs [the symbols Λ and Σ denote isospins $T = 0$ and 1 respectively]. This process is discussed in References 20 - 23.

To give the reader some of the flavour of this work, we mention two specific models (I and II) and show a few of the resultant fits.

The following is a simple parametrisation (I) due to Inami[22]. For the $T_s = 0$ $\bar{K}N$ amplitudes we write

$$A_I^{(0)} = \Lambda_{A1} \frac{\Gamma(1-\bar{\alpha}_\Lambda)\Gamma(1-\alpha(t))}{\Gamma(1-\bar{\alpha}_\Lambda-\alpha(t))} + \Lambda_{A_2} \frac{\Gamma(-\bar{\alpha}_\Lambda)\,\Gamma(1-\alpha(t))}{\Gamma(1-\bar{\alpha}_\Lambda-\alpha(t))} \tag{7.19}$$

$$B_I^{(0)} = \Lambda_{B1} \frac{\Gamma(-\bar{\alpha}_\Lambda)\,\Gamma(1-\alpha(t))}{\Gamma(1-\bar{\alpha}_\Lambda-\alpha(t))} \tag{7.20}$$

and for the $T_s = 1$ $\bar{K}N$ amplitudes we write

$$A_I^{(1)} = \Sigma_{A1} \frac{\Gamma(1-\bar{\alpha}_\Sigma)\Gamma(1-\alpha(t))}{\Gamma(1-\bar{\alpha}_\Sigma-\alpha(t))} \tag{7.21}$$

$$B_I^{(1)} = \Sigma_{B1}\,(t_o - t)\,\frac{\Gamma(1-\bar{\alpha}_\Sigma)\Gamma(1-\alpha(t))}{\Gamma(2-\bar{\alpha}_\Sigma-\alpha(t))} \tag{7.22}$$

Here $\bar{\alpha}_\Lambda = \alpha_\Lambda(s) - \frac{1}{2}$, $\bar{\alpha}_\Sigma = \alpha_\Sigma(s) - \frac{1}{2}$ where Λ, Σ represent the Λ_α-Λ_γ Σ_β-Σ_δ pairs respectively; $\alpha(t)$ is the ρ-f-ω-A_2 trajectory. The factor $(t_o - t)$ is needed to make a change of sign in Σ_{B1} between large t and t = 0. We now make a best fit to the data by adjusting the parameters.

A better fit may be obtained if we add extra subsidiary terms to $A_1^{(0,1)}$ and $B_1^{(0,1)}$ thus increasing the number of free parameters. Such an improved fit (II) has been presented by Berger and Fox[20] who add terms as follows

$$A_{II}^{(0)} = A_I^{(0)} + \Lambda_{A3} \frac{\Gamma(1-\bar{\alpha}_\Lambda)\Gamma(2-\alpha(t))}{\Gamma(2-\bar{\alpha}_\Lambda-\alpha(t))} \tag{7.23}$$

$$B_{II}^{(0)} = B_I^{(0)} + \Lambda_{B2} \frac{\Gamma(1-\bar{\alpha}_\Lambda)\,(2-\alpha(t))}{\Gamma(2-\bar{\alpha}_\Lambda-\alpha(t))} +$$

$$+ \Lambda_{B3} \frac{\Gamma(1-\bar{\alpha}_\Lambda)\Gamma(1-\alpha(t))}{\Gamma(2-\bar{\alpha}_\Lambda-\alpha(t))} \tag{7.24}$$

$$A_{II}^{(1)} = A_I^{(1)} + \Sigma_{A2} \frac{\Gamma(1-\bar{\alpha}_\Sigma)\Gamma(1-\alpha(t))}{\Gamma(2-\bar{\alpha}_\Sigma-\alpha(t))} \tag{7.25}$$

$$B_{II}^{(1)} = B_I^{(1)} + \Sigma_{B3} \frac{\Gamma(2-\bar{\alpha}_\Sigma)\Gamma(2-\alpha(t))}{\Gamma(3-\bar{\alpha}_\Sigma-\alpha(t))}$$

$$+ \Sigma_{B4} \frac{\Gamma(2-\bar{\alpha}_\Sigma)\Gamma(1-\alpha(t))}{\Gamma(3-\bar{\alpha}_\Sigma-\alpha(t))} \tag{7.26}$$

Some results of the models I and II are shown in our Figures. In Figures 7.4(a) and 7.4(b) are shown K^+p backward scattering at 1.79 GeV/c and 2.76 GeV/c respectively.

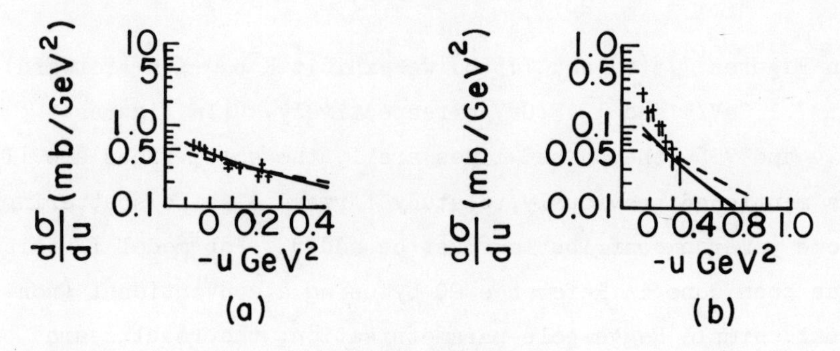

FIGURE 7.4

Backward Scattering at (a) 1.79 GeV/c (b) 2.76 GeV/c. Dashed line, Model I; Solid line Model II; (Reference 20)

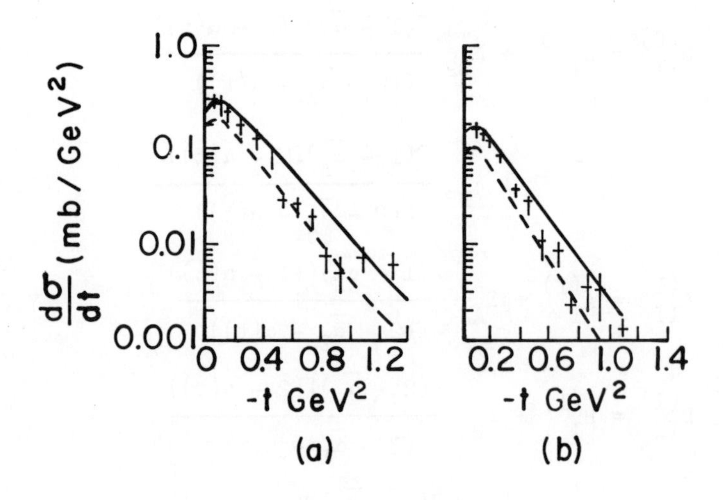

FIGURE 7.5

K^-p Charge Exchange at (a) 7.1 GeV/c (b) 12.3 GeV/c. Dashed line, Model I; Solid line Model II; (Reference 20)

In Figures 7.5(a) and 7.5(b) we exhibit $K^-p \to K^on$ (forward) at 7.1 GeV/c and 12.3 GeV/c respectively. [In Figures 7.4 and 7.5, the dashed lines are I, the solid lines are II]. As mentioned previously, to study forward elastic scattering some pomeron contribution must be added. For model II this has been done in Reference 20 by using a conventional (non-dual) simple Regge-pole parametrisation; the results are given in Figures 7.6(a) and 7.6 (b) for K^-p and K^+p respectively at ~ 10 GeV/c. We have presented only a few representative fits out of those given in the original literature.

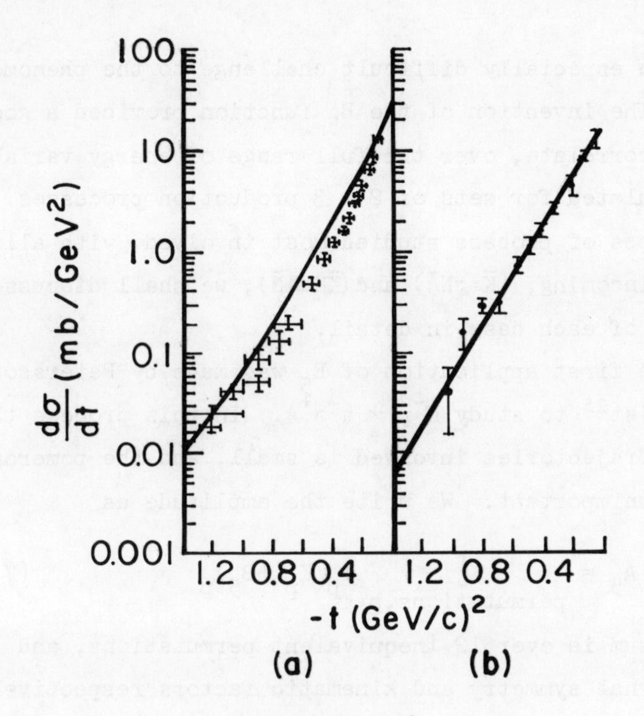

FIGURE 7.6

Elastic Scattering (a) K^-p
(b) K^+p at ~ 10 GeV/c (Reference 20)

A third interesting meson-nucleon process is $\pi N \to \eta N$. Here the pomeron is excluded and there is the further simplification of only one isospin amplitude. In the s- and u-channels we insert $N_\alpha - N_\gamma$, while in the t-channel only the A_2-trajectory survives. The experimental data are not so plentiful for this process, but a simple model due to Miyamura[24] appears to be in adequate agreement [see also Reference 25].

7.4 B_5 PHENOMENOLOGY

Exclusive multiparticle production at high energies

presents an especially difficult challenge to the phenomenologist. The invention of the B_5 function provided a good chance to correlate, over the full range of energy variables, data accumulated for sets of $2 \rightarrow 3$ production processes. The two types of process studied most involved, with all particles incoming, $(\bar{K}\pi\pi N\bar{\Lambda})$ and $(\bar{K}K\pi N\bar{N})$; we shall discuss an example of each case in detail.

The first application of B_5 was made by Petersson and Tornqvist[25] to study $K^- p \rightarrow \pi^- \pi^+ \Lambda$. In this process the number of trajectories involved is small, and the pomeron should be unimportant. We write the amplitude as

$$A_5 = \sum_{\text{permutations},p} I_p \, K_p \, (B_5)_p \tag{7.27}$$

where the sum is over 12 inequivalent permutations, and I, K are internal symmetry and kinematic factors respectively.

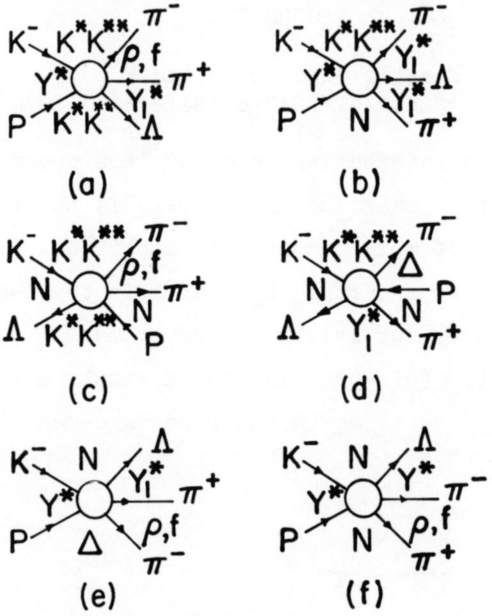

FIGURE 7.7

Diagrams for $K^- p \rightarrow \pi^- \pi^+ \Lambda$

Exclusion of exotics leaves only the 6 permutations shown in Figure 7.7 (a) - (f), where the leading trajectories are also indicated. Petersson and Tornqvist rejected (e) and (f) as negligible because they involve baryon exchanges, and by exchange-degeneracy considerations were led to retain only the configurations (a) and (b). Baryon spin is simply ignored, but because of the abnormal overall parity for the leading resonances one includes a pseudoscalar kinematic factor. The model used, for the process

$$K^-(p_1) + p(p_5) \to \pi^-(-p_2) + \pi^+(-p_3) + \Lambda(-p_4) \quad (7.28)$$

is

$$A = N\, \varepsilon_{\alpha\beta\gamma\delta}\, p_1^{\alpha}\, p_2^{\beta}\, p_3^{\gamma}\, p_4^{\delta}$$

$$[B_5(1 - \alpha_{K*}(s_{12}), 1 - \alpha_\rho(s_{23}), \tfrac{3}{2} - \alpha_\Sigma(s_{34}),$$

$$1 - \alpha_K(s_{45}), \tfrac{3}{2} - \alpha_\Sigma(s_{51})) +$$

$$+ B_5(1 - \alpha_{K*}(s_{12}), \tfrac{3}{2} - \alpha_\Sigma(s_{24}), \tfrac{3}{2} - \alpha_\Sigma(s_{34}),$$

$$1 - \alpha_N(s_{35}), \tfrac{3}{2} - \alpha_\Sigma(s_{51}))] \quad (7.29)$$

and the trajectories are taken from knowledge of two-body processes, with an imaginary part added, if above threshold, as follows

$$\alpha_{K*}(s) = 0.3 + 0.9\, s \quad (7.30)$$

$$\alpha_\rho(s) = 0.48 + 0.9\, s + i\, 0.13 \sqrt{s - 4m_\pi^2} \quad (7.31)$$

$$\alpha_\Sigma(s) = 0.22 + 0.9\, s + i\, 0.13(s - (m_\Lambda + m_\pi)^2) \quad (7.32)$$

$$\alpha_N(s) = -\, 0.30 + 0.9\, s \quad (7.33)$$

The imaginary parts are such that the meson widths are approximately constant and baryon widths increase linearly in mass, as indicated by experiment.

Note that the argument $(1 - \alpha_N(s_{35}))$ in the second B_5 function has incorrect pole positions, but for the phase-space region of interest it is the simplest method to obtain correct Regge behaviour without adding new kinematic factors.

Comparison with experimental data (there is only one remaining free parameter - the overall normalisation, N) is very impressive, as far as general features are concerned. Although naturally the result disagrees in the details, the success for the energy dependence is indicated in Figure 7.8. The predictions for resonance production, angular

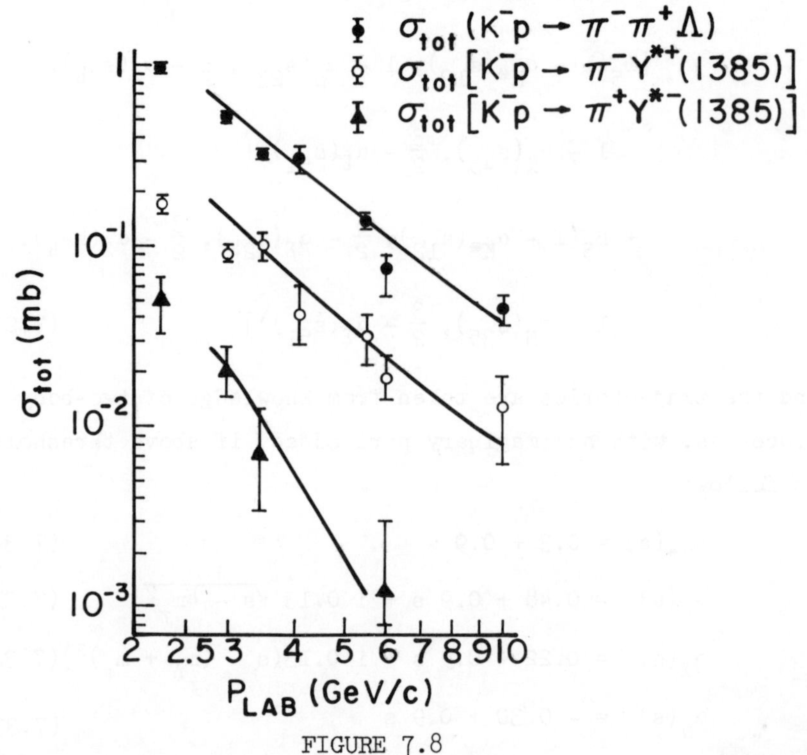

FIGURE 7.8

Energy Dependences in $K^-p \to \pi^-\pi^+\Lambda$
(Reference 25)

distributions and differential cross-sections are all qualitatively acceptable.

The diagrams used by Petersson and Tornqvist are not legal forms of the quark (Harari-Rosner) diagrams[27] we discussed much earlier, and there has been some discussion of this point. If the legal diagrams (c) and (d) are used, the fit worsens; in this connection, note that legal diagram (d) contains Δ_δ-exchange and we have already seen in two body $\pi N \to \pi N$ that Δ_δ does not fit in well for this type of phenomenology.

For the crossed reaction[28] $\pi^+ p \to K^+ \pi^+ \Lambda$ the same model gives reasonable energy dependence and differential cross-sections, and even the normalisation is qualitatively acceptable. Further tests of isospin and crossing symmetry properties can be made by studying $\pi^- n \to \pi^- K^o \Lambda$, $\pi^- p \to \pi^- K^+ \Lambda$, $\pi^+ n \to \pi^+ K^o \Lambda$ [see Reference 29].

Now we turn to the $(\bar{K}K\pi N\bar{N})$ complex. According to the electric charge assignments we can conveniently consider three different classes

$$\text{(I)} \quad K^- K^+ \pi^o p\bar{p} \tag{7.34}$$

$$\text{(II)} \quad \bar{K}^o K^+ \pi^- p\bar{p} \tag{7.35}$$

$$\text{(III)} \quad K^- K^+ \pi^- p\bar{n} \tag{7.36}$$

The pomeron is expected to be most suppressed in Class II, because for I and III it may be coupled to $K\bar{K}$.

For these reasons, the first applications were made to Class II; in particular, Chan, Raitio, Thomas and Tornqvist[30] considered the examples $K^+ p \to K^o \pi^+ p$, $K^- p \to \bar{K}^o \pi^- p$, $\pi^- p \to K^o K^- p$ where good data are available over a range of incident momenta 2.5 to 13 GeV/c. The possible non-exotic graphs are shown in Figure 7.9 (a) - (d). By exchange-

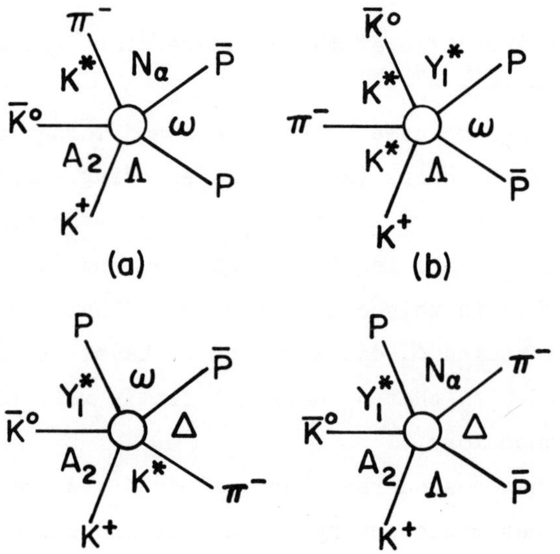

FIGURE 7.9

Diagrams for $\bar{K}^oK^+\pi^-p\bar{p}$

degeneracy arguments these authors decide to retain only (a), (c), (d) and then write the amplitude as:

$$A = N\,\varepsilon_{\alpha\beta\gamma\delta}\,p_\pi{}^\alpha\,p_K{}^\beta\,p_{\bar{K}}{}^\gamma\,p_{\bar{p}}{}^\delta\,[B_5(1-\alpha_{K*},\,1-\alpha_A,\,\frac{1}{2}-\alpha_\Lambda,\,1-\alpha_\omega,\,\frac{1}{2}-\alpha_N)\,+$$

$$+\,B_5(1-\alpha_{K*},\,1-\alpha_A,\,\frac{3}{2}-\alpha_\Sigma,\,1-\alpha_\omega,\,\frac{3}{2}-\alpha_\Delta)\,+$$

$$+\,B_5(\frac{3}{2}-\alpha_\Delta,\,\frac{1}{2}-\alpha_\Lambda,\,1-\alpha_A,\,\frac{3}{2}-\alpha_\Sigma,\,\frac{1}{2}-\alpha_N)] \tag{7.37}$$

As in $\bar{K}\pi\pi N\bar{\Lambda}$ certain baryon exchanges may be shifted by $\frac{1}{2}$ unit to obtain correct Regge behaviour without extra kinematic factors.

Again by one parameter, the overall normalisation, we may hope to accommodate a large amount of data. The results seem to be as successful as the earlier example. The energy dependences of the three processes considered

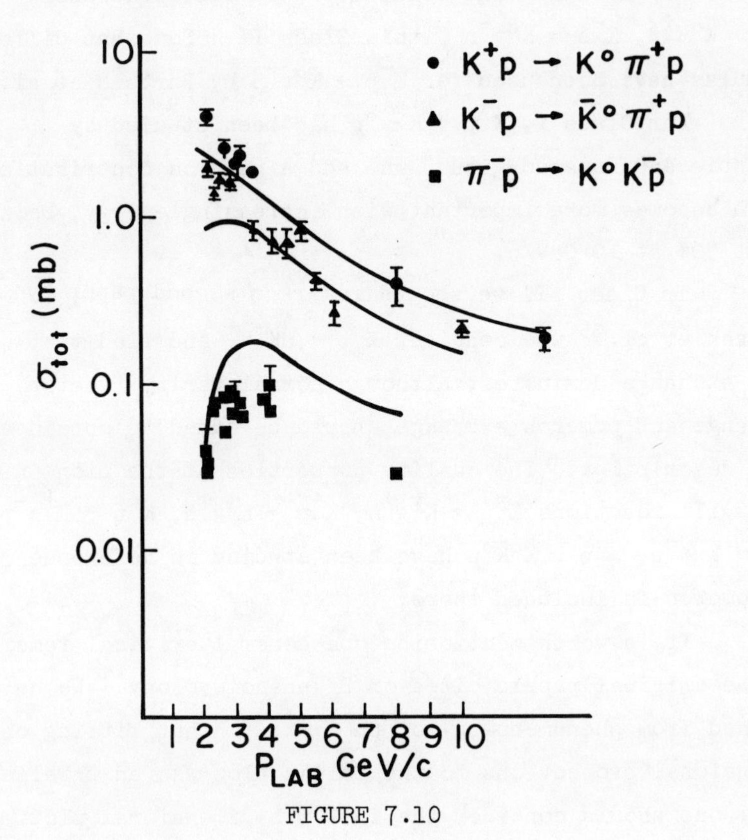

FIGURE 7.10

Energy Dependences in $\bar{K}^oK^+\pi^-p\bar{p}$ (Reference 30)

are shown in Figure 7.10. Note that two of the three overall normalisations are predictions; this is a significant test of crossing symmetry and we see that the much smaller (by a factor 20) π^-p process is correct within a factor two. Similarly fits to resonance production, angular distributions and differential cross-sections are generally acceptable, aside from discrepencies in detail.

The $(\bar{K}K\pi N\bar{N})$ reactions have had several subsequent treatments. The same model has been used[31] to study the resonance production in $K^-p \to \bar{K}^0 n$, $K^+n \to K^0 p$, $\pi^-p \to K^0\Lambda$, $K^-n \to \pi^-\Lambda$. Raitio[32] has studied two further processes $K^+n \to K^0\pi^+n$, $K^-n \to K^0\pi^-n$ within Class II. Somewhat different formulas have been used for $K^-p \to \bar{K}^0\pi^-p$ by Bartsch et al.[33]

For Class I, $K^\pm p \to K^\pm\pi^0 p$ has been studied by Kajantie and Papageorgiou[34] who add a pomeron contribution which becomes more important with increasing energy, being about 50% at 10 GeV/c.

In Class III we should remark a second paper by Bartsch et al.[35] who consider $K^-p \to nK^-\pi^+$ and find that pion exchange dominates, although normal-parity (vector) exchange and pomeron exchange should be added to obtain a good description. The duality properties of the pion in the Class III reactions $K^+p \to K^+\pi^+n$, $K^-p \to K^+\pi^+n$, $K^+n \to K^+\pi^-p$, $K^-n \to K^-\pi^-p$, $\pi^-p \to K^0\bar{K}^0 n$ have been studied in Reference 36; no pomeron is included there.

It is worth mentioning one general critical remark on the original papers cited on B_5 phenomenology. We have learned from phenomenology of $\bar{p}n \to \pi^+\pi^-\pi^-$ that fitting one-dimensional projections to the Dalitz plot was an unreliable test: one should consider the full two-dimensional plot. Similarly in B_5 phenomenology for each incident energy and neglecting spin then one might better fit the final state two-dimensional Dalitz plot instead of several one-dimensional projections.

Finally we note another use of B_5 relevant to phenomenology. Consider the long-standing problem of whether a kinematic Deck bump[37] is equivalent to a resonance parametrisation; before the advent of dual resonance models it was argued[38] that the two were equivalent by duality. The existence of B_5 has allowed a detailed analysis of this question showing that the equivalence holds only for the imaginary part of the amplitude; thus if the amplitude is purely real there is a Deck bump but no resonance (Reference 39) while if the amplitude is purely imaginary then the Deck bump is indeed globally dual to resonances. [For intermediate phases of the amplitude, an intermediate situation exists].

7.5 SINGLE-PARTICLE INCLUSIVE SPECTRA

We now consider inclusive reactions, firstly where only one particle is detected in the final state. We consider a reaction of the form

$$a(p_a) + b(p_b) \rightarrow c(-p_c) + \text{anything} \tag{7.38}$$

where particle c is the detected one. By using a Mueller[40] generalised optical theorem, the inclusive cross-section can be related to a discontinuity in $(p_a + p_b + p_c)^2$ of the forward $a + b + \bar{c} \rightarrow a + b + \bar{c}$ reaction. In the dual resonance model, we use B_6 to describe this $3 \rightarrow 3$ reaction. Of the $\frac{1}{2}(6 - 1)! = 60$ inequivalent permutations of the external lines in B_6, only 18 can contribute to the $(ab\bar{c})$ discontinuity, which has been evaluated in References 41 - 44.

For simplicity, we shall treat in detail only the central region ($\underline{p}_c$ finite in the center of mass, for $s \rightarrow \infty$). [We follow most closely DeTar et al., Reference 43]. The appropriate limits are:

$$S_{ab} \rightarrow +\infty + i\varepsilon \; ; \; S_{\overline{ab}} \rightarrow +\infty - i\varepsilon \tag{7.39}$$

$$M^2 = (p_a + p_b + p_c)^2 \to +\infty \pm i\varepsilon \tag{7.40}$$

$$S_{a\bar{c}},\ S_{b\bar{c}} \to -\infty \tag{7.41}$$

$$\frac{S_{a\bar{c}}\, S_{b\bar{c}}}{S_{ab}} = \kappa,\ \text{fixed.} \tag{7.42}$$

In this central region, only one permutation $(a\bar{c}b\bar{b}c\bar{a})$ survives. We write for this permutation, the representation

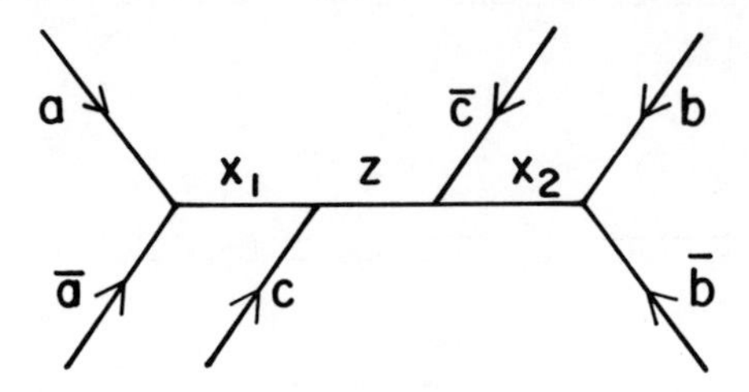

FIGURE 7.11

Six-point Function

appropriate to Figure 7.11 as

$$B_6 = \int_0^1 dx_1\, dz\, dx_2\, x_1^{-\alpha_{a\bar{a}} - 1}\, z^{-\alpha_{a\bar{a}c} - 1}\, x_2^{-\alpha_{b\bar{b}} - 1}$$

$$(1 - x_1)^{-\alpha_{a\bar{c}} - 1}\, (1 - z)^{-\alpha_{a\bar{a}\bar{c}} - 1}\, (1 - x_2)^{-\alpha_{b\bar{c}} - 1}$$

$$(1 - x_1 z - x_2 z + x_1 x_2 z)^{\alpha_{a\bar{a}c}}$$

$$\left[\frac{1 - x_1 z - x_2 z + x_1 x_2 z}{(1 - x_1 z)(1 - x_2 z)}\right]^{\alpha_{a\bar{c}b} - \alpha_{a\bar{c}} - \alpha_{b\bar{c}}} \tag{7.43}$$

Now we put

$$x_1 = 1 - \exp(y_1/\alpha_{a\bar{c}}) \tag{7.44}$$

$$x_2 = 1 - \exp(y_2/\alpha_{b\bar{c}}) \tag{7.45}$$

to find, without approximations

$$B_6 = \int_0^\infty \frac{dy_1}{\alpha_{a\bar{c}}} \int_0^\infty \frac{dy_2}{\alpha_{b\bar{c}}} \int_0^1 dz\,(1 - e^{y_1/\alpha_{a\bar{c}}})^{-\alpha_{a\bar{a}} - 1}$$

$$z^{-\alpha_{a\bar{a}c} - 1}\,(1 - e^{y_2/\alpha_{b\bar{c}}})^{-\alpha_{b\bar{b}} - 1}\, e^{-y_1}\,\cdot$$

$$\cdot\,(1 - z)^{-\alpha_{a\bar{a}c} - 1}\, e^{-y_2}\,\cdot$$

$$\cdot\,[1 + z(1 - e^{y_1/\alpha_{a\bar{c}}} - e^{y_2/\alpha_{b\bar{c}}})]^{\alpha_{a\bar{a}c}}$$

$$\left[\frac{1 + z(1 - e^{y_1/\alpha_{a\bar{c}}} - e^{y_2/\alpha_{b\bar{c}}})}{(1 - z + ze^{y_1/\alpha_{a\bar{c}}})(1 - z + ze^{y_2/\alpha_{b\bar{c}}})}\right]^{\alpha_{a\bar{c}b} - \alpha_{a\bar{c}} - \alpha_{b\bar{c}}} \tag{7.46}$$

In the asymptotic limit we are considering the important region of integration is x_1, $x_2 \simeq 0$ or equivalently y_1, y_2 $\simeq 0$. Now we therefore approximate

$$1 - e^{y_1/\alpha_{a\bar{c}}} \simeq (-y_1/\alpha_{a\bar{c}}) \tag{7.47}$$

$$1 - e^{y_2/\alpha_{b\bar{c}}} \simeq (-\,y_2/\alpha_{b\bar{c}}) \tag{7.48}$$

$$1 + z(1 - e^{y_1/\alpha_{a\bar{c}}} - e^{y_2/\alpha_{b\bar{c}}}) \simeq 1 \tag{7.49}$$

and the final square bracket is

$$\left[\frac{1 + z(1 - e^{y_1/\alpha_{a\bar{c}}} - e^{y_2/\alpha_{b\bar{c}}})}{(1 - z + ze^{y_1/\alpha_{a\bar{c}}})(1 - z + ze^{y_2/\alpha_{b\bar{c}}})}\right]^{\alpha_{a\bar{c}b} - \alpha_{a\bar{c}} - \alpha_{b\bar{c}}}$$

$$\simeq \exp\left[\frac{\alpha_{a\bar{c}b}\, y_1 y_2\, z(1 - z)}{\alpha_{a\bar{c}}\, \alpha_{b\bar{c}}}\right] \tag{7.50}$$

Hence

$$B_6 \simeq (-\alpha_{a\bar{c}})^{\alpha_{a\bar{a}}} (-\alpha_{b\bar{c}})^{\alpha_{b\bar{b}}} \int_0^1 dz\, z^{-\alpha_{a\bar{a}c} - 1} \cdot$$

$$\cdot (1 - z)^{-\alpha_{a\bar{a}\bar{c}} - 1}$$

$$\int_0^\infty dy_1 \int_0^\infty dy_2\, y_1^{-\alpha_{a\bar{a}} - 1} y_2^{-\alpha_{b\bar{b}} - 1} \cdot$$

$$\cdot \exp[-y_1 - y_2 + \frac{y_1 y_2}{x}] \qquad (7.51)$$

where we have defined

$$x = \frac{\bar{\kappa}}{z(1 - z)} \qquad (7.52)$$

$$\bar{\kappa} = \frac{\alpha_{a\bar{c}}\, \alpha_{b\bar{c}}}{\alpha_{a\bar{c}b}} \underset{s \to \infty}{\approx} \kappa = \frac{\alpha_{a\bar{c}}\, \alpha_{b\bar{c}}}{\alpha_{ab}} \simeq p_T^2 + m_c^2 \qquad (7.53)$$

where p_T is the transverse momentum of particle c.

To take now the discontinuity in $\alpha_{a\bar{c}b}$ we notice that our y_1, y_2 integral has a corresponding cut in x and its discontinuity is conveniently found by changing variables to

$$\eta_2 = y_2(1 - \frac{y_1}{x}) \qquad (7.54)$$

whereupon the final two integrals in Equation (7.51) become

$$\int_0^\infty dy_1\, y_1^{-\alpha_{a\bar{a}} - 1} e^{-y_1} \int_0^\infty d\eta_2\, \eta_2^{-\alpha_{b\bar{b}} - 1} e^{-\eta_2} \cdot$$

$$\cdot (1 - \frac{y_2}{x})^{\alpha_{b\bar{b}}} \qquad (7.55)$$

and the x discontinuity is therefore

$$\Gamma(-\alpha_{b\bar{b}}) \sin\pi\, \alpha_{b\bar{b}} \int_x^\infty dy_1\, y_1^{-\alpha_{a\bar{a}} - 1} \left(\frac{y_1}{x} - 1\right)^{\alpha_{b\bar{b}}} e^{-y_1}$$

$$= \frac{\pi\, e^{-x}\, x^{-\alpha_{a\bar{a}}}}{\Gamma(\alpha_{b\bar{b}} + 1)} \int_0^\infty dt\, t^{\alpha_{b\bar{b}}} (1 + t)^{-\alpha_{a\bar{a}} - 1} e^{-xt} \quad (7.56)$$

$$= \pi\, e^{-x}\, x^{-\alpha_{a\bar{a}}}\, \psi(\alpha_{b\bar{b}} + 1, -\alpha_{a\bar{a}} + \alpha_{b\bar{b}} + 1; x) \quad (7.57)$$

where we have made the change of variables $y_1 = x(t + 1)$ and have used the integral representation for the confluent hypergeometric function[45]

$$\psi(a, c; z) = \frac{1}{\Gamma(a)} \int_0^\infty dt\, t^{a-1} (1 + t)^{c-a-1} e^{-zt} \quad (7.58)$$

[Note: the discontinuity must be symmetric in $\alpha_{a\bar{a}}$, $\alpha_{b\bar{b}}$ although this is not manifest here; it can be exhibited explicity by exploiting the relationship between ψ and ϕ confluent hypergeometric functions in the form[45]

$$\psi(a, c; z) = \frac{\Gamma(1 - c)}{\Gamma(a - c + 1)} \phi(a, c; z) + \frac{\Gamma(c - 1)}{\Gamma(z)} z^{1-c} \phi(a - c + 1, 2 - c; z) \quad (7.59)$$

which enables us to re-write the discontinuity as

$$\pi\, e^{-x} \left[x^{-\alpha_{a\bar{a}}} \frac{\Gamma(\alpha_{a\bar{c}} - \alpha_{b\bar{b}})}{\Gamma(\alpha_{a\bar{a}} + 1)} \phi(\alpha_{b\bar{b}} + 1, -\alpha_{a\bar{a}} + \alpha_{b\bar{b}} + 1; x) + x^{-\alpha_{b\bar{b}}} \frac{\Gamma(\alpha_{b\bar{c}} - \alpha_{a\bar{a}})}{\Gamma(\alpha_{b\bar{b}} + 1)} \phi(\alpha_{a\bar{a}} + 1, -\alpha_{b\bar{b}} + \alpha_{a\bar{a}} + 1; x)\right] \quad (7.60)$$

which is manifestly symmetric].

Next we use the asymptotic form[45]

$$\psi(a, c; z) \underset{z\to\infty}{\sim} z^{-a} \quad (7.61)$$

to find that the asymptotic discontinuity is

$$\sim \; x^{-\alpha_{a\bar{a}} - \alpha_{b\bar{b}} - 1} \; e^{-x} \tag{7.62}$$

We define a single-particle distribution function $f^c(p_T)$, after division by the total cross-section $\sigma^{\text{total}}(ab)$, as

$$f^c(p_T) = E_c \frac{d\sigma^c_{ab}}{d^3p} \frac{1}{\sigma^{\text{total}}(ab)} \tag{7.63}$$

and we write

$$\sigma^{\text{total}}(ab) = \frac{\pi}{\Gamma(1 + \alpha_M)} (\alpha_{ab})^{\alpha_M - 1} \tag{7.64}$$

where α_M is the leading exchange trajectory. Identifying $\alpha_M = \alpha_{a\bar{a}} = \alpha_{b\bar{b}}$ and noticing that x is minimised for $z \simeq \frac{1}{2}$ which therefore dominates the z-integral, we find by collecting terms that up to a overall constant (and restoring units of α')

$$f^c(p_T) \approx \Gamma(1 + \alpha_M)\, 2^{\alpha_{a\bar{a}c} + \alpha_{\bar{a}\bar{a}c} - 4\alpha_M} (p_T)^{-2\alpha_M - 2} \cdot$$

$$\cdot \; e^{-4\alpha' p_T^2} \tag{7.65}$$

Remarkably, we see that the transverse momentum is cut off by the gaussian $\exp[-4\alpha' p_T^2]$, which restricts the p_T^2 values to around $(4\alpha')^{-1}$, in qualitative agreement with experiment.

The behaviour of $f^c(p_T)$ for small p_T is affected also by the power term; for further details and the appropriate calculations for the fragmentation regions we refer to the literature [Reference 43].

It is worth mentioning that Huang and Segré[47] have used the B_6 Mueller discontinuity to suggest a new scaling law at fixed-angle in the single-particle inclusive distribution. Defining as the three independent energy variables

$$s = (p_a + p_b)^2 \tag{7.66}$$

$$z = \cos\theta_{ac} \tag{7.67}$$

$$r = |\underline{p}_a|/|\underline{p}_c| \tag{7.68}$$

where $\underline{p}_a$, $\underline{p}_c$ are measured in the center of mass system, then we find that [for $s \to \infty$ and at fixed angle z] to leading order

$$\frac{d\sigma_{ab}^{c}}{d^3p^c} \sim \exp[\alpha' s\, G(r, z)] \tag{7.69}$$

with the universal function G(r,z) given by

$$G(r, z) = (1 + rz)\,\ell n(1 + rz) + (1 - rz)\,\ell n(1 - rz) - $$

$$- (1 + r)\,\ell n(1 + r) - (1 - r)\,\ell n(1 - r) \tag{7.70}$$

for the deep inelastic region; this specific form seems to have some qualitative agreement with data.

Concerning the universal cut-off in p_T we should add the remarks that:

i) Such a cut-off is also predicted by the statistical bootstrap model of Hagedorn,[48-52] but there are some important differences. Firstly the Hagedorn model gives an exponential cut-off $\sim \exp[-(p_T{}^2 + m_c{}^2)^{1/2}/T]$ where T is the limiting temperature $T \approx m_\pi$; the dual resonance model gives a gaussian cut-off. Secondly in the Hagedorn model the temperature T is also what governs the rate of growth of the spectrum: the level density grows as $\rho(m) \sim \exp[m/T]$; in the dual resonance model the cut-off $(4\alpha')^{-1}$ seems to be unconnected to the temperature $T = \frac{1}{2\pi}\frac{\sqrt{6}}{\sqrt{d}}$, governing the rate of growth of the level density (see section 2.6). We can see this in a different way:

the density of levels required to factorise the $3 \to 3$ channel where we are taking the Mueller discontinuity grows only as a power of $(\text{mass})^2$ [not exponentially] so that the level density temperature does not play a role in the above calculation.

ii) The gaussian cut-off in p_T is related to the fixed-angle behaviour $\sim \exp(-as)$ for $|s| \to \infty$ in the four-particle Euler B-function (see section 2.2). In fact, using the universal function $G(r, z)$ mentioned above we may interpolate directly between the two. Taking $r \to 0$ and using $sr^2 \to 4p_T^{\,2}$ we find

$$\exp[\alpha' s\, G(r, z)] \underset{r \to 0}{\to} \exp[-4\alpha'\, p_T^{\,2}] \tag{7.71}$$

On the other hand, taking $r \to 0$ (the elastic limit) we find

$$\exp[\alpha' s\, G(r, z)] \to \exp[-\alpha' s\, f(z)] \tag{7.72}$$

where

$$f(z) = \frac{1 - z}{2}\, \ell n\left(\frac{2}{1 - z}\right) + \frac{1 + z}{2}\, \ell n\left(\frac{2}{1 + z}\right) \tag{7.73}$$

precisely as we found much earlier in studying the fixed-angle property of the Euler B function.

iii) Finally we should remark that although the observed behaviour in p_T of, for example, $p+p \to \pi^o$ + anything at very high energies does indeed fall exponentially (or possibly as a gaussian) for $p_T^{\;\pi^o} \lesssim 2$ GeV, at higher values the fall off is much slower[53]. This fact is an embarrassment to the dual Born amplitude, but it is reasonable to expect that unitarity corrections would fill in the necessary extra cross-section (a very small fraction of the total cross-section) for large transverse momentum.

7.6 MANY-PARTICLE INCLUSIVE SPECTRA

We can find the two-particle inclusive cross-section by taking the appropriate discontinuity in B_8. This has been done in References 54-56. Let us mention only one of the principal results for two-particle correlations in

$$a + b \rightarrow c + d + \text{anything} \tag{7.74}$$

For the central region one finds that the generalisation of the single-particle cut-off $\exp[-4\alpha' p_T^2]$ is a dependence of the form

$$\exp[-4\alpha' ({p_T^c}^2 + {p_T^d}^2 + 2p_T^c\, p_t^d \cos\beta)] \tag{7.75}$$

where β is a function of the relative rapidity y defined by

$$y = \frac{1}{2} \log \frac{(E_c + p_c'')(E_d + p_d'')}{(E_c - p_c'')(E_d - p_d'')} \tag{7.76}$$

and the azimuthal angle ϕ between the momenta $\underline{p}_c$ and $\underline{p}_d$. Here $E_{c,d}$ and $p_{c,d}''$ are the energy and longitudinal momentum of c, d respectively (in the center of mass frame). The precise form of cos β depends on the limit taken. If we take $|p_T^c| >> |p_T^d| >> m_c, m_d$ then

$$\cos\beta = \cosh\frac{y}{2} \cos\frac{\phi}{2} + \frac{1}{4}(\cosh y - \cos\phi) \cdot \cdot\, \ell n\left[\frac{\cosh\frac{y}{2} - \cos\frac{\phi}{2}}{\cosh\frac{y}{2} + \cos\frac{\phi}{2}}\right] \tag{7.77}$$

On the other hand, if we take $|p_T^c| = |p_T^d| >> m_c, m_d$ then

$$\cos\beta = \cosh\frac{y}{2} - \frac{1}{4}(1 - \cos\phi) + \frac{1}{4}(\cosh y - \cos\phi) \cdot \cdot\, \ell n[(\cosh y - \cos\phi)/\{2(1 + \cosh\frac{y}{2})^2\}] \tag{7.78}$$

These formulae have the consequence that for fixed p_T^c, p_T^d and y the relative probability is always highest when $\phi = \pi$ as expected. For more details on this, and the properties of the correlation function, see Reference 56. At present, the experimental data are not sufficiently analysed to allow a check of the B_8 predictions.

More generally we may study n-particle inclusive spectra by considering a discontinuity of B_{2n+4}. This has been mentioned briefly by Virasoro[41], and further analysed by Buras[57].

7.7 SUMMARY

Perhaps the three most impressive results of the phenomenological work that we have described may be listed as (i) the explanation of the principal features of the $\bar{p}n \to \pi^+\pi^-\pi^-$ Dalitz plot; (ii) the B_5 phenomenology applied to $K^-p \to \pi^-\pi^+\Lambda$ and similar processes; (iii) the cut-off in transverse momentum for high energy collisions.

We should emphasise that all the phenomenological work has been done with the (suitably mutilated) Euler B function model. Qualitatively similar results are expected from the (suitably mutilated) Neveu-Schwarz and symmetric-group models. These improved models have not yet been tested phenomenologically; it is quite possible that, when they are, some more useful hints may emerge.

To summarise, it is satisfying to see the good qualitative agreement with the gross features of the experimental data, although this does not surprise us since we have built into the model most of the general properties we believe to be true for the strong-interaction

S-matrix - narrow resonances, linear Regge trajectories, crossing symmetry, absence of exotics and exchange degeneracy.

REFERENCES

1. P. Anninos, L. Gray, P. Hagerty, T. Kalogeropoulos, S. Zenone, R. Bizzarri, G. Ciapetti, M. Gaspero, I. Laasko, S. Lichtman and G.C. Moneti, Phys. Rev. Letters 20, 402 (1968).
2. In preparing subsection 7.2, I have benefited from discussions with T. Kalogeropoulos.
3. C. Lovelace, Physics Letters 28B, 264 (1968).
4. Also in Reference 3 is the proposal for a $\pi\pi \to \pi\pi$ amplitude that we have discussed in subsection 2.8. Note that we here consider applications only to hadronic processes where direct information is available, and do not treat meson-meson processes such as $\pi\pi \to \pi\pi$, $\pi K \to \pi K$, $K\bar{K} \to K\bar{K}$, $\pi\eta \to \pi\eta$. For a detailed treatment of meson-meson scattering see J. L. Petersen, Physics Reports 2C, 155 (1971) and references cited therein. Also we do not consider partially non-strong processes such as the three-pion decays of K and η [see Lovelace, Reference 3].
5. G. Altarelli and H. R. Rubinstein, Phys. Rev. 183, 1469 (1969).
6. H. R. Rubinstein, E. J. Squires and M. Chaichian, Physics Letters 30B, 189 (1969).
7. C. Boldrighini and A. Pugliese, Lettere al Nuovo Cimento 2, 239 (1969).
8. G. P. Gopal, R. Migneron and A. Rothery, Phys. Rev. D3, 2262 (1971).
9. The data of Figure 7.1 have been replotted for direct comparison.
10. S. Pokorski, R. O. Raitio and G. H. Thomas, Nuovo Cimento 7A, 828 (1972).

11. R. Odorico, Physics Letters 33B, 489 (1970).
12. R. Jengo and E. Remiddi, Lettere al Nuovo Cimento 1, 637 (1969).
13. J. F. L. Hopkinson and R. G. Roberts, Lettere al Nuovo Cimento 2, 466 (1969).
14. M. Chaichian and J. F. L. Hopkinson, Lettere al Nuovo Cimento 4, 616 (1970).
15. K. Igi, Phys. Letters 28B, 330 (1968).
16. M. A. Virasoro, Phys. Rev. 184, 1621 (1969).
17. Y. Hara, Phys. Rev. 182, 1906 (1969).
18. R. F. Amann, Lettere al Nuovo Cimento 2, 87 (1969).
19. S. Fenster and K. C. Wali, Phys. Rev. D1, 1409 (1970).
20. E. L. Berger and G. C. Fox, Phys. Rev. 188, 2120 (1969).
21. K. Igi and J. K. Storrow, Nuovo Cimento 62A, 972 (1969).
22. T. Inami, Nuovo Cimento 63A, 987 (1969).
23. K. P. Pretzel and K. Igi, Nuovo Cimento 63A, 609 (1969).
24. O. Miyamura, Progress in Theoretical Physics 42, 305 (1969).
25. M. L. Blackmon and K. C. Wali, Phys. Rev. D2,258 (1970).
26. B. Petersson and N. A. Tornqvist, Nucl. Phys. B13, 629 (1969).
27. H. Harari, Phys. Rev. Letters 22, 562 (1969).
J. L. Rosner, Phys. Rev. Letters 22, 689 (1969).
28. N. A. Tornqvist, Nucl. Phys. B18, 530 (1970).
29. P. Hoyer, B. Petersson and N. A. Tornqvist, Nucl. Phys. B22, 497 (1970).

30. H. M. Chan, R. O. Raitio, G. H. Thomas and N. A. Tornqvist, Nucl. Phys. B19, 173 (1970). See also: V. Waluch, S. Flatté, J. H. Friedman, and D. Sivers Phys. Rev. D5, 4 (1972).

31. B. Petersson and G. H. Thomas, Nucl. Phys. B20, 451 (1970).

32. R. O. Raitio, Nucl. Phys. B21, 427 (1970).

33. J. Bartsch et al., Nucl. Phys. B20, 63 (1970).

34. K. Kajantie and S. Papageorgiou, Nucl. Phys. B22, 31 (1970).

35. J. Bartsch et al., Nucl. Phys. B23, 1 (1970).

36. P. Hoyer, B. Petersson, A. T. Lea, J. E. Paton and G. H. Thomas, Nucl. Phys. B32, 285 (1971).

37. R. T. Deck, Phys. Rev. Letter 13, 169 (1964).

38. G. F. Chew and A. Pignotti, Phys. Rev. Letters 20, 1078 (1968).

39. P. H. Frampton and N. A. Tornqvist, Lettere al Nuovo Cimento 4, 233 (1972).

40. A. H. Mueller, Phys. Rev. D2, 2363 (1970).

41. M. A. Virasoro, Phys. Rev. D3, 2834 (1971).

42. D. Gordon and G. Veneziano, Phys. Rev. D3, 2116 (1971).

43. C. DeTar, K. Kang, C. I. Tan and J. H. Weis, Phys. Rev. D4, 425 (1971).

44. K. J. Biebl, D. Bebel and D. Ebert, Fortschr. Phys. 20, 555 (1972).

45. A. Erdélyi et al., Higher Transcendental Functions, Bateman Manuscript Project, McGraw-Hill (1953).

46. See, for example, M. Banner et al., Phys. Letters 41B, 547 (1972).

47. K. Huang and G. Segré , Phys. Rev. Letters 27, 1095 (1971).

48. R. Hagedorn, Nuovo Cimento Suppl. 3, 147 (1965).
49. R. Hagedorn, Nuovo Cimento 52A, 1336 (1967).
50. R. Hagedorn, Nuovo Cimento 56A, 1027 (1968).
51. R. Hagedorn, Nuovo Cimento Suppl 3, 147 (1965).
52. R. Hagedorn and G. Ranft, Nuovo Cimento Suppl 6, 169 (1968).
53. B. Alper et al., Phys. Letters 44B, 521 (1973).
M. Banner et al., Phys. Letters 44B, 537 (1973).
F. W. Büsser et al., Phys. Letters 46B, 471 (1973).
54. B. Hasslacher, C. S. Hsue and D. K. Sinclair, Phys. Rev. D4, 3089 (1971).
55. C. L. Jen, K. Kang, P. Shen and C. I. Tan, Phys. Rev. Letters 27, 458 (1971).
56. C. L. Jen, K. Kang, P. Shen and C. I. Tan, Ann. Phys. (New York) 72, 548 (1972).
57. A. J. Buras, Nuovo Cimento 12A, 863 (1972).

APPENDIX

ZERO-SLOPE LIMIT

In the limit of small slope of the Regge trajectories, the dual resonance models reduce to field theories where a small number of states couple through a simple lagrangian. This limit is the subject of the present Appendix.

A.1 THE $\lambda\phi^3$ LIMIT

The simplest zero-slope limit[1] is that of the conventional dual model, where we retain only the ground state.

Let us normalise the S-matrix for N external ground-state particles by

$$\langle p_1\, p_2 \cdots p_L \,|S|\, p_{L+1} \cdots p_N \rangle =$$

$$= i(2\pi)^4\, \delta^4\Big(\sum_{i=1}^{N} p_i\Big) \prod_{i=1}^{N} \left[(2\pi)^3\, 2p_i^{\,o}\right]^{\frac{1}{2}} \cdot$$

$$\cdot\, T_N(p_1\, p_2 \cdots p_N) \qquad \text{(A.1)}$$

and then write T_N as a sum over inequivalent planar

amplitudes

$$T_N = \sum_{P\{q_i\}} F_N(p_{q_1}, p_{q_2}, \cdots, p_{q_N}) \tag{A.2}$$

where F_N has resonance poles only in planar channels. We then write

$$F_N(p_1, p_2, \cdots, p_N) = \frac{g^{N-2}}{2^{N-3}} (\alpha')^{\frac{1}{2}(N-4)} \cdot \cdot B_N(p_1, p_2, \cdots, p_N) \tag{A.3}$$

in which B_N is the generalised Euler B function model, and powers of α' have been added to make the definition dimensionally consistent. [The S-matrix element has dimension $M^{-3/2\ N}$].

Consider first the $N = 4$ case. Here we have

$$T_4 = \frac{1}{2} g^2[B(-\alpha_s, -\alpha_t) + B(-\alpha_t, -\alpha_u) + + B(-\alpha_u, -\alpha_s)] \tag{A.4}$$

Now consider the $\alpha' \to 0$ limit of

$$B(-\alpha_s, -\alpha_t) = \frac{\Gamma(-\alpha_s)\Gamma(-\alpha_t)}{\Gamma(-\alpha_s - \alpha_t)} \tag{A.5}$$

$$= - \frac{\alpha_s + \alpha_t}{\alpha_s \alpha_t} \frac{\Gamma(1 - \alpha'(s - M^2))\Gamma(1 - \alpha'(t - M^2))}{\Gamma(1 - \alpha'(s + t - 2M^2))} \tag{A.6}$$

$$\underset{\alpha' \to 0}{\to} \frac{1}{\alpha'} \left(\frac{1}{M^2 - s} + \frac{1}{M^2 - t}\right) (1 + O(\alpha')) \tag{A.7}$$

Define now a coupling constant λ by

$$\lim_{\alpha' \to 0} \left(\frac{g}{\sqrt{\alpha'}} \right) = \lambda \tag{A.8}$$

Then treating the other terms in T_4 similarly we find

$$T_4 = \lambda^2\left(\frac{1}{M^2 - s} + \frac{1}{M^2 - t} + \frac{1}{M^2 - u}\right)(1 + O(\alpha')) \quad (A.9)$$

This corresponds precisely to the Born amplitude of the field theory with lagrangian

$$L = \frac{1}{2}[(\partial_\mu\phi)^2 - M^2\phi^2] + \frac{1}{6}\lambda\phi^3 \quad (A.10)$$

where the factor $\frac{1}{6}$ arises due to Wick's procedure.
We can see the same result in the integral representation by

$$\frac{1}{2}g^2 \int_0^1 dx\, x^{-\alpha_s - 1}(1 - x)^{-\alpha_t - 1} \approx$$

$$\approx \frac{1}{2}\lambda^2\alpha'\left[\int_0^\varepsilon dx\, x^{-\alpha_s - 1} + \int_{1-\varepsilon}^1 dx\,(1 - x)^{-\alpha_t - 1}\right] \quad (A.11)$$

$$= \frac{1}{2}\lambda^2\alpha'\left[-\frac{\varepsilon^{-\alpha_s}}{\alpha_s} - \frac{\varepsilon^{-\alpha_t}}{\alpha_t}\right] \quad (A.12)$$

$$\underset{\alpha'\to 0}{\to} \frac{1}{2}\lambda^2\left[\frac{1}{M^2 - s} + \frac{1}{M^2 - t}\right] \quad (A.13)$$

For $N > 4$, we need to realise that each planar B_N gives rise, in the $\alpha' \to 0$ limit, to a certain number[2]

$$\nu_N = \frac{(2N-4)!}{(N-1)!\,(N-2)!} \quad (A.14)$$

of different Feynman tree amplitudes. On the other hand, there are 2^{N-3} cyclically inequivalent orderings of the external particles associated with each Feynman tree.

When we consider a particular "corner" of the integration region in B_N (e.g. all $x_i \to 0$) we find the tree amplitude multiplied by $(\alpha')^{-N-3}$. Combining these observations with the definition of F_N which contains a factor

$$\frac{g^{N-2}}{2^{N-3}}(\alpha')^{\frac{1}{2}(N-4)} = \lambda^{N-2}\left(\frac{\alpha'}{2}\right)^{N-3} \quad (A.15)$$

we see that the $\alpha' \to 0$ limit of T_N coincides again with the N-point Born amplitude of our $\lambda\phi^3$ lagrangian.

It is amusing that we can understand, at least heuristically, the zero-slope limit in the string picture[3]. Recall that the classical action for the string was

$$S = \frac{1}{\alpha'} \int_{\tau_i}^{\tau_f} d\tau \int_0^{\pi} d\sigma \sqrt{(\dot{x}\cdot x')^2 - \dot{x}^2 x'^2} \tag{A.16}$$

Let us choose the orthogonal gauge $\dot{x}\cdot x' = 0$, and re-define $\sigma' = \varepsilon\sigma/\pi$. We then have

$$S = \int_{\tau_i}^{\tau_f} d\tau \int_0^{\varepsilon} \frac{d\sigma'}{\alpha'} \sqrt{\left(\frac{dx_i}{d\sigma'}\right)^2} \sqrt{-\dot{x}^2} \tag{A.17}$$

and putting $\varepsilon\left(\frac{dx}{d\sigma}\right) = \ell$, with dimension length and letting $\alpha', \ell \to 0$ such that $\ell/\alpha' \to m$ remains finite we arrive at

$$S = \underset{\alpha'\to 0}{\to} m \int_{\tau_i}^{\tau_f} d\tau \sqrt{-\dot{x}} \tag{A.18}$$

which is the action for a point particle of mass m. The one-dimensional string whose action is the area of its world-sheet collapses in the zero-slope limit to a point particle whose action is the length of its world-line.

A.2 YANG-MILLS FIELD THEORY

A more interesting zero-slope limit can be obtained[4] by starting from an N-point function having unequal intercepts. Let us add a conserved fifth-component to the external four-momenta, with the values $\pm$ n c where c = constant and n = integer.

Then

$$B_N = \int \prod_{i=1}^{N} dz_i \, (dV_3)^{-1} \prod_{i \neq j} (z_i - z_j)^{-\alpha' \hat{p}_i \cdot \hat{p}_j} \tag{A.19}$$

where

$$\hat{p}_{i\alpha} = (p_{i\mu}, p_i{}^5) \tag{A.20}$$

If we take alternating signs $\pm$ c for the fifth component this amounts to taking (for external pions) $\alpha_\rho(o) = 1$ and

$$\alpha_\pi(o) = 1 - \alpha' c^2 \tag{A.21}$$

in the odd G-parity channels. Another possibility is to sum over all possible assignments $\pm$ c to the fifth components, compatible with conservation; this leads to additional trajectories of intercepts

$$\left.\begin{aligned} \alpha_+(o) &= 1 - (2n)^2 \alpha' c^2 \\ \alpha_-(o) &= 1 - (2n+1)^2 \alpha' c^2 \end{aligned}\right\} n = 0, 1, 2, \cdots \tag{A.22}$$

in the even and odd channels respectively.

In the first method of assignment a scalar daughter (σ) of the ρ is introduced, with coupling proportional to c. We would like to examine a zero-slope limit retaining only ρ and π states (<u>not</u> σ); this is possible only with the second method of assignment - by summing over the possible signs $\pm$ c, the σ-particle completely decouples from all amplitudes involving π or ρ as external particles.

Let us first consider the $\pi^+\pi^- \to \pi^+\pi^-$ amplitude with the fifth component assignments $\pm$ c as indicated by Figures A.1(a), (b), (c). Then bearing in mind that

$$\alpha' c = 1 + \alpha' m^2 \tag{A.23}$$

where m is the pion mass, and putting in isospin by the

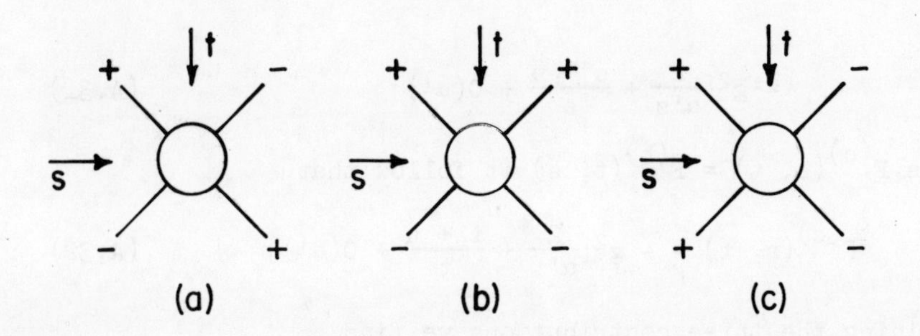

FIGURE A.1

Diagrams for $\pi^+\pi^- \to \pi^+\pi^-$

conventional multiplicative trace factor we arrive at

$$F^{(a)}(s, t) = -g^2 \frac{\Gamma(-1 - \alpha's)\Gamma(-1 - \alpha't)}{\Gamma(-2 - \alpha's - \alpha't)} \tag{A.24}$$

$$F^{(b)}(s, t) = -g^2 \frac{\Gamma(-1 - \alpha's)\Gamma(+3 - \alpha'(t - 4m^2))}{\Gamma(2 - \alpha'(s + t - 4m^2))} \tag{A.25}$$

$$F^{(c)}(s, t) = -g^2 \frac{\Gamma(3 - \alpha'(s - 4m^2))\Gamma(-1 - \alpha't)}{\Gamma(2 - \alpha'(s + t - 4m^2))} \tag{A.26}$$

To find the $\alpha' \to 0$ limit we may re-write (using $\Gamma(1 + z) = z\Gamma(z)$)

$$F^{(a)} = \frac{g^2(2 + \alpha's + \alpha't)(1 + \alpha's + \alpha't)(\alpha's + \alpha't)}{(1 + \alpha's)\,\alpha's\,(1 + \alpha't)\,\alpha't} \cdot$$

$$\cdot \frac{\Gamma(1 - \alpha's)\Gamma(1 - \alpha't)}{\Gamma(1 - \alpha's - \alpha't)} \tag{A.27}$$

$$= g^2\frac{2}{\alpha'}\left(\frac{s + t}{st}\right)\left(1 + \frac{1}{2}\alpha'(s + t) + O(\alpha'^2)\right) \tag{A.28}$$

$$= g^2\left[\frac{2}{\alpha'}\left(\frac{1}{s} + \frac{1}{t}\right) + \frac{(s+t)^2}{st} + O(\alpha')\right] \tag{A.29}$$

Similarly

$$F^{(b)} = \frac{-g^2(2 - \alpha'(t - 4m^2))(1 - \alpha'(t - 4m^2))}{(-1 - \alpha's)(-\alpha's)(1 - \alpha's - \alpha'(t - 4m^2))} \cdot$$

$$\cdot \frac{\Gamma(1 - \alpha's)\Gamma(1 - \alpha'(t - 4m^2))}{\Gamma(1 - \alpha's - \alpha'(t - 4m^2))} \tag{A.30}$$

$$=-g^2[\frac{2}{\alpha' s} + \frac{s + u}{s} + O(\alpha')] \qquad (A.31)$$

Since $F^{(c)}(s, t) = F^{(b)}(t, s)$ it follow that

$$F^{(c)}(s, t) = - g^2[\frac{2}{\alpha' t} + \frac{t + u}{t} + O(\alpha')] \qquad (A.32)$$

Combining the three contributions we find

$$F(s, t) = F^{(a)}(s, t) + F^{(b)}(s, t) + F^{(c)}(s, t) \qquad (A.33)$$

$$=-g^2[\frac{u - t}{s} + \frac{u - s}{t}] + O(\alpha') \qquad (A.34)$$

Taking into account the other possible isospin combinations, we find that the $\alpha' \to 0$ limit is identical to the Born amplitude derived from the interaction

$$L_{\rho\pi\pi} = \frac{1}{2} g \; \pi^a \; \partial_\mu \; \pi^b \; \rho_\mu{}^c \; \varepsilon_{abc} \qquad (A.35)$$

By dimensionality considerations, we should also expect possible $\rho^2\pi^2$, ρ^3 and ρ^4 interactions (Notice that π^4 is absent). To find these couplings we consider the $\pi\pi \to \rho\rho$

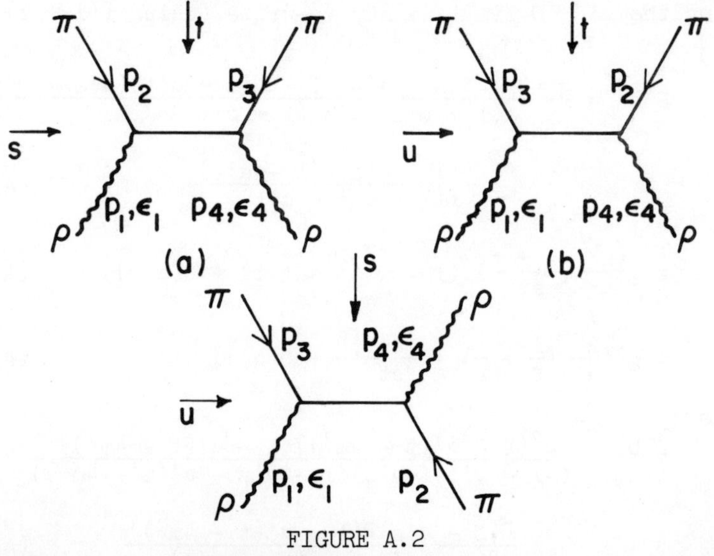

FIGURE A.2
Diagrams for $\pi\pi \to \rho\rho$

amplitudes indicated in Figure A.2. To derive these amplitudes we start from the operatorial vertex[5] for coupling four external parent states, namely

$$F_4 = \int_0^1 dx\, x^{-\alpha_s - 1} (1 - x)^{-\alpha_t - 1}$$

$$\langle 0|\, \exp[a_1^{(1)} \cdot (p_2 + xp_3) + a_2^{(1)} \cdot (p_3 + (1-x)p_4) +$$

$$+ a_3^{(1)} \cdot (p_4 + xp_1) + a_4^{(1)} \cdot (p_1 + (1 - x)p_2) -$$

$$- (1 - x)\, (a_1^{(1)} \cdot a_2^{(1)} + a_3^{(1)} \cdot a_4^{(1)}) -$$

$$- x(a_2^{(1)} \cdot a_3^{(1)} + a_4^{(1)} \cdot a_1^{(1)}) -$$

$$- x(1 - x)(a_2^{(1)} \cdot a_4^{(1)} + a_1^{(1)} \cdot a_3^{(1)}] \quad (A.36)$$

and take

$$|\pi\rangle = |0\rangle \quad (A.37)$$

$$|\rho\rangle = \sqrt{2\alpha'}\; a_\mu^{(1)+}\; |0\rangle\; \varepsilon_\mu \quad (A.38)$$

Then we immediately find, by contracting F_4 with the appropriate combinations of π and ρ states, the amplitudes

$$F'^{(a)} = -g^2[2\alpha'(\varepsilon_1 p_2)(\varepsilon_4 p_1)\; B(-\alpha_s, -\alpha_t) +$$

$$+ 2\alpha'(\varepsilon_1\, p_2)(\varepsilon_4\, p_2)\; B(-\alpha_s, 1 - \alpha_t) +$$

$$+ 2\alpha'(\varepsilon_1 p_3)(\varepsilon_4 p_1)\; B(1 - \alpha_s, -\alpha_t) +$$

$$+ 2\alpha'(\varepsilon_1\, p_3)(\varepsilon_4 p_2)\; B(1 - \alpha_s, 1 - \alpha_t) -$$

$$- (\varepsilon_1\, \varepsilon_4) B(1 - \alpha_s, -\alpha_t)] \quad (A.39)$$

$$F'^{(b)} = F'^{(a)}\; (p_2 \leftrightarrow p_3,\; s \leftrightarrow u) \quad (A.40)$$

$$F'^{(c)} = - g^2[2\alpha'(\varepsilon_1 p_2)(\varepsilon_4 p_3)\, B(-\alpha_s, -\alpha_t) +$$
$$+ 2\alpha'(\varepsilon_1 p_2)(\varepsilon_4 p_1)\, B(1 - \alpha_s, - \alpha_u) +$$
$$+ 2\alpha'(\varepsilon_1 p_4)(\varepsilon_4 p_3)\, B(1 - \alpha_s, -\alpha_u) +$$
$$+ 2\alpha'(\varepsilon_1 p_4)(\varepsilon_4 p_1)\, B(2 - \alpha_s, -\alpha_u) -$$
$$- (\varepsilon_1\, \varepsilon_4)\, B(1 - \alpha_s, 1 - \alpha_u)] \qquad (A.41)$$

To find the $\alpha' \to 0$ limit is now a simple exercise in the repeated use of $\Gamma(1 + z) = z\Gamma(z)$. The result is

$$F'^{(a)} = - g^2[2(\varepsilon_1 p_2)(\varepsilon_4 p_3)(\frac{1}{s - m^2} + \frac{1}{t}) -$$
$$- 2(\varepsilon_1 p_3)(\varepsilon_4 p_2)\, \frac{1}{t} +$$
$$+ (\varepsilon_1\varepsilon_4)(1 + \frac{s - m^2}{t})]\, (1 + O(\alpha')) \qquad (A.42)$$

$$F'^{(b)} = -g^2[2(\varepsilon_1 p_3)(\varepsilon_4 p_2)(\frac{1}{u - m^2} + \frac{1}{t})-$$
$$- 2(\varepsilon_1 p_2)(\varepsilon_4 p_3)\, \frac{1}{t} +$$
$$+ (\varepsilon_1\varepsilon_4)(1 + \frac{u - m^2}{t})]\, (1 + O(\alpha')) \qquad (A.43)$$

$$F'^{(c)} = - g^2[- \frac{2(\varepsilon_1 p_2)(\varepsilon_4 p_3)}{s - m^2} -$$
$$- \frac{2(\varepsilon_1 p_3)(\varepsilon_4 p_2)}{u - m^2} - (\varepsilon_1\varepsilon_4)](1 + O(\alpha'))$$
$$(A.44)$$

These amplitudes correspond to the $\pi^2\rho^2$ and ρ^3 interactions defined by the lagrangians

$$L_{\pi^2\rho^2} = \frac{1}{2}\, g^2[(\rho_a\rho^a)(\pi^b\pi_b) - (\rho_a\pi^a)^2] \qquad (A.45)$$

$$L_{\rho^3} = \frac{1}{2}\, g(\partial_\mu\rho_\nu{}^a - \partial_\nu\rho_\mu{}^a)\, \rho_\mu{}^b\rho_\nu{}^c\varepsilon_{abc} \qquad (A.46)$$

It remains to extract the ρ^4 contact term from the $\rho\rho \to \rho\rho$ amplitude. This can be done by the same method and the result is

$$L_{\rho^4} = \frac{1}{4} g^2[(\rho^2)^2 - (\rho_\mu{}^a \rho_\nu{}^a)(\rho_\mu{}^b \rho_\nu{}^b)] \tag{A.47}$$

Combining the four interaction terms for $\rho\pi^2, \rho^3, \rho^2\pi^2$ ρ^4 together with the free lagrangian

$$L = \frac{1}{2}[(\partial_\mu \pi^a)^2 - m^2 \pi^a \pi_a] + [\frac{1}{2}(\partial_\mu \rho_\nu{}^a - \partial_\nu \rho_\mu{}^a)]^2 + + L_{\rho\pi\pi} + L_{\pi^2\rho^2} + L_{\rho^3} + L_{\rho^4} \tag{A.48}$$

we find a beautiful result, namely that L is precisely the locally gauge invariant Yang-Mills lagrangian[6]. That is, it can be re-written

$$L = \frac{1}{2}(D_\mu \pi^a)^2 - \frac{1}{2} m^2(\pi^a \pi_a) + \frac{1}{4} \tilde{F}_{\mu\nu}{}^a \tilde{F}_{\mu\nu}{}^a \tag{A.49}$$

where

$$D_\mu \pi^a = \partial_\mu \pi^a - g\, \varepsilon_{abc}\, \pi^b \rho_\mu{}^c \tag{A.50}$$

$$\tilde{F}_{\mu\nu}{}^a = \partial_\mu \rho_\nu{}^a - \partial_\nu \rho_\mu{}^a + \varepsilon_{abc}\, \rho_\mu{}^b \rho_\nu{}^c \tag{A.51}$$

This lagrangian is invariant under a unitary gauge transformation

$$\pi'(x) = U(\theta)\, \pi(x) \tag{A.52}$$

$$U(\theta) = \exp[-\, i\, \underline{T}\cdot\underline{\theta}(x)] \tag{A.53}$$

with $\underline{T}$ the generators of isospin, given by

$$(T_a)_{bc} = -\, i\, \varepsilon_{abc} \tag{A.54}$$

The transformation on the gauge field ρ^μ, and the form of $\tilde{F}_{\mu\nu}{}^a$ are such that $D_\mu \pi^a$ and $\tilde{F}_{\mu\nu}{}^a$ transform covariantly, that is

$$D_\mu \pi^a \to U(\theta)\, (D_\mu \pi^a) \tag{A.55}$$

$$\tilde{F}_{\mu\nu}{}^{a} \to U(\theta)(\tilde{F}_{\mu\nu}{}^{a}) \tag{A.56}$$

and hence such that L is invariant. For infinitesimal θ the ρ_μ transformation is

$$\rho_\mu{}^a \to \rho_\mu{}^a - \frac{1}{g}\,\partial_\mu\theta^a + \varepsilon_{abc}\,\theta^b\,\rho_\mu{}^c \tag{A.57}$$

corresponding to

$$\pi^a \to \pi^a + \varepsilon_{abc}\,\theta^b\,\pi^c \tag{A.58}$$

[For the reader unfamiliar with gauge invariance, we outline the derivation. We need

$$D_\mu\pi \to U(D_\mu\pi) \tag{A.59}$$

$$= U(\partial_\mu - ig\,\underline{\rho}\cdot\underline{T})\,\pi \tag{A.60}$$

$$= D_\mu{}'\,\pi' \tag{A.61}$$

$$= \partial_\mu\,\pi' - i\,g\,\underline{\rho}'\cdot\underline{T}\,\pi' \tag{A.62}$$

$$= \partial_\mu(U\pi) - i\,g\,\underline{\rho}'\cdot\underline{T}\,(U\pi) \tag{A.63}$$

and hence

$$\underline{\rho}'\cdot\underline{T} = U(\underline{\rho}\cdot\underline{T})\,U^{-1} - \frac{i}{g}\,(\partial_\mu U)\,U^{-1} \tag{A.64}$$

for

$$U_{ij} = \delta_{ij} - \varepsilon_{ijk}\,\theta_k \tag{A.65}$$

we then obtain

$$((\underline{\rho} + \underline{\delta\rho})\cdot\underline{T})_{ab} = (\underline{\rho}\cdot\underline{T})_{ab} + \varepsilon_{pqr}\,(T^r)_{ab}\,\theta^p\,\rho_\mu{}^q - \frac{i}{g}\,(\underline{T}^r)_{ab}\,\partial_\mu\theta^r \tag{A.66}$$

which implies

$$\delta\rho_\mu{}^a = -\frac{1}{g}\,\partial_\mu\theta^a + \varepsilon_{abc}\,\theta^b\rho_\mu{}^c \tag{A.67}$$

as required. This ensures covariance of $(D_\mu\,\pi)$. Now consider

$$\delta(\tilde{F}_{\mu\nu}{}^a) = \delta[\partial_\mu\rho_\nu{}^a - \partial_\nu\rho_\mu{}^a + g\,\varepsilon_{abc}\,\rho_\mu{}^b\rho_\nu{}^c] \tag{A.68}$$

under the change $\delta\rho_\mu{}^a$. We find

$$\delta(\partial_\mu \rho_\nu{}^a - \partial_\nu \rho_\mu{}^a) = \varepsilon_{abc} [(\partial_\mu \theta^b)\, \rho_\nu{}^c - (\partial_\nu \theta^b)\, \rho_\mu{}^c] + $$

$$+ \varepsilon_{abc}\, \theta^b [\partial_\mu \rho_\nu{}^c - \partial_\nu \rho_\mu{}^c] \quad \text{(A.69)}$$

and only slightly more complicated is

$$\delta(g\, \varepsilon_{abc}\, \rho_\mu{}^b \rho_\nu{}^c) = - \varepsilon_{abc} [(\partial_\mu \theta^b)\, \rho_\nu{}^c - (\partial_\nu \theta^b)\, \rho_\mu{}^c] +$$

$$+ \varepsilon_{abc}\, \theta^b [\partial_\mu \rho_\nu{}^c - \partial_\nu \rho_\mu{}^c] +$$

$$+ g(\varepsilon_{abc}\, \varepsilon_{cde} - \varepsilon_{aec}\, \varepsilon_{cdb}) \cdot$$

$$\cdot\, \rho_\mu{}^b \rho_\nu{}^e \theta^d \quad \text{(A.70)}$$

Now use

$$\varepsilon_{abc}\, \varepsilon_{cde} - \varepsilon_{aec}\, \varepsilon_{cdb} = ([T^a, T^d])_{be} \quad \text{(A.71)}$$

$$= \varepsilon_{adc}\, \varepsilon_{cbe} \quad \text{(A.72)}$$

to confirm that

$$\delta \tilde{F}_{\mu\nu}{}^a = \varepsilon_{abc}\, \theta^b\, [\partial_\mu \rho_\nu{}^c - \partial_\nu \rho_\mu{}^c + \varepsilon_{cde}\, \rho_\mu{}^d \rho_\nu{}^e] \quad \text{(A.73)}$$

$$= \varepsilon_{abc}\, \theta^b\, \tilde{F}_{\mu\nu}{}^c \quad \text{(A.74)}$$

as required.]

Concerning this Yang-Mills limit we should add the remarks:

i) Although the isospin factor is added in a rather trivial way (multiplicatively) in the dual model, it becomes an essential ingredient for the local gauge invariance of the limiting field theory.

ii) The zero-slope limit of the Neveu-Schwarz dual pion model is similar to that discussed for the conventional model in this subsection. One important difference, however, is that we find[7] an additional term $(\underline{\pi}\cdot\underline{\pi})^2$ in the lagrangian. This gives rise to a new stable ground-state and, through the Higgs mechanism, the hitherto-massless gauge field aquires mass.

iii) Starting from a dual model with $\alpha_\rho(o) < 1$ one can arrive at a massive Yang-Mills theory in the zero-slope limit as shown by Gervais and Neveu in Reference 8; this article makes a beautiful application of the zero-slope limit by using it to discover, for the Yang-Mills theory, a new (non-hermitian) gauge which is neither the usual U-gauge nor the R-gauge and in which both unitarity and renormalisability are manifest.

A.3 REGGE SLOPE EXPANSION

So far we have considered the precise limit $\alpha' \to 0$, but it is natural to ask: what happens when α' is small but non-zero? To answer this question, in the conventional model, amounts to taking a Taylor series expansion in α' of the amplitudes.

For the four-particle Euler B function, for example, we may expand[9] (putting $x = -\alpha_s$, $y = -\alpha_t$)

$$B(x, y) = \frac{x + y}{xy} E(x, y) \tag{A.75}$$

$$E(x, y) = \frac{\Gamma(1 + x)\Gamma(1 + y)}{\Gamma(1 + y + y)} \tag{A.76}$$

$$= 1 - xy \{\psi^{(1)} + \frac{1}{2!} \psi^{(2)}(x + y) + \frac{1}{3!} \psi^{(3)}(x^2 + y^2) - $$

$$- \frac{1}{2} [\psi^{(1)2} - \frac{1}{2!} \psi^{(3)}] xy + \frac{1}{4!} \psi^{(4)}(x^3 + y^3) -$$

$$- \frac{1}{2}[\psi^{(1)}\psi^{(2)} - \frac{1}{3!} \psi^{(4)}]xy(x + y)\} + O(\alpha'^6) \tag{A.77}$$

whereupon

$$B(x, y) = (\frac{1}{x} + \frac{1}{y}) - \psi^{(1)}(x + y) -$$

$$- \frac{1}{2!}\psi^{(2)}(x^2 + y^2) - \cdots \qquad (A.78)$$

Such expansions converge in the region $|x|, |y| < 1$ indicated in Figure A.3. In these equations

$$\psi^{(n)} = \frac{d^n}{dz^n}\Gamma(z)\Big|_{z=1} \qquad (A.79)$$

$$= (-1)^{n+1} n!\, \zeta(n + 1) \qquad (A.80)$$

where $\zeta(z)$ is the Riemann zeta function, and $\psi(z)$ is the diagamma function $\psi(z) = \Gamma'(z)/\Gamma(z)$.

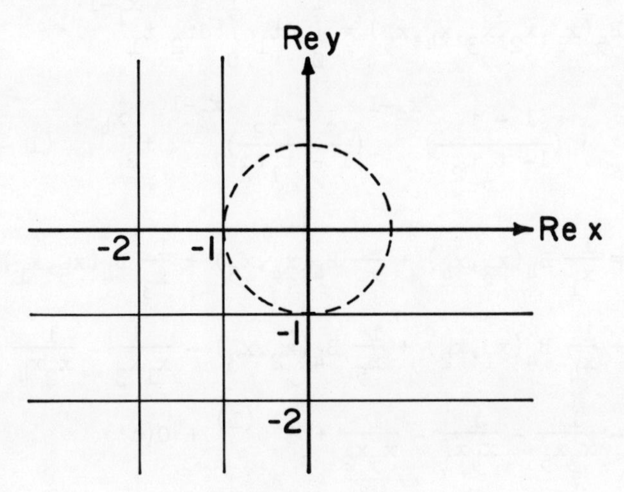

FIGURE A.3
Region of Convergence

Writing the four-meson amplitude as

$$T_4 = \frac{1}{2} g^2[B(-\alpha_s, -\alpha_t) + B(-\alpha_t, -\alpha_u) +$$

$$+ B(-\alpha_u, -\alpha_s)] \qquad (A.81)$$

we find that the first non-singular term in T_4 is given by

$$g^2\alpha'[3m^2 - (s + t + u)]\psi^{(1)} =$$

$$= g^2\alpha'[-3m^2 + \sum_{\substack{i,j=1 \\ i\neq j}}^{4} p_i\cdot p_j + \frac{3}{2}\sum_{i=1}^{4} p_i^2]\psi^{(1)} \tag{A.82}$$

which arises from a langrangian (with $\lambda = g/\sqrt{\alpha'}$)

$$-\psi^{(1)}\lambda^2\alpha'^2[\frac{1}{4}(\partial_\mu\phi)^2\,\phi^2 + \frac{1}{4}(\partial_\mu{}^2\phi)\,\phi^3 + \frac{1}{8}m^2\phi^4] \tag{A.83}$$

By dimensionality considerations we can expect a term $\sim \lambda^3(\alpha')^2\,\phi^5$ at the same order. To find it we must make a Taylor expansion of the five-point function $B_5(x_1, x_2, x_3, x_4, x_5)$ where $x_i = -\alpha_{i,i+1}$ as follows:

$$B_5(x_1,x_2,x_3,x_4,x_5) = \int_0^1 dt_1 \int_0^1 dt_2\, t_1^{x_1-1} \cdot$$

$$\cdot \left(\frac{1-t_1}{1-t_1t_2}\right)^{x_2-1}\left(\frac{1-t_2}{1-t_1t_2}\right)^{x_3-1} t_2^{x_4-1}\,(1-t_1t_2)^{x_5-2} \tag{A.84}$$

$$= \frac{1}{x_1}B_4(x_3,x_4) + \frac{1}{x_2}B_4(x_4,x_5) + \frac{1}{x_3}B_4(x_5,x_1) +$$

$$+ \frac{1}{x_4}B_4(x_1,x_2) + \frac{1}{x_5}B_4(x_2,x_3) - \frac{1}{x_1x_3} - \frac{1}{x_2x_4} -$$

$$- \frac{1}{x_3x_5} - \frac{1}{x_4x_1} - \frac{1}{x_5x_2} + 3\,\psi^{(1)} + O(\alpha') \tag{A.85}$$

This contact term can be reproduced, taking into account the $\lambda^3(\alpha'/2)^2$ normalisation factor discussed earlier, if we take as the full lagrangian.

$$L = \frac{1}{2}[(\partial_\mu\phi)^2 - m^2\phi^2] + \frac{1}{6}\lambda\phi^3 - \psi^{(1)}\lambda^2\alpha'^2[\frac{1}{4}(\partial_\mu\phi)^2 +$$

$$+ \frac{1}{4}(\partial_\mu{}^2\phi)\,\phi^3 + \frac{1}{8}m^2\phi^4] + \frac{9}{5!}\lambda^3(\alpha')^2\,\psi^{(1)}\phi^5 + O(\alpha'^3) \tag{A.86}$$

In general, we expect that the N-point function B_N will require a contact term $\sim \lambda^{N-2}\alpha'^{N-3}\phi^N$ together with derivative terms of order $\lambda^{N-2}\alpha'^{N-3+r}$ $(r = 1, 2, 3, \cdots)$.

Of course, one would like to be able to re-sum the infinite expansion thus obtained, to exhibit a closed-form for the full lagrangian but the mathematical complexity has (so far) obstructed this possibility.

We may regard these expansions, at least from the mathematical viewpoint if not from the degree of physical profundity, as analogous to the semi-classical expansion in h (h = Planck's constant) of quantum mechanics or to the post-Newtonian expansion in $\frac{1}{c}$ (c = velocity of light) of relativity theory. In the dual resonance model the constant α' and the associated fundamental length (putting $\alpha' = (2m_\rho^2)^{-1}$)

$$\ell = \frac{\hbar\sqrt{\alpha'}}{c} = \frac{\hbar}{\sqrt{2}\, m_\rho c} \tag{A.87}$$

$$= 1.8 \times 10^{-14} \text{ cm.} \tag{A.88}$$

$$= 0.18 \text{ fermi} \tag{A.89}$$

which characterises the spatial extension of hadrons play a similarly central role; at the same time, it is not surprising that in the limit $\alpha' \to 0$ where the hadron becomes purely pointlike we arrive at a local lagrangian field theory, as we have discussed in this Appendix.

REFERENCES

1. J. Scherk, Nucl. Phys. B31, 222 (1971).
See also: R. Sawyer, in High Energy Physics, edited by K. T. Mahanthappa, W. D. Walker and W. E. Brittin (Colorado Associated Univ. Press, Boulder, 1970) p. 419.
2. N. Nakanishi, Prog. Theor. Phys. 48, 355 (1972).
3. I am grateful to P. Ramond for a discussion of this point.
4. A. Neveu and J. Scherk, Nucl. Phys. B36, 155 (1972).
5. D. J. Gross and J. H. Schwarz, Nucl. Phys. B23, 33 (1970).
6. C. N. Yang and R. L. Mills, Phys. Rev. 96, 191 (1954).
7. R. F. Cahalan and P. H. Frampton, Phys. Lett. 50B, 475 (1974).
8. J. L. Gervais and A. Neveu, Nucl. Phys. B46, 381 (1972).
9. P. H. Frampton and K. C. Wali, Phys. Rev. D8, 1879 (1973).

SUPPLEMENT

SUPERSTRINGS

S.1 INTRODUCTION

Here we shall discuss in turn the most significant developments in string and superstring theories from 1974 up to the present (1986). First and foremost superstring theories are now used to describe, instead of hadrons, a model for quantum gravity possibly unified with the other elementary particle interactions. This interpretation takes advantage of the massless spin-one and spin-two string states as gauge bosons and the graviton respectively; such states were, of course, an embarrassment in dual resonance models for hadronic interactions.

The first theoretical development beyond 1974 was the modification of the Ramond-Neveu-Schwarz (RNS) string to obtain ten-dimensional supersymmetry, as well as just two-dimensional supersymmetry on the worldsheet. The modification gives the "superstring" theory developed by Green and Schwarz in 1980-82. It has two big advantages over the RNS string: (i) it automatically removes the spin-zero tachyon (which was a "displaced pion," and hence always kept, in the hadronic version); (ii) it renders the loop expansion probably finite, and hence these are the first candidates for finite quantum gravity theories.

The development which sparked the renaissance of widespread interest in superstrings was associated with chiral anomaly cancellation in ten spacetime dimensions. One topic of intense research during the preceding years was higher-dimensional spacetimes. The chiral anomalies for higher-dimensional Yang-Mills theories were first calculated, then those for higher dimensional gravity. It appeared that the superstrings without any Yang-Mills gauge group in ten dimensions were anomaly free, while those with such a gauge group were (apparently) inconsistent. Survival of chiral fermions in four dimensions requires a gauge group. It was therefore a shot in the arm for superstrings when Green and Schwarz discovered in 1984 that the open superstring theory could cancel all anomalies for gauge group O(32).

The anomaly-cancellation mechanism works equally for the groups O(32) or E(8) x E(8), but the Chan-Paton procedure does not work for E(8) x E(8). A few months later, still in 1984, Gross, Harvey, Martinec and Rohm invented the heterotic string, which accommodated either of the two gauge groups. The E(8) x E(8) heterotic model seems the most promising model phenomenologically.

The phenomenology of superstrings requires the compactification of six dimensions. The superstring can be formulated consistently only on special types of manifold, the most studied being Ricci-flat, Kahler manifolds for six dimensions (Calabi-Yau manifolds). There is a very wide choice of compact manifolds, some of which may give promising phenomenologies in four dimensions. It is presently unknown how the superstring theory which starts off with only one parameter compactifies, or, in other words, what is the correct vacuum state of the superstring.

S.2 GRAVITATION

As a fundamental theory of strong interactions the dual resonance models seem fatally flawed for several reasons: (i) the spectrum contains massless spin-one and spin-two particles; (ii) the spacetime dimension is unrealistic; (iii) the incorporation of electroweak currents coupling to pointlike constituents of the hadrons seems impossible. This last problem became the principal claim to success of quantum chromodynamics[1] after the discovery of asymptotic freedom[2] in 1973. At present, QCD is generally accepted as the theory of strong interactions because not only does asymptotic freedom allow the successful use of perturbation theory but also the lattice version introduced by Wilson[3] gives strong support[4] for the assumption that QCD implies quark confinement. Thus, QCD offers a fundamental theory of the strong (color) forces between quarks while dual resonance models are evidently not a fundamental theory of strong interactions. Nevertheless, the dual model may offer an approximation to QCD in certain regimes, e.g. the long-distance interquark potential in heavy quarkonia and the large N limit (N = number of colors) of a generalized QCD.

In the zero-slope limit $\alpha' \to 0$ it was shown in "Dual Resonance Models" (see page 401) that for the Neveu-Schwarz model with Chan-Paton factors one can arrive at Yang-Mills theory. In this limit the massless spin-one states of the open string become gauge bosons. Since the closed string sector contains a massless spin-two state, it was natural to check its correspondence to the graviton couplings of general relativity. This was first pursued by Yoneya[5,6] and then emphasized by Scherk and Schwarz[7,8]. Incidentally, the latter authors later reconsidered applying strings as a generalization of the QCD of quarks and gluons[9].

Let us illustrate the zero-slope limit of the graviton sector[5] in the Shapiro-Virasoro model[10,11]. In an operator formalism[12,13] the scattering amplitude for n gravitons and 2 scalar tachyons can be written as

$$\varepsilon^{(1)}_{\mu_1\nu_1}\,\varepsilon^{(2)}_{\mu_2\nu_2}\cdots\varepsilon^{(n)}_{\mu_n\nu_n}\,T_{\mu_1\nu_1\cdots\mu_n\nu_n}$$

$$= \frac{g^n(\alpha')^{\frac{1}{2}(n-2)}}{2^{n-1}} \lim_{\substack{z_1\to 0\\ z_{n+1}\to 1\\ z_{n+2}\to\infty}} \int \prod_{i=2}^{n} \frac{d^2 z_i}{|z_i|^2}\,|z_{i+1}-z_i|^2$$

$$\langle 0|T\,\{V^{(0)}(z_1,\bar z_1,p_1)\,V^{(0)}(z_{n+2},\bar z_{n+2},p_2)$$

$$\prod_{i=1}^{n}\varepsilon^{(i)}_{\mu_1\nu_1}V^{(2)}_{\mu_i\nu_i}(z_{i+1},\bar z_{i+1},k_i)\}|0\rangle \tag{S.1}$$

where $k_i^2 = 0$, $p_1^2 = p_2^2 = -2/\alpha'$ and

$$V^{(0)}(z,\bar z,p) = :\exp\left(i\sqrt{\alpha'}\,p\cdot Q(z,\bar z)\right): \tag{S.2}$$

$$V^{(2)}_{\mu\nu}(z,\bar z,k) = :\frac{1}{2}\left[z\frac{dQ_\mu}{dz}\bar z\frac{d\bar Q_\nu}{d\bar z} + z\frac{dQ_\nu}{dz}\bar z\frac{d\bar Q_\mu}{d\bar z}\right]\exp(i\sqrt{\alpha'}\,k\cdot Q): \tag{S.3}$$

$$Q_\mu(z) = q_{o\mu} + i\,p_{o\mu}\,\ell nz + \sum_{n=1}^{\infty}(n)^{-\frac{1}{2}}(a_{n\mu}z^n + a^\dagger_{n\mu}z^{-n}) \tag{S.4}$$

$$\bar{Q}_\mu(\bar{z}) = \bar{q}_{o\mu} + i\,\bar{p}_{o\mu}\,\ell n\bar{z} + \sum_{n=1}^{\infty}(n)^{-1/2}(b_{n\mu}\bar{z}^{n} + b^{\dagger}_{n\mu}\bar{z}^{-n}) \; . \quad (S.5)$$

The operators $a_{n\mu}$ and $b_{n\mu}$ are mutually commuting sets and α' is the slope parameter. To take the $\alpha' \to 0$ limit, one can conveniently leave the tachyon at nonzero mass μ^2 by the artefact of adding a momentum component; this is sufficient to illustrate the result and is easier than dealing with only the massless states. Strictly speaking, only massless states survive the $\alpha' \to 0$ limit but since, in any case, this theory contains tachyons it seems esoteric to argue[7] about what is more appropriate.

With the notation for a Virasoro function[11]

$$\Gamma\begin{pmatrix} a & b & c \\ d & e & f \end{pmatrix} \equiv \frac{\Gamma\left(a - \tfrac{1}{2}\alpha'(u-\mu^2)\right)\Gamma\left(b - \tfrac{1}{2}\alpha'(t-\mu^2)\right)\Gamma\left(c - \tfrac{1}{2}\alpha' s\right)}{\Gamma\left(d + \tfrac{1}{2}\alpha'(u-\mu^2)\right)\Gamma\left(e + \tfrac{1}{2}\alpha'(t-\mu^2)\right)\Gamma\left(f + \tfrac{1}{2}\alpha' s\right)} \quad (S.6)$$

[here $s = (p_1 + p_2)^2$, $t = (p_1 + k_1)^2$, $u\ (p_1 + k_2)^2$] the amplitude for n = 2 gravitons and 2 scalar tachyons is found to be a series of Virasoro terms[5]

$$\frac{2}{g^2\pi}\,\varepsilon^1_{\mu_1\nu_1}\varepsilon^2_{\mu_2\nu_2}T_{\mu_1\nu_1,\mu_2\nu_2} = (\varepsilon_1\cdot\varepsilon_2)^2\,\Gamma\begin{pmatrix} 1 & 1 & -1 \\ 0 & 0 & 2 \end{pmatrix}$$

$$+ 2\alpha'(\varepsilon_1\cdot p_1)(\varepsilon_2\cdot p_1)(\varepsilon_1\cdot\varepsilon_2)\,\Gamma\begin{pmatrix} 1 & 1 & -1 \\ 0 & 0 & 2 \end{pmatrix}$$

$$+ 2\alpha'(\varepsilon_1\cdot p_2)(\varepsilon_2\cdot p_1)(\varepsilon_1\cdot\varepsilon_2)\left[\Gamma\begin{pmatrix} 1 & 1 & -1 \\ 0 & 0 & 2 \end{pmatrix} - \Gamma\begin{pmatrix} 1 & 1 & 0 \\ 1 & 0 & 2 \end{pmatrix}\right]$$

$$+ 2\alpha'(\varepsilon_1\cdot p_1)(\varepsilon_2\cdot p_2)(\varepsilon_1\cdot\varepsilon_2)\left[\Gamma\begin{pmatrix} 1 & 1 & -1 \\ 0 & 0 & 2 \end{pmatrix} - \Gamma\begin{pmatrix} 1 & 1 & 0 \\ 0 & 1 & 2 \end{pmatrix}\right]$$

$$+ 2\alpha'(\varepsilon_1\cdot p_2)(\varepsilon_2\cdot p_2)(\varepsilon_1\cdot\varepsilon_2)\,\Gamma\begin{pmatrix} 1 & 1 & -1 \\ 0 & 0 & 2 \end{pmatrix}$$

$$+ \alpha'^2(\varepsilon_1\cdot p_1)^2(\varepsilon_2\cdot p_1)^2\,\Gamma\begin{pmatrix} 1 & 1 & -1 \\ 0 & 0 & 2 \end{pmatrix} +$$

$$+ \alpha'^2(\varepsilon_1\cdot p_2)^2(\varepsilon_2\cdot p_2)^2\Gamma\begin{pmatrix}1 & 1 & -1\\ 0 & 0 & 2\end{pmatrix}$$

$$+ 2\alpha'^2(\varepsilon_1\cdot p_1)^2(\varepsilon_2\cdot p_1)(\varepsilon_2\cdot p_2)\left[\Gamma\begin{pmatrix}1 & 1 & -1\\ 0 & 0 & 2\end{pmatrix} - \Gamma\begin{pmatrix}1 & 1 & 0\\ 1 & 0 & 2\end{pmatrix}\right]$$

$$+ 2\alpha'^2(\varepsilon_1\cdot p_1)(\varepsilon_2\cdot p_1)^2(\varepsilon_1\cdot p_2)\left[\Gamma\begin{pmatrix}1 & 1 & -1\\ 0 & 0 & 2\end{pmatrix} - \Gamma\begin{pmatrix}1 & 1 & 0\\ 0 & 1 & 2\end{pmatrix}\right]$$

$$+ 2\alpha'^2(\varepsilon_1\cdot p_2)^2(\varepsilon_2\cdot p_1)(\varepsilon_2\cdot p_2)\left[\Gamma\begin{pmatrix}1 & 1 & -1\\ 0 & 0 & 2\end{pmatrix} - \Gamma\begin{pmatrix}1 & 1 & 0\\ 0 & 1 & 2\end{pmatrix}\right]$$

$$+ 2\alpha'^2(\varepsilon_1\cdot p_1)(\varepsilon_2\cdot p_2)^2(\varepsilon_1\cdot p_2)\left[\Gamma\begin{pmatrix}1 & 1 & -1\\ 0 & 0 & 2\end{pmatrix} - \Gamma\begin{pmatrix}1 & 1 & 0\\ 1 & 0 & 2\end{pmatrix}\right]$$

$$+ \alpha'^2(\varepsilon_1\cdot p_1)^2(\varepsilon_2\cdot p_2)^2\Gamma\begin{pmatrix}2 & 0 & -1\\ -1 & 1 & 2\end{pmatrix}$$

$$+ \alpha'^2(\varepsilon_1\cdot p_2)^2(\varepsilon_2\cdot p_1)^2\Gamma\begin{pmatrix}0 & 2 & -1\\ 1 & -1 & 2\end{pmatrix}$$

$$+ \alpha'^2(\varepsilon_1\cdot p_1)(\varepsilon_2\cdot p_2)(\varepsilon_1\cdot p_2)(\varepsilon_2\cdot p_1)\left[2\Gamma\begin{pmatrix}1 & 1 & -1\\ 0 & 0 & 2\end{pmatrix} + \Gamma\begin{pmatrix}0 & 2 & -1\\ 1 & -1 & 2\end{pmatrix}\right.$$

$$\left. + \Gamma\begin{pmatrix}2 & 0 & -1\\ -1 & 1 & 2\end{pmatrix} - 2\Gamma\begin{pmatrix}1 & 0 & 0\\ 0 & 1 & 1\end{pmatrix} - 2\Gamma\begin{pmatrix}0 & 1 & 0\\ 1 & 0 & 1\end{pmatrix} + \Gamma\begin{pmatrix}0 & 0 & 1\\ 1 & 1 & 0\end{pmatrix}\right]. \qquad (S.7)$$

The graviton polarization tensor has been separated into $\varepsilon^i_{\mu_i\nu_i} = \varepsilon^i_{\mu_i}\,\varepsilon^i_{\nu_i}$ with $\varepsilon^i\cdot k^i = 0$.

It is straightforward to take the $\alpha' \to 0$ limit of Eq. (S.7) keeping $(g^2\alpha')$ fixed. The result is

$$\varepsilon^1_{\mu_1\nu_1}\varepsilon^2_{\mu_2\nu_2}T_{\mu_1\nu_1,\mu_2\nu_2} = \frac{\pi g^2\alpha'}{4}\left\{\frac{1}{s}\left[(\varepsilon_1\cdot\varepsilon_2)^2(t-\mu^2)(u-\mu^2)\right.\right.$$

$$+ 4(\varepsilon_1\cdot p_2)(\varepsilon_2\cdot p_1)(\varepsilon_1\cdot\varepsilon_2)(t-\mu^2) + 4(\varepsilon_1\cdot p_1)(\varepsilon_2\cdot p_2)(\varepsilon_1\cdot\varepsilon_2)(u-\mu^2)$$

$$- 4(\varepsilon_1\cdot p_1)^2(\varepsilon_2\cdot p_2)^2 - 4(\varepsilon_1\cdot p_2)^2(\varepsilon_2\cdot p_1)^2$$

$$\left. + 8(\varepsilon_1\cdot p_1)(\varepsilon_1\cdot p_2)(\varepsilon_2\cdot p_1)(\varepsilon_2\cdot p_2)\right] -$$

$$- 4(\varepsilon_1 \cdot p_1)^2(\varepsilon_2 \cdot p_2)^2 \frac{1}{t - \mu^2} - 4(\varepsilon_1 \cdot p_2)^2(\varepsilon_2 \cdot p_1)^2 \frac{1}{u - \mu^2}\}. \qquad (S.8)$$

This zero-slope limit is to be compared with the tree approximation to the lagrangian of general relativity coupled to a scalar tachyon

$$\tilde{L} = \frac{1}{K^2} \sqrt{-g}\, R + \frac{1}{2} \sqrt{-g}\, (g^{\mu\nu} \phi_{;\mu} \phi_{;\nu} - \mu^2 \phi^2) \ . \qquad (S.9)$$

With Gupta-Feynman quantization[14] we put $g_{\mu\nu} = \eta_{\mu\nu} + \kappa h_{\mu\nu}$ and expand in $\kappa = (16\pi\, G)^{1/2}$ where G is Newton's constant. The four relevant tree amplitudes are shown in Figs. S.1(a) through S.1(d). The corresponding amplitudes are given by

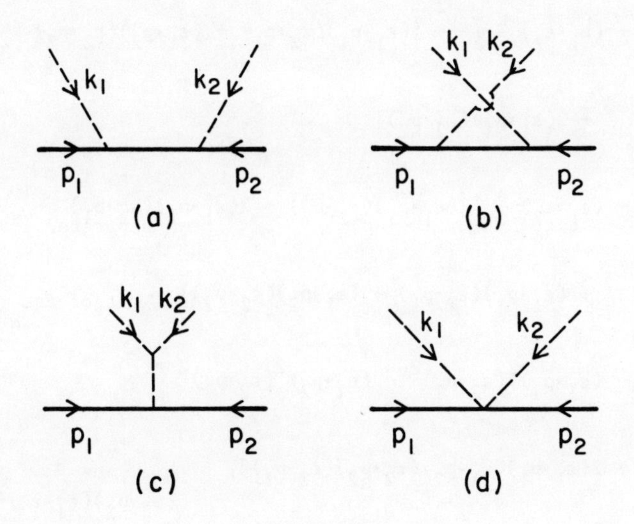

FIGURE S.1

Tree amplitudes for scalar-graviton scattering

$$A^{(a)} = -\frac{2\kappa^2}{t-\mu^2}(\varepsilon_1\cdot p_1)^2(\varepsilon_2\cdot p_2)^2, \tag{S.10}$$

$$A^{(b)} = -\frac{2\kappa^2}{u-\mu^2}(\varepsilon_2\cdot p_1)^2(\varepsilon_1\cdot p_2)^2, \tag{S.11}$$

$$\begin{aligned}A^{(c)} = \frac{2\kappa^2}{s}\{&(\varepsilon_1\cdot\varepsilon_2)^2[-\frac{s^2}{4}+\frac{(t-\mu^2)(u-\mu^2)}{4}]\\
&+(\varepsilon_1\cdot\varepsilon_2)\frac{t-\mu^2}{2}[(\varepsilon_1\cdot p_1)(\varepsilon_2\cdot p_1)+(\varepsilon_1\cdot p_2)(\varepsilon_2\cdot p_2)\\
&+2(\varepsilon_1\cdot p_2)(\varepsilon_2\cdot p_1)]\\
&+(\varepsilon_1\cdot\varepsilon_2)\frac{u-\mu^2}{2}[(\varepsilon_1\cdot p_1)(\varepsilon_2\cdot p_1)+(\varepsilon_1\cdot p_2)(\varepsilon_2\cdot p_2)\\
&+2(\varepsilon_1\cdot p_1)(\varepsilon_2\cdot p_2)]\\
&+(\varepsilon_1\cdot\varepsilon_2)\frac{s}{2}[2(\varepsilon_1\cdot p_1)(\varepsilon_2\cdot p_2)+2(\varepsilon_1\cdot p_2)(\varepsilon_2\cdot p_1)\\
&+(\varepsilon_1\cdot p_1)(\varepsilon_2\cdot p_1)+(\varepsilon_1\cdot p_2)(\varepsilon_2\cdot p_2)]\\
&-(\varepsilon_1\cdot p_1)^2(\varepsilon_2\cdot p_2)^2-(\varepsilon_1\cdot p_2)^2(\varepsilon_2\cdot p_1)^2\\
&+2(\varepsilon_1\cdot p_1)(\varepsilon_1\cdot p_2)(\varepsilon_2\cdot p_2)(\varepsilon_2\cdot p_1)\},\end{aligned} \tag{S.12}$$

$$\begin{aligned}A^{(d)} = 2\kappa^2[&(\varepsilon_1\cdot\varepsilon_2)\frac{s}{4}-(\varepsilon_1\cdot\varepsilon_2)(\varepsilon_1\cdot p_1)(\varepsilon_2\cdot p_2)\\
&-(\varepsilon_1\cdot\varepsilon_2)(\varepsilon_1\cdot p_2)(\varepsilon_2\cdot p_1)].\end{aligned} \tag{S.13}$$

Adding these four terms gives complete agreement with Eq. (S.8) when we make the identification $\pi g^2 \alpha' = 2\kappa^2$. Using general arguments about the S-matrix theory of massless spin-two particles[15] one expects the agreement to hold for all tree amplitudes.

In a model with open and closed strings, the graviton exchange diagram is $\sim \kappa^2 \sim G \sim \alpha' g_{YM}^4$ if the Yang-Mills gauge bosons are open strings with couplings g_{YM}. In an arbitrary spacetime dimension D the gravitational attraction is not inverse square but is generally $\sim r^{2-D}$ so that $[G] = M^{2-D}$. Also $[g_{YM}] = M^{\frac{1}{2}(4-D)}$. Hence one has up to numerical coefficients

$$G = g_{YM}^4 (\alpha')^{\frac{1}{2}(6-D)} . \tag{S.14}$$

We do not know G or g_{YM} in $D \neq 4$ and hence should focus on $D = 4$ to estimate

$$\alpha' = G\, g_{YM}^{-4} \approx 10^{-34}\ \mathrm{GeV}^{-2} \tag{S.15}$$

which amounts to a change in the slope parameter (or string tension) by 34 orders of magnitude from its value for dual models of strong interactions.

Incidentally, in the heterotic string, the relation $G \sim g_{YM}^4$ is replaced by $G \sim g_{YM}^2$ since the Yang-Mills gauge bosons are in the same sector of the spectrum as the graviton. So then

$$\alpha' = G\, g_{YM}^{-2} \approx 10^{-36}\ \mathrm{GeV}^{-2} \tag{S.16}$$

replaces Eq. (S.15).

In any case, the fact that the graviton of closed string theory couples like that in general relativity suggests that the Einstein theory is the low-energy limit of a string or superstring theory. The presence of massless spin-one gauge fields suggests the unification of gravity with the Yang-Mills theories of elementary particle interactions; this requires as a third ingredient the chiral fermion fields describing quarks and leptons.

The first aim then is a consistent quantum gravity model. But the second aim of a Theory of Everything is essential because experimental evidence germane to quantum gravity will not be available in the foreseeable future.

S.3 TEN-DIMENSIONAL SUPERSYMMETRY

The Ramond-Neveu-Schwarz (RNS) model[16,17] posseses a two-dimensional supersymmetry of the string worldsheet which can be shown to be a local symmetry[18]. Superspace techniques were introduced in two dimensions for the RNS model[19,20] and for the symmetric group models[21,22]. The global and local two-dimensional supersymmetry in the RNS model was recognized a few years before the ten-dimensional global and local supersymmetry for (a subsector of) the theory[23,24]. String theories which incorporate ten-dimensional supersymmetry are known as "superstrings."

The possibility of supersymmetry in four dimensions was first studied by Russian authors[25,26], then emphasized by Wess and Zumino[27] who were clearly inspired[28] by the properties of the RNS string model. The superspace version for four dimensions was proposed by Salam and Strathdee[29].

Globally supersymmetric field theories with gauge symmetry appeared as interesting candidates for phenomenological applications to strong and electroweak interactions and their grand unification. One major problem was (and still is!) that the 24 known particles--12 flavors of quarks and leptons and 12 gauge bosons--are predicted to have superpartners called squarks, sleptons and gauginos respectively. These superpartners have never been seen and probably don't exist at energies accessible to accelerators (why should they all lie just over the horizon?). It also appeared that global supersymmetry might solve the gauge hierarchy problem concerning the presence of light Higgs scalars by arranging that these scalars are the superpartners of light chiral fermions. Also, the quadratic divergence for $\lambda\phi^4$ is cancelled by global supersymmetry. Nevertheless, although non-renormalization theorems can avoid re-tuning in quantum corrections, the tuning of parameters in the classical lagrangian is not, in general, avoided.

Global supersymmetry softens the ultra-violet divergences of a field theory in four dimensions but in no case does it render a non-renormalizable theory renormalizable. In a renormalizable theory, global supersymmetry reduces, in general, the number of renormalization constants and in special cases even renders the theory completely finite[30]. This last circumstance may be the most remarkable aspect of global supersymmetry, especially if it can be connected to a superstring ancestry in a convincing way.

Much work was done on globally supersymmetric field theory [Reviews are in Refs. 31-33] but because no great success was forthcoming in the strong and electroweak sectors it was natural to retreat to gravitation. Local gauging of supersymmetry leads inevitably to inclusion of gravity and hence supergravity [34, reviewed in Ref. 35]. Here the primary aim was to secure a finite theory of quantum gravity, but no four-dimensional supergravity avoids uncontrollable divergences and the situation for higher dimensions is even worse. Still, the study of supergravity in higher dimensions, especially ten[36,37], has been--as will become clear--at least useful in the evolution of superstring theory.

The initial studies of supersymmetry and supergravity in four or more dimensions were carried out with no regard to the RNS dual resonance model which had, at least in part, provided the original inspiration. This was the case until 1977[23] when Gliozzi, Olive and

Scherk made an important observation concerning the spectrum of the RNS model. The complete RNS model is not ten-dimensionally supersymmetric, as can be seen most simply by noticing that the tachyon is a spin-zero particle with no spin-half superpartner. This tachyon was the pion in the strong interaction model and hence rather central in those considerations. But when we drop the hadronic interpretation and adopt the more fundamental interpretation of the previous subsection we may try to remove this tachyon.

Let us examine the RNS model spectrum. The NS sector is created by operators

$$\text{bose} \qquad a_n^{\mu\dagger} \qquad n = \text{integer} \tag{S.17}$$

$$\text{fermi} \qquad b_r^{\mu\dagger} \qquad r = \text{half-integer} . \tag{S.18}$$

The physical states $|\phi\rangle$ satisfy

$$L_n|\phi\rangle = G_r|\phi\rangle = (L_o - \tfrac{1}{2})|\phi\rangle = 0 \tag{S.19}$$

and the mass-shell condition is

$$\alpha' m^2 = n = -\tfrac{1}{2},\ 0,\ +\tfrac{1}{2},\ 1,\ \ldots . \tag{S.20}$$

The NS bosons fall into two sets according to their eigenvalue of the G-parity

$$G = -(-1)^{\sum_r b_r^\dagger b_r} . \tag{S.21}$$

The half-integer values of $\alpha' m^2$, including the tachyon, have $G = -1$; the integer values of $\alpha' m^2$ have $G = +1$. The ground state of the even G sector is the massless vector

$$b_{1/2}^{\mu\dagger}|0\rangle\ \varepsilon_\mu(p) \qquad (p^2 = 0) . \tag{S.22}$$

The R fermionic sector of the RNS model is generated by operators

$$\text{bose} \qquad a_n^{\mu\dagger} \qquad n = \text{integer} \tag{S.23}$$

$$\text{fermi} \qquad d_n^{\mu\dagger} \qquad n = \text{integer} \tag{S.24}$$

with physical states $|\phi\rangle$ satisfying

$$F_n|\phi\rangle = L_n|\phi\rangle = 0 \ . \tag{S.25}$$

The case $F_o|\phi\rangle = 0$ is the Dirac-Ramond equation while $L_o|\phi\rangle = 0$ is the mass-shell condition giving $\alpha' m^2 = 0, 1, 2, \ldots$ with no tachyon.

The ten-dimensionally supersymmetric string theory--a superstring--is formed by taking only the even G bosons while for the Ramond fermions we take the Majorana-Weyl projection. At the massless level this leaves eight components each for spin-one and spin-half, appropriate to a ten-dimensional supersymmetric gauge multiplet.

The authors of Ref. 23 made it very plausible that the entire spectrum is ten-dimensionally supersymmetric by counting the number of bosons $d_{NS}(n)$ and fermions $d_R(n)$ at the level $\alpha' m^2 = n = 0, 1, 2, 3, \ldots$ in this truncated theory. For the bosons we have

$$f_{NS}(q) = \sum_{n=0}^{\infty} d_{NS}(n) q^{2n} \tag{S.26}$$

$$= \frac{1}{q} \, \mathrm{Tr}\left(\frac{1}{2}\,(1 + G) q^{2R}\right) \tag{S.27}$$

with

$$R = \sum_{n=0}^{\infty} n a_n^{\dagger} \cdot a_n + \sum_{r=1/2}^{\infty} r \, b_r^{\dagger} \cdot b_r \tag{S.28}$$

giving

$$f_{NS}(q) = \prod_{n=1}^{\infty} (1 - q^{2n})^{-8} \frac{1}{2q} [\prod_{n=1}^{\infty} (1 + q^{2n-1})^8 - \prod_{n=1}^{\infty} (1 - q^{2n-1})^8] . \tag{S.29}$$

On the other hand, the spectrum of projected fermions has degeneracy

$$f_R(q) = \sum_{n=0}^{\infty} d_R(n) q^{2n} \tag{S.30}$$

$$= 8 \ \mathrm{tr}(q^{2R}) \tag{S.31}$$

since the ground state is an 8-fold Majorana-Weyl spinor and where

$$R = \sum_{n=1}^{\infty} n a_n^{\dagger} \cdot a_n + \sum_{n=1}^{\infty} n \ d_n^{\dagger} \cdot d_n . \tag{S.32}$$

This gives

$$f_R(q) = 8 \prod_{n=1}^{\infty} (1 - q^{2n})^{-8} (1 + q^{2n})^8 . \tag{S.33}$$

Because of a remarkable identity discovered by Jacobi in 1829 (see e.g. Ref. 38)

$$f_R(q) = f_{NS}(q) . \tag{S.34}$$

This is a necessary condition for supersymmetry. The proof that the spectrum is supersymmetric was constructed by Green and Schwarz[24] who exhibited the supersymmetry operator which transforms the bosons into fermions using the vertex operator for fermion emission.

In the closed string sector of this model the massless states correspond to N = 1, D = 10 supergravity with components: the graviton g_{MN} (35 states), the gravitino ψ_M (56), the antisymmetric tensor field A_{MN} (28), a spin-half χ (8) and a dilaton ϕ (1). These 128 states fall into 64 each of bosons and fermions. The field A_{MN} will play a central role in our discussion of anomaly cancellation.

In 1982, Green and Schwarz invented two further superstring theories which involve only closed stings[39]. The massless sector of these theories corresponds to N = 2, D = 10 supergravity. The two N = 2 supersymmetry operators can be of opposite handedness (non-chiral) or the same handedness (chiral). The corresponding massless states can be easily found using the O(8) group by producing the left and right sectors as follows:

$$\text{non-chiral} \qquad \begin{pmatrix} 8_v \\ 8_s \end{pmatrix} \times \begin{pmatrix} 8_v \\ 8_c \end{pmatrix}$$

giving 128 bosons: $g_{MN}(35)$, $A_M(8)$, $A_{MN}(28)$, $A_{MNP}(56)$, $\phi(1)$ and 128 fermions: $\psi^i_M(2 \times 56)$, $\chi^i(2 \times 8)$

$$\text{chiral} \qquad \begin{pmatrix} 8_v \\ 8_s \end{pmatrix} \times \begin{pmatrix} 8_v \\ 8_s \end{pmatrix}$$

giving 128 bosons: $g_{MN}(35)$, $A^i_{MN}(2 \times 28)$, $A_{MNPQ}(35)$, $\phi^i(2 \times 1)$ and 128 fermions counted as above, except now the two ψ^i_M and the two χ^i have the same rather than opposite handedness. The nonchiral N = 2, D = 10 supergravity can be obtained by reducing N = 1, D = 11 supergravity while the chiral version is specific to D = 10.

In summary then, by 1982 three superstring theories had been established. The open superstring allowed introduction of a Yang-Mills gauge group by the Chan-Paton procedure, although consistency[40] required that the group be either an SO(n) or an Sp(2n) group; as will be discussed below, anomaly cancellation singles out the choice SO(32). The two N = 2 theories, nonchiral and chiral, have no Yang-Mills gauge symmetry in ten dimensions.

S.4 NON-STRING KALUZA-KLEIN THEORY

During the period 1978-84 there was considerable activity in higher-dimensional gauge field theories, beginning with Ref. 41. The successes and limitations of this approach help explain the extreme attractiveness of the superstring theory which subsumed the Kaluza-Klein approach in 1984. Because the earlier attempts were non-renormalizable and non-unified, they were motivated, for some of us, by the expectation of an underlying string theory, although during the period mentioned it was never obvious how the string theory would eventually merge with the Kaluza-Klein theory.

A principal difficulty of, and restriction on, all Kaluza-Klein theories is to obtain chiral fermions on M_4, four dimensional Minkowski space. This leads to two simple, but important, remarks:

(i) The initial dimension must be even $d = 2n$, with a spatial rotation group $O(2n - 1)$, n = integer. In $O(2n)$ rotational invariance implies that parity (P) is conserved because P is not a discrete operation but merely part of the rotation group e.g. for $d = 3$ and $O(2)$ parity is a 180° rotation. P is here defined by changing sign of all spatial components. In such a case, no chirality is possible.

Equivalently, for $d = 4$, the four gamma matrices γ_μ can be combined to give the pseudoscalar γ^5 which allows the decomposition

$$\psi = \left(\frac{1 + \gamma_5}{2}\right)\psi + \left(\frac{1 - \gamma_5}{2}\right)\psi = \psi_R + \psi_L \; . \tag{S.35}$$

In $d = 5$, by contrast, $\gamma^0, \gamma^1, \gamma^2, \gamma^3, \gamma^4$ form a complete set and there is no "γ^6".

(ii) There must be explicit gauge fields in the starting dimension $d = 2n > 4$. If we start with a chiral fermion ψ_L in $d = 2n > 4$ (eigenvalue of γ_{2n+1} equal to -1) then compactification to $d = 4$ with a compact manifold C_{2n-4} gives chiralities as follows

$$M_{2n} \to M_4 \times C_{2n-4}$$

$$\gamma_{2n+1} = \gamma_5 \times \gamma_{2n-3} \qquad (S.36)$$

$$- = \begin{cases} + & \times & - \\ - & \times & + \end{cases}$$

so that both chiralities appear symmetrically on M_4. Thus the original Kaluza-Klein program starting with just higher dimensional gravity fails. One must introduce an asymmetry between the chiralities by arranging explicit gauge fields in a topologically nontrivial vacuum configuration on C_{2n-4}; this then allows survival of chirality on M_4. [For a proof see E. Witten, Proceedings of the 1983 Shelter Island Conference, MIT Press (1985) p. 227.]

The necessity of explicit Yang-Mills fields in these theories was disappointing from the viewpoint of unification. One did, however, still hope to explain fermion quantum numbers, including the existence of families, by new anomaly cancellation conditions. This is because such gauge theories with chiral fermions have anomalies which must be zero in order to define the quantum theory and keep unitarity. The nonrenormalizability of such higher dimensional theories is irrelevant to the necessity for anomaly cancellation.

The leading contribution to the gauge anomaly in dimension $d = 2n$ is from a polygonal Feynman graph with $(n + 1)$ sides e.g. the triangle in $d = 4$ through the hexagon in $d = 10$ (Fig. S.2). These anomalies may be computed using Fujikawa's method, or by differential geometry. One can use [42, Reprint 1] a brute force method of studying the linear divergence of the Feynman diagram (Fig. S.3). Shifting integration variables appropriately one finds that the $5! = 120$ crossed diagrams combine in 60 pairs and enable one to compute

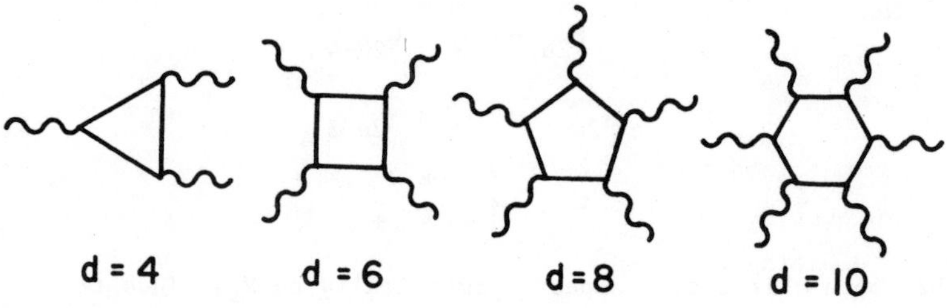

d = 4 d = 6 d = 8 d = 10

FIGURE S.2

Leading contributions to gauge anomaly

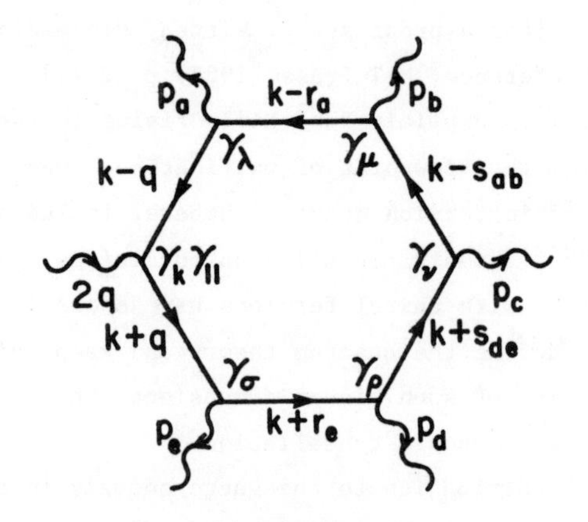

FIGURE S.3

Hexagon diagram

$$2q_{\kappa}V_{\kappa\lambda\mu\nu\rho\sigma}(p_a p_b p_c p_d p_e)$$

$$= 2^6\ X_6(ST)\ \varepsilon_{\lambda\mu\nu\rho\sigma\alpha\beta\gamma\delta\varepsilon} p_a^{\alpha} p_b^{\beta} p_c^{\gamma} p_d^{\delta} p_e^{\varepsilon} \tag{S.37}$$

with

$$X_6 = \frac{-1}{2^{10}\pi^5 5!} \tag{S.38}$$

and

$$(ST) = STr(\Lambda^a \Lambda^b \Lambda^c \Lambda^d \Lambda^e \Lambda^f) \tag{S.39}$$

is the symmetrized trace of generators in the fermion representation. We must decompose this trace into irreducible tensors formed from the defining representation

$$\begin{aligned} STr(\Lambda^a \Lambda^b \Lambda^c \Lambda^d \Lambda^e \Lambda^f) &= A_6 STr(\lambda^a \lambda^b \lambda^c \lambda^d \lambda^e \lambda^f) \\ &+ A_6^{4,2} S\big(tr(\lambda^a \lambda^b \lambda^c \lambda^d) tr(\lambda^e \lambda^f)\big) \\ &+ A_6^{3,3} S\big(tr(\lambda^a \lambda^b \lambda^c) tr(\lambda^d \lambda^e \lambda^f)\big) \\ &+ A_6^{2,2,2} S(tr(\lambda^a \lambda^b) tr(\lambda^c \lambda^d) tr(\lambda^e \lambda^f)\big). \end{aligned} \tag{S.40}$$

For SU(N) and k-rank antisymmetric tensors the relevant formulas for these hexagon anomalies are

$$A_6(N, k) = \sum_{p=0}^{k=1} (-1)^{k-p+1} (k - p)^5 \binom{N}{p} \tag{S.41}$$

$$A_6^{4,2}(N, k) = 15\ A_4(N - 2, k - 1) \tag{S.42}$$

where

$$A_4(N, k) = \sum_{p=0}^{k=1} (-1)^{k-p+1} (k - p)^3 \binom{N}{p} \tag{S.43}$$

$$A_6^{2,2,2}(N, k) = 15 \binom{N-6}{k-3} \tag{S.44}$$

$$A_6^{3,3}(N, k) = \frac{2}{3} A_6^{4,2}(N, k) + \frac{4}{3} A_6^{2,2,2}(N, k)\ . \tag{S.45}$$

For example:

$$A_6(N,\ 0) = 1 \tag{S.46}$$

$$A_6(N,\ 2) = (N - 32)\ . \tag{S.47}$$

The k = 2 representation of SU(N) behaves like the adjoint of O(N). Only the disconnected (non-leading) anomalies may be compensated by Chern-Simons terms. O(32) is the <u>only</u> nonexceptional group for which the adjoint has leading hexagon anomaly $A_6 = 0$.

One immediate point is that if we are in d = 10 with only left-handed, and no right-handed, Weyl spinors then, without adding Chern-Simons terms, there cannot be anomaly cancellation for any nontrivial representation of any gauge group G. This can be seen by going into the Cartan subalgebra (diagonal) with real eigenvalues λ_i and noticing that

$$\sum_i \lambda_i^6 > 0\ . \tag{S.48}$$

The only possible cancellation in these nonstring cases is between left and right helicities (in d = 4k + 2). Thus, all supersymmetric Yang-Mills theories in d = 10 (or any d = 4k + 2) are anomalous. Because of the zero-slope limit, it appeared in 1983 that open superstrings were anomalous although it was said[43] in a footnote: "The only possible loophole would be if the infinite sum over massive fermions gives a nonvanishing anomaly precisely cancelling the anomaly of the massless fermions. We regard this as extremely unlikely." For O(32), however, this is precisely what happens from this viewpoint.

We now discuss those combinations of the representations [k] (k-rank antisymmetric) of SU(N) which have, in d = 2n, no gauge, gravity, or mixed chiral anomalies. For d = 6, the first non-trivial solution[44] ("non-trivial" means able to give chirality in d = 4 after compactification) is for SU(6) where there is only one possibility. For higher SU(N) there is one solution for SU(7), two for SU(8,9), three for SU(10,11), and so on. In general, for d = 2n and SU(N) with only antisymmetric representations the number of such independent solutions is

$$[\tfrac{1}{2}(N - n - 1)] \qquad (S.49)$$

where [x] is the integer part of x.

There is a systematic procedure[45] for constructing these solutions: consider the superalgebra SU(N/M) with $(N - M) \geq (n + 1)$. Choose first a combination of $[k]^{(N-M)}$ which is real (n = even) or pure imaginary (n = odd). Now promote these Young tableaus $[k]^{(N-M)}$ to super Young tableaus $[k]^{(N/M)}$ defined by

$$[k]^{(N/M)} = \sum_{p=0}^{m} (-1)^p \left([k-p]^{(N)}, \{p\}^{(M)}\right) \qquad (S.50)$$

where $\{p\}^{(M)}$ is a symmetric p-rank representation of SU(M) and the sign $(-1)^p$ determines the helicity of the state. The combination arrived at has no chiral anomalies of gauge or gravitational type.

It is convenient to introduce the principal representations[46,47]

$$P_{(2n)}(N, k) = [k]^{(N/N-2k+1)} + (-1)^n [k-1]^{(N/N-2k+1)} \qquad (S.51)$$

and by taking

$$[\tfrac{1}{2}(n + 3)] \leq k \leq k_{max} = \begin{cases} [N/2] & n = \text{odd} \\ [\,1/2\,(N-1)] & n = \text{even} \end{cases} \qquad (S.52)$$

we obtain the full number $[\frac{1}{2}(N - n - 1)]$ of independent solutions as conjectured by others. Note, in particular, that the number of independent choices is small.

Now let us consider briefly the compactification to four dimensions of such non-string Kaluza-Klein theories.

We seek solutions of the Einstein-Yang-Mills equations in d = 2n with less than the maximal symmetry (M_{2n}). Actually, we shall discuss the manifolds of the form $M_4 \times S_{2n-4}$ (sphere).

Such classical solutions require that the energy-momentum tensor T_{MN} recieve contributions from a nonzero vacuum value of the gauge field e.g. a monopole configuration for $n = 3$ (S_2), an instanton configuration for $n = 4$ (S_4), etc. This drives the compactification. At the same time, and quite attractively, this precisely enables survival of chiral fermions on M_4.

We shall require classical stability[48] of the vacuum under small perturbations--surely a necessary condition for stability in the complete quantum theory.

Consider first the case $d = 6$ with manifold $M_4 \times S_2$ and monopole-induced compactification.[49] The solution sought is therefore of the form

$$g_{MN}dz^M dz^N = g_{mn}(x)dx^m\, dx^n + g_{\mu\nu}(y)dy^\mu\, dy^\nu \tag{S.53}$$

$$g_{\mu\nu}(y)dy^\mu\, dy^\nu = a^2(d\theta^2 + \sin^2\theta\, d\phi^2) \tag{S.54}$$

$$A_M dz^M = A_\mu(y)dy^\mu \tag{S.55}$$

$$= \frac{n}{2g}\,(\cos\theta \mp 1)d\phi \qquad \begin{matrix} 0 \leqslant \theta < \pi \\ 0 < \theta \leqslant \pi \end{matrix} \quad . \tag{S.56}$$

Here, n = monopole charge. It is worth pointing out that the monopole is not in the space: the field on S_2 is as if there were a monopole at its center, see Eq. (S.56).

To investigate which fields remain massless on M_4, we expand in monopole harmonics $Y_{q,LM}(\theta,\phi)$ which are eigenfunctions of $\underline{J}^2 = (\underline{L}^2 + \underline{q}^2)$ with $q = ny/2$ where y is the U(1) charge of the field considered. For example, a massless (in d = 6) scalar field is expanded as

$$\Phi(x, y) = \sum_{JM} \phi_{JM}(x) Y_{q,JM}(\theta, \phi) . \tag{S.57}$$

The only massless component (recall $J \geq |q|$) is for $q = 0$. If $n \neq 0$, this means $y = 0$. Recall that

$$\Box_z = \Box_x + \Box_y \tag{S.58}$$

so the eigenvalue of $\Box_y$ must be zero (s-wave, independent of θ and ϕ).

More interesting is a 4-component Weyl spinor Ψ_L in $d = 6$ with its expansion

$$\begin{pmatrix} \Psi_{L1} \\ \Psi_{L2} \\ \Psi_{L3} \\ \Psi_{L4} \end{pmatrix} = \sum_{JM} \begin{pmatrix} \Psi^1_{L,JM}(x)\ Y_{q-1/2,JM}(\theta, \phi) \\ \Psi^2_{L,JM}(x)\ Y_{-q+1/2,JM}(\theta, \phi) \\ \Psi^1_{R,JM}(x)\ Y_{q+1/2,JM}(\theta, \phi) \\ \Psi^2_{R,JM}(x)\ Y_{-q-1/2,JM}(\theta, \phi) \end{pmatrix} . \tag{S.59}$$

For a given q, consider $J = |q| - 1/2$. Only the upper two ($q > 0$) or lower two ($q < 0$) components are nonzero. The corresponding multiplicity is $(2J + 1) = 2|q| = |ny|$. Hence we obtain $|ny|$ Weyl spinors massless on M_4 with the same ($q > 0$) or opposite ($q < 0$) helicity compared to the helicity in $d = 6$.

Using this rule, we can easily analyze the anomaly-free combinations.

It turns out that these $d = 6$ examples are not very successful phenomenologically, so I will go immediately to instanton-induced compactification in $d = 8$ with manifold $M_4 \times S_4$. This is similar to the $d = 6$ case except that now we write[50]

$$A^a_\mu(y)dy^\mu = \frac{2}{g}\,\frac{\eta_{a\mu\nu}y^\nu}{y^2 + b^2}\,dy^\mu \tag{S.60}$$

where $\mu = 5, 6, 7, 8$ and $a = 1, 2, 3$ for the generators of SU(2)g. The full gauge group is $SU(N) \supset SU(2)g$. Again we take Weyl spinors in antisymmetric representations $[k]^{(N)}$ of SU(N). These contain only $t = 0$ and $t = 1/2$ representations of SU(2)g. Writing

$$S_4 = \frac{SO(5)}{SO(4)} = \frac{SO(5)}{SU(2)_A \times SU(2)_B} \tag{S.61}$$

the invariance group of the instanton is $SU(2)_{B+g}$ and we must analyze an 8-component chiral spinor Ψ_L in $d = 8$ by expansion into irreducible representations of $SO(5) \supset SU(2)_A \times SU(2)_{B+g}$. We find, by a generalization of the procedure done explicitly above for $d = 6$, that

$$[k]^{(N)}_L \to \underset{SO(5)\ \text{singlet}}{[k-1]^{(N-2)}_L} \quad . \tag{S.62}$$

This is a very simple rule. In fact, it extends to complete superalgebra tableaus

$$[k]^{(N/M)}_L \to [k-1]^{(n-2/M)} \quad . \tag{S.63}$$

Now we ask: can we find completely anomaly-free fermions in $d = 8$ that give interesting chiral fermions on M_4? This is not a trivial question since the anomaly conditions are so restrictive. Nevertheless, the answer is affirmative.[46,47]

To arrive at SU(N) in $d = 4$ we start from $SU(N + 2)$ in $d = 8$. Examples are:

$$N = 11 \qquad [5]^{(13/3)}_L + 2[4]^{(13/5)}_L + 3[3]^{(13/7)}_L$$

$$\to \qquad (11^4 + \overline{11}^3 + \overline{11}^2 + \overline{11}) \text{ of } SU(11) \tag{S.64}$$

which is a three-family model.[51]

$$N = 9 \qquad [4]_L^{(11/3)} + 3[3]_L^{(11/5)}$$

$$\rightarrow \quad 9^3 + 9(\bar{9}) \text{ of } SU(9). \qquad (S.65)$$

This also has three families[52] since on reduction to SU(5)

$SU(9) \rightarrow SU(5)$

$$[3]^{(9)} \quad : \quad 84 = \overline{10} + 4(10) + 6(5) + 4(1) \qquad (S.66)$$

$$9[8]^{(9)} \quad : \quad 9(\bar{9}) = 9(\bar{5} + 4(1)) \qquad (S.67)$$

so that adding, and dropping real representations that are expected to develop superheavy Dirac masses, leaves

$$3(10 + \bar{5}) \text{ of } SU(5). \qquad (S.68)$$

The SO(5) isometry group is passive: all the massless chiral fermions on M_4 are SO(5) singlets. SO(5) nonsinglets are superheavy, near the Planck scale.

Finally, concerning nonstring Kaluza-Klein theory we should remark that the quantum fluctuations are expected to be large (the theory is non-renormalizable), and the only features of the classical solutions expected to survive are those massless particles protected by an unbroken symmetry like gauge vectors and chiral fermions. We should not take seriously any massless scalars which, since unprotected here by supersymmetry, will become superheavy due to quantum corrections.

Rather, we have shown how to obtain phenomenologically an acceptable mass spectrum for spin-1 and spin-1/2 in the effective theory on M_4.

Two objections to this approach are

1) the theory is non-unified; the gauge coupling is independent of the gravitation coupling;

2) the quantum corrections are out of control.

Both objections are overcome by superstrings, to which we now turn.

S.5 ANOMALY CANCELLATION

If we consider the naive zero-slope limit of the open superstring, the resultant point field theory has chiral fermions in the adjoint of the gauge group G and, as mentioned above, is anomalous for all G.[42,43] In an already classic paper, Green and Schwarz [Ref. 53, Reprint No. 2] found a loophole in this argument for G = 0(32). Let us therefore focus on the anomaly cancellation in this case.

For the adjoint of 0(N), put k = 2 in the SU(N) formulae already given to find

$$A_6(N,\ 2) = N - 32 \tag{S.69}$$

$$A_6^{4,2}(N,\ 2) = 15 \tag{S.70}$$

$$A_6^{2,2,2}(N,\ 2) = 0 \tag{S.71}$$

$$A_6^{3,3}(N,\ 2) = 10\ . \tag{S.72}$$

But now we must realize that $\mathrm{tr}(\lambda\lambda\lambda) = 0$ in 0(N) if $N \neq 6$ and hence we have

$$\mathrm{STr}(\lambda\lambda\lambda\lambda\lambda\lambda) = (N - 32)\mathrm{Str}(\lambda\lambda\lambda\lambda\lambda\lambda) + 15S\big(\mathrm{tr}(\lambda\lambda\lambda\lambda)\mathrm{tr}(\lambda\lambda)\big) \tag{S.73}$$

for Λ of 0(N) in the adjoint representation. We see that 0(32) supersymmetric Yang-Mills theory is not enough.

The field theory limit of the superstring has the bosonic terms

$$S = \int d^{10}x \sqrt{g} \left[-\frac{1}{2\kappa^2} R - \frac{1}{\kappa^2} \frac{1}{\phi^2} \partial_\mu \phi \partial_\mu \phi \right.$$

$$\left. - \frac{1}{4g^2 \phi} F^2 - \frac{3\kappa^2}{2g^4} \frac{1}{\phi^2} H^2 \right] \tag{S.74}$$

where

$$F = dA + A^2 \tag{S.75}$$

$$H = dB - \omega^o_{3Y} - \omega^o_{3L} \cdot \tag{S.76}$$

Note that the ten dimensional spacetime dictates that the dimensions are $[g] \sim L^3$, $[\kappa] \sim L^4$ so

$$\frac{\kappa}{g^2} \sim \frac{1}{L^2} \sim \frac{1}{\alpha'} \sim T \tag{S.77}$$

where α' = Regge slope and T = string tension. In fact, g^2 is absorbable into $\langle\phi\rangle$ leaving no dimensionless parameters at all, just T.

The presence of the Yang-Mills Chern-Simons form ω^o_{3Y} in Eq. (51) was discovered in N = 1 supergravity plus N = 1 super-Yang-Mills in d = 10 by two different groups[36,37] in 1982. One has

$$\omega^o_{3Y} = tr(AF - \frac{1}{3} A^3) \tag{S.78}$$

$$d\omega^o_{3Y} = tr\ F^2 \tag{S.79}$$

$$\omega^o_{3L} = tr(\omega R - \frac{1}{3} \omega^3) \tag{S.80}$$

$$d\omega^o_{3L} = tr\ R^2 \tag{S.81}$$

where the last two will be important in the gravitational anomalies.

Although $B_{\mu\nu}$ is in the supergravity multiplet, and hence a "gauge singlet," it does transform under Yang-Mills gauge transformation as dictated by local supergravity. Under

$$\delta A = d\Lambda + [A, \Lambda] \tag{S.82}$$

where Λ = gauge function, then

$$\delta\omega^{o}_{3Y} = d\omega^{1}_{2Y} . \tag{S.83}$$

Corresponding to

$$\mathrm{TrF}^6 = (N - 32)\mathrm{trF}^6 + 15\ \mathrm{trF}^4\mathrm{trF}^2 \tag{S.84}$$

the consistent anomaly is

$$G_1 \sim \int (\tfrac{1}{3}\ \omega^{1}_{2Y}\mathrm{trF}^4 + \tfrac{2}{3}\ \omega^{1}_{6Y}\mathrm{trF}^2) . \tag{S.85}$$

If we add to the effective action

$$S_1 = \int (B\mathrm{trF}^4 + \tfrac{2}{3}\ \omega^{o}_{3Y}\omega^{o}_{7Y}) \tag{S.86}$$

with $\delta B = (-\ \omega^{1}_{2Y} - \omega^{1}_{2L})$ we find that

$$\delta S_1 = \int \delta B\ \mathrm{trF}^4 - \tfrac{2}{3}\left((d\omega^{1}_{2Y})\omega^{o}_{7Y} - \tfrac{2}{3}\ \omega^{o}_{3Y}(d\omega^{1}_{6Y})\right) \tag{S.87}$$

$$= \int (\delta B\ \mathrm{trF}^4 + \tfrac{2}{3}\ \omega^{1}_{2Y}\mathrm{trF}^4 - \tfrac{2}{3}\ \omega^{1}_{6Y}\mathrm{trF}^2) \tag{S.88}$$

$$= -\ G_1 - \int \omega^{1}_{2L}\mathrm{trF}^4 . \tag{S.89}$$

The miracle continues for O(32) since one finds that one $(3/2)_L$ gravitino plus one $(1/2)_R$--the content of the supergravity multiplet--plus n $(1/2)_L$ gives a gravitational anomaly[54] of the form

$$\frac{n - 496}{7560}\ \mathrm{trR}^6 + [\tfrac{1}{8} + \frac{n - 496}{5760}]\ \mathrm{trR}^4\ \mathrm{trR}^2$$

$$+ [\tfrac{1}{32} + \frac{n - 496}{13824}]\ (\mathrm{trR}^2)^3 \tag{S.90}$$

so we must have n = 496 to cancel the leading gravitational anomaly = dimension O(32). By adding

$$S_2 = - \int \left(\frac{1}{32} B(trR^2)^2 + \frac{1}{8} B\, trR^4 + \frac{1}{12} \omega^o_{3L}\omega^o_{7L}\right) \quad (S.91)$$

and using $\delta\omega^o_{3L} = - d\omega^1_{2L}$, $\delta\omega^o_{7L} = - d\omega^1_{6L}$ we find

$$\delta S_2 = - G_2 + \int \left(\frac{1}{32} \omega^1_{2Y}(trR^2)^2 + \frac{1}{8} \omega^1_{2Y} trR^4\right) \quad (S.92)$$

where

$$G_2 = - \int \left(\frac{1}{32} \omega^1_{2L}(trR^2)^2 + \frac{1}{24} \omega^1_{2L} trR^4 + \frac{1}{12} \omega^1_{6L} trR^2\right) \quad (S.93)$$

is the consistent anomaly corresponding to Eq. (S.90).

The mixed gauge-gravity anomalies require two further numerical coincidences, making a total of four, in order that O(32) be an anomaly-free superstring.

For O(32) the mixed anomaly is

$$- trR^2\left[trF^4 + \frac{\hat{A}}{8} (trF^2)^2\right] + \hat{B} trF^2\left[\frac{1}{8} trR^4 + \frac{5}{32} (trR)^2\right] \quad (S.94)$$

where $\hat{A} = \hat{B} = 1$. For general O(N) one would have in this formula

$$\hat{A} = \frac{24}{N - 8} \text{ and } \hat{B} = \frac{N - 2}{30} . \quad (S.95)$$

The corresponding consistent anomaly is

$$G_3 = \int \left(\frac{1}{3} \omega^1_{2L} trF^4 + \frac{2}{3} \omega^1_{6Y} trR^2 + \frac{\hat{A}}{24} \omega^1_{2L}(trF^2)^2 + \frac{\hat{A}}{12} \omega^1_{2Y} trR^2 trF^2\right.$$
$$- \frac{\hat{B}}{24} \omega^1_{2Y} trR^4 - \frac{\hat{B}}{12} \omega^1_{6L} trF^2$$
$$\left.- \frac{5\hat{B}}{96} \omega^1_{2Y}(trR^2)^2 + \frac{5\hat{B}}{48} \omega^1_{2L} trR^2 trF^2\right). \quad (S.96)$$

We now add to the effective action

$$S_3 = \int \left(\frac{K}{8} \mathrm{BtrF}^2 \mathrm{trR}^2 + \frac{L}{48} \overset{o}{\omega}_{3L} \overset{o}{\omega}_{3Y} \mathrm{trR}^2 \right.$$

$$\left. - \frac{M}{24} \overset{o}{\omega}_{3Y} \overset{o}{\omega}_{3L} \mathrm{trF}^2 - \frac{2N}{3} \overset{o}{\omega}_{3L} \overset{o}{\omega}_{7Y} + \frac{P}{12} \overset{o}{\omega}_{3Y} \overset{o}{\omega}_{7L} \right) \qquad (S.97)$$

where K, L, M, N, P are initially at our disposal. Now we wish to arrange that

$$\delta S_3 = - G_3 - (\delta S_1 + G_1) - (\delta S_2 + G_2) \qquad (S.98)$$

and by looking at coefficients of particular terms we find:

$$\omega^1_{2L}(\mathrm{trF}^2)^2: \qquad M = \hat{A} \qquad (S.99)$$

$$\omega^1_{2Y}\mathrm{trF}^2\mathrm{trR}^2: \qquad M = 3 - 2\hat{A} \text{ hence } \hat{A} = 1 \qquad (S.100)$$

$$\omega^1_{2L}\mathrm{trF}^2\mathrm{trR}^2: \qquad L = 6 - 5\hat{B} \qquad (S.101)$$

$$\omega^1_{2Y}(\mathrm{trR}^2)^2: \qquad 2L = 5\hat{B} - 3 \text{ and hence } \hat{B} = 1. \qquad (S.102)$$

Hence we soon find that $\hat{A} = \hat{B} = 1$ both requiring O(32) uniquely.

A similar cancellation of all gauge and gravitational anomalies occurs if the group E(8) x E(8) is used. Note that E(8) x E(8) has the same rank (16) and dimension (496) as O(32). However, one cannot use the Chan-Paton procedure to introduce E(8) x E(8) on open superstrings because the matrix form of the E(8) generators does not have the requisite factorization properties. Construction of the closed heterotic string with E(8) x E(8) gauge group will be described later.

In the O(32) open superstring theory consider the amplitude with six external O(32) gauge bosons. The three contributions are depicted in Fig. S.4. The nonplanar orientable diagram (the final figure of Fig. S.4) is non-anomalous because the hexagon anomaly corresponding to the first figure in Fig. S.5 is cancelled by the massless tensor

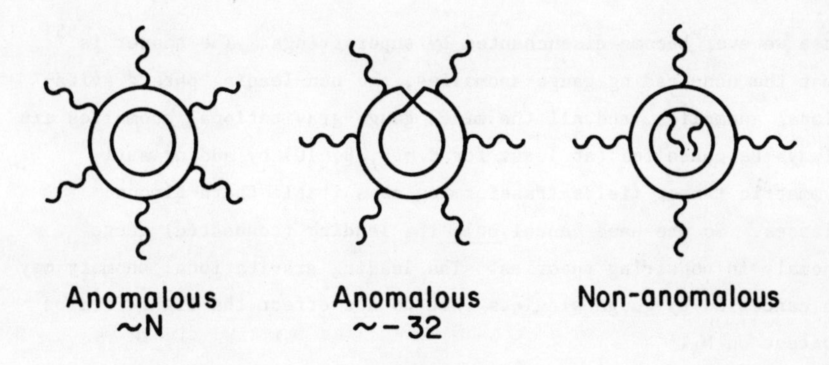

FIGURE S.4
Contributions to six-gauge-boson amplitude

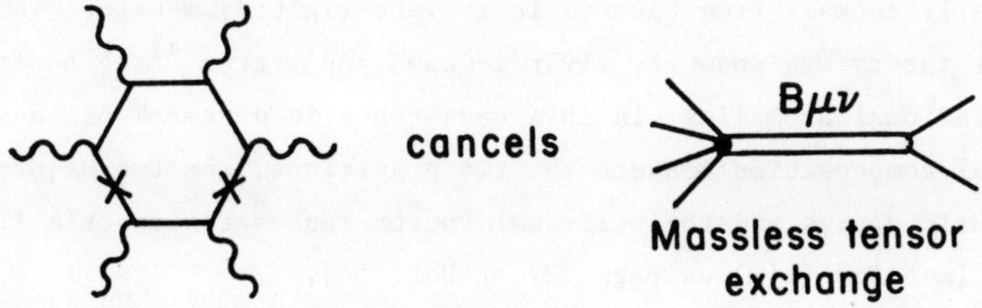

cancels

$B\mu\nu$

Massless tensor exchange

FIGURE S.5
Mechanism of anomaly cancellation

($B_{\mu\nu}$) exchange shown also in Fig. S.5. The $B_{\mu\nu}$ is in the same supermultiplet as the graviton and is a massless closed-string state. This diagram is really a "tree" rather than a "one-loop" diagram. The appearance of Planck's constant is complicated by the fact that closed strings have an intrinsic "h". Hence, the prejudice from quantum field theory that quantum corrections are loop effects is not true for the open superstring loop expansion.

We may ask how the mechanism of cancelling non-leading anomalies by Chern-Simons terms impacts on nonstring Kaluza-Klein theory, in case we ever become disenchanted by superstrings. The answer is[55] that the non-leading gauge anomalies, the non-leading pure gravitational anomalies, and all the mixed gauge-gravitational anomalies can always be cancelled (at least for d = 6, 8, 10) by adding antisymmetric tensor fields transforming as suitable Chern-Simons classes. So one need cancel only the leading (connected) gauge anomaly in nonstring theories. The leading gravitational anomaly may be cancelled by gauge singlets that do not effect the light fermion content on M_4.

For the open superstring the only allowed gauge group is O(32). The N = 2 closed superstrings[39] have no gauge group, but one needs to check gravitational anomalies. The N = 2 nonchiral theory is trivially anomaly free because it is left-right symmetric. The N = 2 chiral theory was shown by Alvarez-Gaumé and Witten[54] to be free of gravitational anomalies; in this case there is a remarkable and nontrivial compensation between the two gravitinos, the two Majorana-Weyl spin-half states and the self-dual fourth-rank antisymmetric field A_{MNPQ} [see Eq. (119) on page 327 of Ref. 54].

S.6 ONE-LOOP FINITENESS; MODULAR INVARIANCE FOR CLOSED STRINGS

We have discussed so far three superstring theories: the O(32) open superstring and the N = 2 nonchiral and chiral closed superstrings. The discussion of finiteness is most complicated for the first theory (open superstrings) because individual amplitudes are infinite, and finiteness is achieved only by cancellation between different amplitudes. For the closed superstrings theories there is just one diagram for each number of loops (the strings are orientable), and finiteness is a consequence of modular invariance. This last property is important to the heterotic string constructed in the next subsection.

Our plan here is to discuss one loop finiteness first for the O(32) open superstring, then move on to the one-loop torus diagram of the N = 2 closed superstrings. We shall indicate explicitly (it has been proved rigorously) that these theories are one-loop finite for up to five external ground-states. Comparably definitive results for more than one loop are unavailable at present. Nevertheless, we shall argue at the end of this subsection that it is much more likely that these theories are finite than, say, $N \leq 8$ supergravity theories in $D = 4$ or any supergravity theories in $D > 4$.

We shall use light-cone gauge. For the bosonic modes the action in light-cone variables is $[\alpha = 1, 2; i = 1, 2, \ldots, (D - 2)]$

$$S = \frac{1}{4\pi\alpha'} \int_0^\pi d\sigma \int d\tau \, \partial_\alpha X^i \partial^\alpha X^i . \tag{S.103}$$

The corresponding equations of motion are

$$\left(\frac{\partial^2}{\partial\sigma^2} - \frac{\partial^2}{\partial\tau}\right) X^i (\sigma, \tau) = 0 \tag{S.104}$$

with boundary conditions for an open string

$$\frac{\partial}{\partial\sigma} X^i = 0 \text{ at } \sigma = 0 \text{ and } \sigma = \pi. \tag{S.105}$$

The general solution is

$$X^i(\sigma, \tau) = x^i + p^i\tau + i \sum_{n\neq 0} \frac{1}{n} \alpha_n^i \cos n\sigma \, e^{-in\tau} . \tag{S.106}$$

Upon quantization, one imposes

$$[\alpha_m^i, \alpha_n^j] = m\delta_{m+n,o}\delta^{ij} . \tag{S.107}$$

Or defining, for $n \geq 1$, $\alpha_n^i = \sqrt{n}\; a_n^i$ and $\alpha_{-n}^i = \sqrt{n}\, a_n^{i+}$

$$[a^i_m, a^{j+}_n] = \delta_{mn}\delta^{ij} \quad . \tag{S.108}$$

Constraints give a mass-shell condition

$$\alpha'(\text{mass})^2 = N - 1 \tag{S.109}$$

$$N = \sum_{n=1}^{\infty} \alpha^i_{-n}\alpha^i_n = \sum_{n=1}^{\infty} n a^{i+}_n a^i_n \quad . \tag{S.110}$$

For superstrings, the $X^i(\sigma, \tau)$ are joined by spinor variables S^a which satisfy the subsidiary conditions in light-cone gauge

$$(h^-)^{ab}S^b = (\gamma^+)^{ab}S^b = 0 \tag{S.111}$$

where $h^- = \frac{1}{2}(1 - \gamma^{11})$. As discussed in Ref. 56, these conditions together with a Majorana condition and the Dirac equation reduce S^a to only <u>eight</u> real components. Ab initio there are 64 complex components--2 spinors on the 2-spacetime of the world sheet and 32-spinors on the 10-spacetime manifold; these 128 real components are halved four times by the Weyl condition, the Majorana condition, the light-cone (γ^+) constraint and finally the Dirac equation. At the endpoint $\sigma = 0$ we expand

$$S^a = \sum_{n=-\infty}^{\infty} S^a_n e^{-in\tau} \tag{S.112}$$

and quantize by

$$\{S^a_m, \bar{S}^b_n\} = (\gamma^+h^-)^{ab}\delta_{m+n,o} \tag{S.113}$$

$$[\alpha^i_m, S^a_n] = 0. \tag{S.114}$$

The superstring mass shell condition is

$$\alpha'(\text{mass})^2 = N = \sum_{n=1}^{\infty} (n a^{i+}_n a^i_n + \frac{n}{2} \bar{S}_{-n} \gamma^- S_n) \quad . \tag{S.115}$$

The ground state of the open superstring is comprised of 16 massless states: 8 form a vector

$$|i\rangle \qquad 1 \leq i \leq 8 \tag{S.116}$$

$$\langle i|j\rangle = \delta_{ij} \quad . \tag{S.117}$$

Since $[S_o, N] = 0$ we may act with S_o to obtain a massless Majorana-Weyl spinor

$$|a\rangle = \frac{i}{8} (\gamma_i S_o)^a \; |i\rangle \quad 1 \leq a \leq 8 \; . \tag{S.118}$$

The S^a_o satisfy

$$S^a_o \, \bar{S}^b_o = \frac{1}{2} \left(\gamma^+ h^-\right)^{ab} + \frac{1}{4} \left(\gamma^{ij+} h^-\right)^{ab} R^{ij}_o \tag{S.119}$$

where γ^{ij+} is the fully antisymmetrized produce of $\gamma^i \gamma^j \gamma^+$ and

$$R^{ij}_o = \frac{1}{8} \bar{S}_o \gamma^{ij-} S_o \tag{S.120}$$

are generators of O(8) in the transverse space.

The excited massive superstring states may be assembled by acting on the ground state with the creation operators α^i_{-n} and S^a_{-n}. For example, at $N = 1$, there are 128 boson states

$$\alpha^i_{-1}|j\rangle \quad \text{and} \quad S^a_{-1}|b\rangle \tag{S.121}$$

and 128 fermion states

$$\alpha^i_{-1}|a\rangle \quad \text{and} \quad S^a_{-1}|i\rangle \; . \tag{S.122}$$

The reader should verify there are 1152 of each at $N = 2$.

It is sometimes useful in computations to define

$$\psi_n^a = \frac{1}{2^{3/4}} \gamma^o \gamma^- S_n^a = \frac{1}{2^{1/4}} S_n^a \tag{S.123}$$

satisfying the simple anticommutation relations

$$\{\psi_n^a, \psi_n^{b+}\} = S^{ab} . \tag{S.124}$$

Since

$$\frac{1}{2} \overline{S}_{-n}\gamma^- S_n = \psi_n^+ \psi_n \tag{S.125}$$

it follows that e.g.

$$\text{tr } w^{1/2n\overline{S}_{-n}\gamma^- S_n} = \text{tr } w^{n\psi_n^+\psi_n} = (1 - w^n)^8 \tag{S.126}$$

$$\text{tr } w^{N_F} = \prod_{n=1}^{\infty} (1 - w^n)^8 \tag{S.127}$$

where

$$N_F = \sum_{n=1}^{\infty} \frac{1}{2} n\overline{S}_{-n}\gamma^- S_n . \tag{S.128}$$

In order to calculate superstring graphs with external massless gauge bosons we shall need the vertex for emitting such a state. In the bosonic string, the massless vector state is

$$|i\rangle = \alpha_{-1}^i |0\rangle . \tag{S.129}$$

If $\zeta^\mu(k)$ is the polarization vector with $k\cdot k = k\cdot\zeta = 0$ then the appropriate vertex is

$$V = g\zeta\cdot P(\tau)e^{ik\cdot X(\tau)} \tag{S.130}$$

$$P^\mu(\tau) = \frac{\partial}{\partial\tau} X^\mu(\tau)\Big|_{\sigma=0} . \tag{S.131}$$

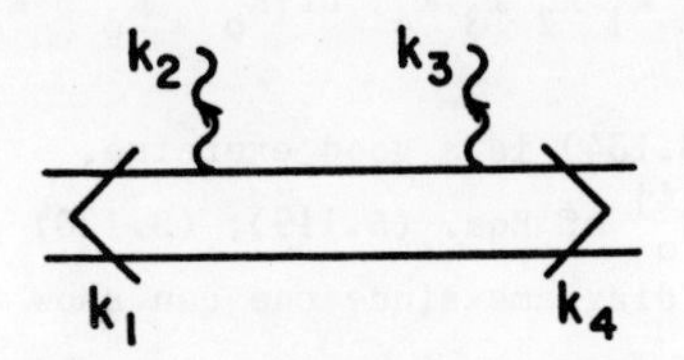

FIGURE S.6

Open superstring tree amplitude

In light-cone gauge, with $k^+ = 0$ and $\zeta^+ = 0$ this is just $g\zeta^i P^i V_o(k)$. For the superstring the relevant vertex is

$$V(k, \zeta) = g\zeta^i(P^i + k^j R^{ij})V_o(k) \qquad (S.132)$$

where $R^{ij} = \frac{1}{8}\bar{S}\gamma^{ij-}S$. Here there is no ordering ambiguity because $k\cdot k = k\cdot\zeta = 0$. Although light-cone gauge itself is not restrictive, the fact that we are putting $k_+ = 0$ means that we may deal only with $M \leqslant 10$ external lines, because for $M > 11$ there are ten independent ten-momenta which may be contracted with the tenth rank antisymmetric tensor to form a lorentz invariant quantity--a quantity which vanishes generically if $k_+ = 0$.

For the open superstring amplitude of Fig. S.6 the S_o modes are very crucial. The relevant matrix element for four external massless gauge bosons is

$$\langle k_1, \zeta_1|V(k_2, \zeta_2)\ \Delta\ V(k_3, \zeta_3)|k_4, \zeta_4\rangle \qquad (S.133)$$

with $V(k, \zeta)$ given by Eq. (S.132). The result is the crossing symmetric amplitude

$$A_4 = K_4\,\frac{\Gamma(-\frac{1}{2}s)\Gamma(-\frac{1}{2}t)}{\Gamma(-\frac{1}{2}s - \frac{1}{2}t)} \qquad (S.134)$$

with the kinematic factor given by

$$K_4 = \zeta_1^{i_1} \zeta_2^{i_2} \zeta_3^{i_3} \zeta_4^{i_4} \, k_1^{j_1} k_2^{j_2} k_3^{j_3} k_4^{j_4} \, \mathrm{tr}\left(R_o^{i_1 j_1} R_o^{i_2 j_2} R_o^{i_3 j_3} R_o^{i_4 j_4}\right) . \qquad (S.135)$$

Derivation of Eq. (S.134) is a good exercise.

The operators R_o^{ij} of Eqs. (S.119), (S.120) play an important role in superstring loop diagrams since one can show that in the S_o^a space

$$\mathrm{Tr}(1) = \mathrm{Tr}(R_o) = \mathrm{Tr}(R_o R_o) = \mathrm{Tr}(R_o R_o R_o) = 0. \qquad (S.136)$$

These vanishing traces imply immediately the superstring renormalization theorems that the loop diagrams with two (Fig. S.7) or three (Fig. S.8) external gauge vectors vanish. As indicated in the figures, this holds for both the annulus and Moebius strip diagrams.

On the other hand, since $\mathrm{tr}(R_o R_o R_o R_o)$ does not vanish, the diagrams with four external gauge bosons do not vanish (Fig. S.9). Actually these diagrams diverge due to soft dilaton emission into the vacuum.

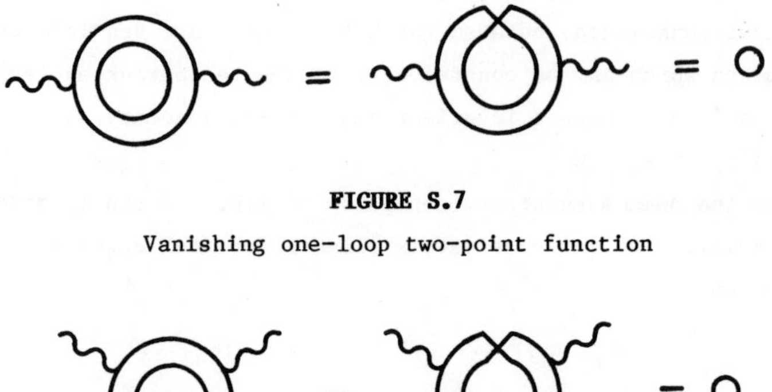

= = 0

FIGURE S.7

Vanishing one-loop two-point function

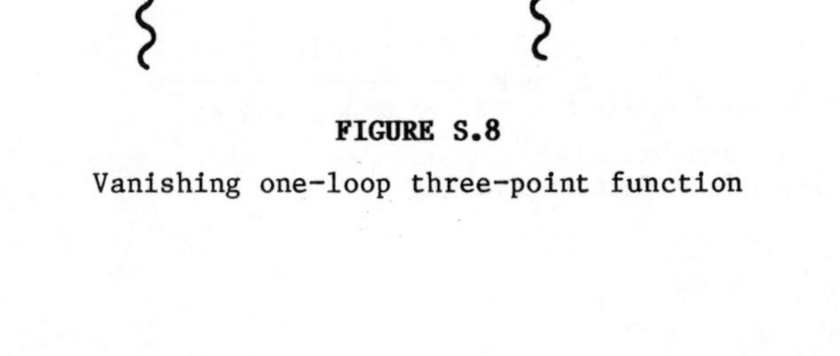

= = 0

FIGURE S.8

Vanishing one-loop three-point function

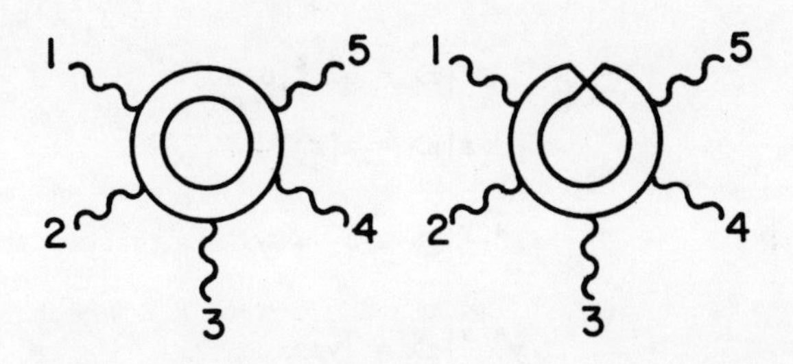

FIGURE S.9
Pentagon diagrams

In "Dual Resonance Models" there was no discussion of string loop diagrams[57] and thus we shall give some of the basic details here. Once one has computed a one-loop graph for the bosonic string, the essential stringiness of the loop has been done, and one-loop superstring graphs are only slightly different.

Therefore we begin by considering the bosonic string d = 26 with M external ground state tachyons emitted from the outer boundary of an annular string loop. We must compute the amplitude

$$\int d^{26}p \; \mathrm{tr}\left(\Delta V(k_1)\Delta V(k_2)\cdots\Delta V(k_M)\right) \tag{S.137}$$

and hence we need the trace

$$\mathrm{Tr}\left[x_1^N V(k_1) x_2^N K V(k_2)\cdots x_M^N V(k_M)\right) \tag{S.138}$$

where

$$V(k) = g{:}\; e^{ik\cdot X(0)}{:} \tag{S.139a}$$

$$N = \sum_{n=1}^{\infty} n a_n^{i+} a_n^i \; . \tag{S.139b}$$

To evaluate Eq. (S.138) one uses coherent states[58] which are eigenstates of the annihilation operator. The basic formulas are

$$|z\rangle = e^{a^+ z}|0\rangle \tag{S.140}$$

$$a|z\rangle = z|z\rangle \tag{S.141}$$

$$e^{a^+ y}|z\rangle = |z + y\rangle \tag{S.142}$$

$$y^{a^+ a}|z\rangle = |yz\rangle \tag{S.143}$$

$$1 = \frac{1}{\pi}\int d^2\lambda\, e^{-|\lambda|^2}|\lambda\rangle\langle\lambda| \tag{S.144}$$

$$\mathrm{tr}A = \frac{1}{\pi}\int d^2\lambda\, e^{-|\lambda|^2}\langle\lambda|A|\lambda\rangle \tag{S.145}$$

where a, y, λ are complex numbers. Hence the trace of Eq. (S.138) can be written

$$g^M \prod_{n=1}^{\infty}\prod_{i=1}^{D-2} T_{ni} \tag{S.146}$$

where

$$T_{ni} = \int \frac{d^2\lambda}{\pi} e^{-|\lambda|^2}\langle\lambda|x_1^{na^+a}\exp\left(\frac{k_1^i a^+}{\sqrt{n}}\right)\exp\left(-\frac{k_1^i a}{\sqrt{n}}\right)x_2^{na^+a}\cdots$$

$$\cdots x_M^{na^+a}\exp\left(\frac{k_M^i a^+}{\sqrt{n}}\right)\exp\left(-\frac{k_M^i a}{\sqrt{n}}\right)|\lambda\rangle \tag{S.147}$$

where $a = a_n^i$. Now define $q_I = k_I/\sqrt{n}$, $u_I = x_I^n$ and

$$O_I = u_I^{a^+a}e^{q_I a^+}e^{-q_I a} \tag{S.148}$$

then

$$O_I|z\rangle = e^{-q_I z}|u_I(z + q_I)\rangle \tag{S.149}$$

and

$$O_L O_{L+1} \cdots O_M |z\rangle = e^{-\beta_L z} |z_L\rangle \tag{S.150}$$

defines z_L and β_L. These satisfy the recursion relations

$$z_L = u_L(z_{L+1} + q_L) \tag{S.151}$$

$$\beta_L = q_L z_{L+1} + \beta_{L+1} \ . \tag{S.152}$$

In particular, defining

$$z_1 = Wz + C \tag{S.153}$$

$$\beta_1 = - Bz - D \tag{S.154}$$

the required trace is

$$\frac{1}{\pi} \int d^2z \, e^{-|z|^2} \, e^{-\beta_1} \langle z|z_1\rangle$$

$$= \frac{1}{\pi} \int d^2z \exp(-|z|^2 - \beta_1 + z^* z_1) \tag{S.155}$$

$$= \frac{1}{\pi} \int d^2z \exp[(1 - W)|z|^2 + Bz + Cz^* + D]. \tag{S.156}$$

Putting $z = x + iy$ the Gaussian integrals result in

$$\frac{1}{1 - W} \exp\left(\frac{- D(W - 1) + BC}{1 - W}\right)$$

$$= \frac{1}{1 - W} \exp\left(\frac{- \sum_{I,J=1}^{M} q_I^i q_J^i C_{IJ}}{1 - W}\right) \tag{S.157}$$

where

$$C_{IJ} = \begin{cases} (u_1 u_2 \cdots u_I)(u_{J+1} \cdots u_M) & (I < J) \\ u_{J+1} u_{j+2} \cdots u_I & (I > J) \ . \end{cases} \tag{S.158}$$

Here we used the solutions of Eqs. (S.151) and (S.152), namely

$$W = \prod_{I=1}^{M} u_I \tag{S.159}$$

$$C = \sum_{I=1}^{M} q_I(u_1 u_2 \cdots u_I) \tag{S.160}$$

$$B = -\sum_{k=1}^{M} q_k \sum_{I=k+1}^{M+1} u_I \tag{S.161}$$

$$D = -\sum_{1 \leq K < J \leq M} q_K q_J (u_{K+1} \cdots u_J) \tag{S.162}$$

with $u_{M+1} \equiv 1$. Taking the product over modes and dimensions our trace becomes

$$g^M \left(f(w)\right)^{2-D} \prod_{n=1}^{\infty} \prod_{I<J} \exp\left(\frac{-k_I \cdot k_J \left(C_{JI}^n + (w/C_{JI})^n - 2w^n\right)}{n(1 - w^n)}\right) \tag{S.163}$$

where $w = x_1 x_2 \cdots x_m$ and

$$f(w) = \prod_{p=1}^{\infty} (1 - w^p). \tag{S.164}$$

We may further massage Eq. (S.163) by doing the logarithmic sums over n inside the exponent giving

$$g^M f(w)^{2-D} \prod_{I<J} \left[\prod_{p=1}^{\infty} \frac{(1 - w^{p-1} C_{JI})(1 - w^p/C_{JI})}{(1 - w^p)^2} \right]^{k_I \cdot k_j} . \tag{S.165}$$

The momentum integral is

$$\int d^D p \prod_{i=1}^{M} x_I^{p_I^2/2} \tag{S.166}$$

where

$$p_I = p - \sum_{J=1}^{I-1} k_J = p + \sum_{J=1}^{M} k_J \quad . \tag{S.167}$$

The integrand may be written

$$\exp\left[\sum_{I=1}^{M} \ln x_I \left(\frac{1}{2} p^2 + p \sum_{J=1}^{M} k_J\right) - \frac{1}{2} \sum_{I<J} k_I \cdot k_J \ln C_{JI}\right] \tag{S.168}$$

and gaussian integration then gives, for Eq. (S.166),

$$\left(- \frac{2\pi}{\ln w}\right)^{D/2} \prod_{I<J} \left[C_{IJ}^{-1/2} \exp\left(\frac{\ln^2 C_{IJ}}{2 \ln w}\right)\right]^{k_I \cdot k_J} \quad . \tag{S.169}$$

Incidentally, the gaussian integral we are using is just

$$\int_{-\infty}^{\infty} dx \; e^{-a^2 x^2 \pm bx} = \frac{\sqrt{\pi}}{a} \exp\left(\frac{b^2}{4a^2}\right) \quad . \tag{S.170}$$

Collecting results, the bosonic annulus with M external spinless tachyons has amplitude

$$A_M = g^M \int \prod_{I=1}^{M} dx_I \; w^{-2} f(w)^{-24} \left(- \frac{2\pi}{\ln w}\right)^{13} \prod_{I<J} (\psi_{IJ})^{k_I \cdot k_J} \tag{S.171}$$

where

$$\psi_{IJ} = \psi(C_{JI}, w) \tag{S.172}$$

$$\psi(x, w) = \frac{1}{\sqrt{x}} \exp\left(\frac{\ln^2 x}{2 \ln w}\right) \prod_{n=1}^{\infty} \frac{(1 - w^{n-1} x)(1 - w^n/x)}{(1 - w^n)^2} \quad . \tag{S.173}$$

It is now useful to transform to the disc variables

$$\nu_I = \frac{\ln(x_1 x_2 \cdots x_I)}{\ln w} \tag{S.174}$$

$$q = \exp\left(\frac{2\pi^2}{\ln w}\right) \tag{S.175}$$

so

$$d\nu_I = \frac{1}{\ln w} \sum_{J=1}^{I} \frac{dx_J}{x_J} + \frac{\ln(x_1 \cdots x_J)}{2\pi^2} \frac{dq}{q} \tag{S.176}$$

$$\frac{dq}{q} = - \frac{2\pi^2}{(\ln w)^2} \sum_{I=1}^{M} \frac{dx_I}{x_I} \tag{S.177}$$

giving

$$\prod_{I=1}^{M} dx_I = \frac{1}{2\pi^2} w(-\ln w)^{M+1} \frac{dq}{q} \prod_{I=1}^{M-1} d\nu_I \theta(\nu_{I+1} - \nu_J) \quad . \tag{S.178}$$

In these disc variables the amplitude looks like

$$A_M = g^M \int_0^1 \prod_{I=1}^{M-1} d\nu_I \ \theta(\nu_{I+1} - \nu_I)$$

$$\int \frac{dq}{q^3} \left(- \frac{2\pi^2}{\ln q}\right)^M f(q^2)^{-24} \prod_{I<J} (\psi_{IJ})^{k_I \cdot k_J} \quad . \tag{S.179}$$

As $q \to 0$, the hole at the center of the annulus shrinks to a point, and one picks up the closed string pole corresponding to the massless scalar dilaton coupling to the vacuum. Using

$$\psi_{IJ} \sim (\ln q)^{-1} \tag{S.180}$$

$$\sum_{I<J} k_I \cdot k_J = + M \tag{S.181}$$

(since $k_I^2 = + 2$) we see the integrand of Eq. (S.179) is meromorphic at $q = 0$. Actually it is a higher order pole here because of tachyons; also, the $q^2 \to 1$ singularity of $f(q^2)^{-24}$ is severe.

Rather than continuing to discuss the unpleasant singularities of Eq. (S.179) we go straight to the open superstring. Consider the annulus with M = 4. Firstly we find a factor (the propagator is $(\frac{1}{2} p^2 + N_B + N_F)^{-1}$)

$$\mathrm{tr}\ (w^{N_F}) = f(w)^{+8} \tag{S.182}$$

which precisely cancels the $f(w)^{-8}$ from the bosonic modes. Hence supersymmetry removes the worst divergence immediately. Taking the gauge group O(N) the result is[59]

$$A_4^{(\mathrm{Annulus})} = NK_4 \int_0^1 \frac{dq}{q} \int_0^1 \prod_{I=1}^{3} d\nu_I\ \theta(\nu_{I+1} - \nu_I) \prod_{I<J} (\psi_{IJ})^{k_I \cdot k_J} \tag{S.183}$$

where N is the group factor coming from the empty boundary and

$$K_4 = 16\pi^3 g^4\ \mathrm{Tr}(\lambda^{a_1}\lambda^{a_2}\lambda^{a_3}\lambda^{a_4})t^{i_1 j_1 i_2 j_2 i_3 j_3 i_4 j_4}$$

$$\zeta_1^{i_1}\zeta_2^{i_2}\zeta_3^{i_3}\zeta_4^{i_4} k_1^{j_1} k_2^{j_2} k_3^{j_3} k_4^{j_4} \tag{S.184}$$

$$t^{i_1 j_1 i_2 j_2 i_3 j_3 i_4 j_4} = \mathrm{tr}(R_o^{i_1 j_1} R_o^{i_2 j_2} R_o^{i_3 j_3} R_o^{i_4 j_4}) \quad . \tag{S.185}$$

Note especially that the $(\ell nq)^{-M-1}$ from the Jacobian cancels the $(\ell nq)^5$ from the momentum integral $\int d^{10}p$. Also, since $\sum_{I<J} k_I \cdot k_J = 0$ for massless vector external states, the integrand is meromorphic at q = 0. It is now a simple dilaton pole since tachyons are also eliminated.

The M = 4 Moebius strip (Fig. S.10) has no N since there is no empty boundary. The twisting operator on the propagator is

$$\Omega = (-1)^{N_B + N_F} \tag{S.186}$$

so it makes $x_1 \rightarrow -x_1$ and $w \rightarrow -w$. The integration range of ν_1 becomes

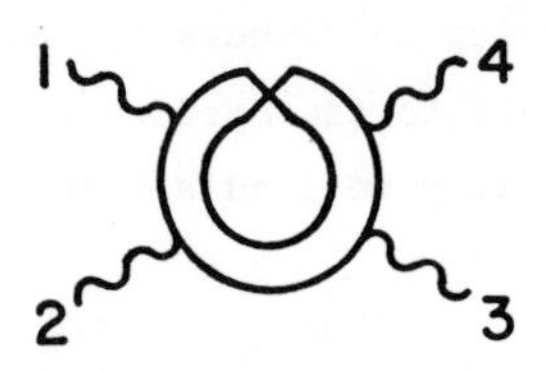

FIGURE S.10

Four-point Moebius strip diagram

(0, 2) instead of (0, 1) and there is an overall - sign for O(N) (it would be + for Sp(2N) and 0 for SU(N); see Ref. 40). Thus

$$A_4^{(\text{Moebius})} = -K_4 \int_0^1 \frac{dq}{q} \int_0^2 \prod_{I=1}^{3} d\nu_I \, \theta(\nu_{I+1} - \nu_I) \prod_{I<J} (\psi_{IJ,N})^{k_I \cdot k_J} . \tag{S.187}$$

In fact, the annulus and the Moebius strip are so similar that we may write[59]

$$A_4^{(\text{Annulus})} + A_4^{(\text{Moebius})} = \int_0^1 \frac{dq}{q} \, [NF(q^2) - 8F(-\sqrt{q})] \tag{S.188}$$

and changing variables to $\lambda = q^2$ and $\lambda = \sqrt{q}$ respectively in the two terms, this can be written for O(32) only as

$$16 \int_{-1}^{+1} \frac{d\lambda}{\lambda} F(\lambda) \qquad \text{(principal part)} . \tag{S.189}$$

We should say that no consistent regularization has been worked through in detail for superstrings. Dimensional regularization is presumably not useful since d = 10 seems fixed. Thus, the above procedure is ambiguous: for example, if we put $\lambda = q$ instead of q^2 for the annulus the result would diverge. Nevertheless, the principal part prescription is the way to isolate O(32) and will be used in what follows.

Consider now the M = 5 pentagon (Fig. S.9). The divergence cancellation for $M \geq 6$ will become clear from M = 5. Really the M = 4 case is too close to the nonrenormalization theorems for the general case to be clear.

In the vector emission vertex, Eq. (S.132), both p^i and R^{ij} have both zero modes (n = 0) and nonzero modes ($n \neq 0$). P is linear in α_n while R is bilinear in S_n. We need at least eight S_o factors for the trace to be nonvanishing.

Already for M = 4 we needed

$$t^{i_1j_1i_2j_2i_3j_3i_4j_4} = \mathrm{tr}\,(R_o^{i_1j_1}R_o^{i_2j_2}R_o^{i_3j_3}R_o^{i_4j_4}) \qquad ((S.190)$$

$$= a_1\varepsilon^{i_1j_1i_2j_2i_3j_3i_4j_4}$$

$$+ a_2[(\delta^{i_1i_2}\delta^{j_1j_2} - \delta^{i_1j_2}\delta^{i_2j_1})$$

$$(\delta^{i_3i_4}\delta^{j_3j_4} - \delta^{i_3j_4}\delta^{i_4j_3})$$

$$+ (13)(24) + (14)(23)]$$

$$+ a_3[\delta^{j_1i_2}\delta^{j_2i_3}\delta^{j_3i_4}\delta^{j_4i_1} + \cdots]_{48\ \text{terms}} \ . \qquad (S.191)$$

The coefficients a_i of these three irreducible tensors can be found by special choices of indices e.g.

$$a_1 = \mathrm{tr}(R_o^{12}R_o^{34}R_o^{56}R_o^{78}) = -\frac{1}{32}\,\mathrm{tr}(\gamma^- \gamma^{78}\gamma^{56}\gamma^{34}\gamma^{12}h_-\gamma^+) \tag{S.192}$$

$$= -\frac{1}{2} \tag{S.193}$$

$$a_2 = \mathrm{tr}(R_o^{12}R_o^{12}R_o^{34}R_o^{34}) = -\frac{1}{32}\,\mathrm{tr}(\gamma^- \gamma^{34}\gamma^{34}\gamma^{12}\gamma^{12}h_-\gamma^+) \tag{S.194}$$

$$= -\frac{1}{2} \tag{S.195}$$

$$a_3 = \mathrm{tr}(R_o^{12}R_o^{23}R_o^{34}R_o^{41}) = (+1)_{\text{bosons}} + \left(-\frac{1}{2}\right)_{\text{fermions}} = +\frac{1}{2}\ . \tag{S.196}$$

For $M = 5$ we consider the pseudotensor and tensor pieces of

$$\mathrm{tr}\left(R_o^{i_1 j_1} R_o^{i_2 j_2} R_o^{i_3 j_3} R_o^{i_4 j_4} R_o^{i_5 j_5}\right)\ . \tag{S.197}$$

The pseudotensor part can be written

$$c_1\left[\delta^{i_1 j_2}\varepsilon^{j_1 i_2 j_3 i_4 j_4 i_5 j_5} + \cdots\right]_{20\ \text{terms}}$$

$$+\ c_2\left[\delta^{i_1 j_3}\varepsilon^{j_1 i_3 i_2 j_2 i_4 j_4 i_5 j_5} +\right]_{20\ \text{terms}} \tag{S.198}$$

where the terms group into those with nearest neighbor and next-nearest neighbor vertices on ϕ respectively. By considering particular indices one finds $c_1 = 3/20$ and $c_2 = 1/20$. Actually, there are Gram determinental constraints which reduce the 40 tensors in Eq. (S.198) to 31. These constraints are the analog of

$$\delta^{\mu\alpha}\varepsilon^{\beta\gamma\delta\varepsilon} + \delta^{\mu\beta}\varepsilon^{\gamma\delta\varepsilon\alpha} + \delta^{\mu\gamma}\varepsilon^{\delta\varepsilon\alpha\beta} + \delta^{\mu\delta}\varepsilon^{\varepsilon\alpha\beta\gamma} + \delta^{\mu\varepsilon}\varepsilon^{\alpha\beta\gamma\delta} = 0 \tag{S.199}$$

in $d = 4$. But we may keep an overcomplete basis if we assign special indices and do not use tensor identities (i.e. do not compare the non-unique coefficients of specific tensors).

The tensor pieces have 524 terms of the types

$$[(\delta\delta)(\delta\delta\delta)]_{160 \text{ terms}} + [(\delta\delta\delta\delta\delta)]_{384 \text{ terms}} \quad . \qquad (S.200)$$

We may separate these further into irreducible parts by giving typical groupings:

$$a_1[(12)(345)]_{80 \text{ terms}} + a_2[(13)(245)]_{80 \text{ terms}}$$
$$+ b_1[(12345)]_{32 \text{ terms}} + b_2[(13524)]_{32 \text{ terms}}$$
$$+ b_3[(13254)]_{160 \text{ terms}} + b_4[(12354)]_{160 \text{ terms}} \quad . \qquad (S.201)$$

Looking at particular values of the indices gives $a_1 = a_2 = -b_2 = b_3 = -b_4 = -\frac{1}{4}$ and $b_1 = \frac{3}{4}$. Note that the only bosonic contribution to this supertrace is in b_1.

Thus the contribution to A_P (P = planar) is[60]

$$A_P^{(R_o^5)} = \zeta_1^{i_1}\zeta_2^{i_2}\zeta_3^{i_3}\zeta_4^{i_4}\zeta_5^{i_5} k_1^{j_1}k_2^{j_2}k_3^{j_3}k_4^{j_4}k_5^{j_5}$$
$$\mathrm{tr}(R_o^{i_1j_1}R_o^{i_2j_2}R_o^{i_3j_3}R_o^{i_4j_4}R_o^{i_5j_5})$$
$$\int \prod_{I=1}^{5} dx_I \frac{1}{w}\left(-\frac{2\pi}{\ell n w}\right) \prod_{1\leq I<J\leq 5} (\psi_{IJ})^{k_I\cdot k_J} \quad . \qquad (S.202)$$

To go from A_P to A_N (N = non-orientable Moebius strip) we replace w by $-w$ and ψ_{IJ} by $\psi_{N,IJ}$. This is the only piece of A_5 with ten S_o factors; the rest have eight S_o factors. Consider first the terms with one S_n and one S_{-n} $(n \neq 0)$. There are eight nonvanishing terms in the factor e.g.

$$\frac{1}{64}\,\mathrm{Tr}\{[S^a_b + \sum_{k\geq 1}(\overline{S}^a_k + \overline{S}^a_{-k})](\gamma^{i_1 j_1 -})^{ab}[S^b_o + \sum_{\ell\geq 1}(S^b_\ell + S^b_{-\ell})]$$

$$[\overline{S}^c_o + \sum_{m\geq 1}\overline{S}^c_m x_2^{-m} + \overline{S}^c_{-m}x_2^m)](\gamma^{i_2 j_2 -})^{cd}[S^d_o + \sum_{p\geq 1}(S^d_p x_2^{-p} + S^d_{-p}x_2^p)]$$

$$w^{N_F} R_o^{i_3 j_3} R_o^{i_4 j_4} R_o^{i_5 j_5}\}. \tag{S.203}$$

Note that $\mathrm{tr}(\overline{S}_k \gamma^{i_1 j_1 -} S_{-k}) = 0$ because there is no second rank anti-symmetric tensor in O(8). The eight nonvanishing terms in (S.203) are all equal. We must then add the ten inequivalent pairs of vertices and obtain

$$A_P^{(S_o^8 S_n S_{-n})} = \zeta_1^{i_1}\zeta_2^{i_2}\zeta_3^{i_3}\zeta_4^{i_4}\zeta_5^{i_5}\,k_1^{j_1}k_2^{j_2}k_3^{j_3}k_4^{j_4}k_5^{j_5}\,\frac{1}{64}$$

$$\int\prod_{I=1}^{5}dx_I w^{-1}\left(-\frac{2\pi}{\ell n w}\right)^5\prod_{1\leq I<J\leq 5}(\psi_{IJ})^{k_I\cdot k_J}$$

$$\sum_{\ell\geq 1}\frac{1}{1-w^\ell}\{[(\frac{w}{x_2})^\ell - x_2^\ell]$$

$$t^{i_1 j_1 i_2 j_2 ij}\,\mathrm{tr}(R_o^{ij} R_o^{i_3 j_3} R_o^{i_4 j_4} R_o^{i_5 j_5})$$

$$+ [(\frac{w}{x_2 x_3})^\ell - (x_2 x_3)^\ell]t^{i_1 j_1 i_3 j_3 ij}\,\mathrm{tr}(R_o^{ij} R_o^{i_2 j_2} R_o^{i_4 j_4} R_o^{i_5 j_5})$$

$$+ \text{cyclic permutations}\}_{2\text{x}5\ \text{terms}} \tag{S.204}$$

where

$$t^{i_1 j_1 i_2 j_2 ij} = \mathrm{tr}\left(\gamma^{i_1 j_1}\gamma^{i_2 j_2}\gamma^{ij}\right) \tag{S.205}$$

$$\begin{aligned} = 32[&\delta_{ij_1}\delta_{i_2 j}\delta_{i_1 j_2} - \delta_{ij_1}\delta_{i_2 j}\delta_{j_i j_2} \\ &- \delta_{ij_1}\delta_{i_1 i_2}\delta_{jj_2} + \delta_{ii_1}\delta_{i_2 j_1}\delta_{jj_2} - \delta_{ii_2}\delta_{i_1 j_2}\delta_{jj_1} \\ &+ \delta_{ii_2}\delta_{i_1 j}\delta_{j_1 j_2} + \delta_{ij_2}\delta_{i_1 i_2}\delta_{jj_1} - \delta_{ij_2}\delta_{i_2 j_1}\delta_{i_1 j}] \,. \end{aligned} \tag{S.206}$$

We obtain $A_N^{(S_o^8 S_n S_{-n})}$ from $A_P^{(S_o^8 S_n S_{-n})}$ by replacing ψ_{IJ} by $\psi_{N,IJ}$, w by $-$ w and x_1 by $- x_1$.

The remaining terms are of the form $(R_o^4 P)$ with

$$P^i = p^i + \sum_{n\geq 1} \sqrt{n}\,(a_n^i + a_n^{i+}) \,. \tag{S.207}$$

The zero mode piece (p^i) modifies the gaussian in the momentum integral already computed in Eqs. (S.166) and (S.169) above. The nonzero modes change the bosonic trace in Eq. (S.157) above. Modifying the operator O_I of Eq. (S.148) to

$$\hat{O}_I^i = P^{i}{}_I(u_I)O_I = \sqrt{n}\,(au_I^{-1} + a^+ u_I)O_I \tag{S.208}$$

giving a modified trace

$$\hat{T} = \frac{1}{\pi}\int d^2 z\,\sqrt{n}\,(z_1 u_1^{-1} + z^* u_1)e^{-|z|^2} e^{-\beta_1} e^{z^* z_1} \tag{S.209}$$

$$= \sqrt{n}\,\left(W u_i^{-1}\frac{\partial}{\partial B} + C u_1^{-1} + u_1\frac{\partial}{\partial C}\right)T \tag{S.210}$$

where T is the old trace (c.f. Eq. (S.157))

$$T = \frac{1}{1-W}\exp\left(-\frac{D(W-1)-BC}{1-W}\right) . \qquad (S.211)$$

It follows that

$$\hat{T} = \frac{1}{1-W}\left(Cu_1^{-1} + Bu_1\right)T . \qquad (S.212)$$

Collecting results we find that[60]

$$A_P^{(R_o^4 P)} = \int \prod_{I=1}^{5} dx_I \, w^{-1}\left(-\frac{2\pi}{\ln w}\right)^5 \prod_{1\leq I<J\leq 5} (\psi_{IJ})^{k_I\cdot k_J}$$

$$\Big[\Big(\zeta_1^{i_1}\sum_{L=1}^{5} k_L^{i_1}\frac{\ln(x_1\cdots x_L)}{-\ln w} K_4(2345)$$

$$+ \zeta_2^{i_2}\sum_{L=2}^{1} k_L^{i_2}\frac{\ln(x_2\cdots x_L)}{-\ln w} K_4(3451)$$

$$+ \text{cyclic permutations}\Big)_{5\ \text{terms}}$$

$$+ \Big\{K_4(2345)\zeta_1^{i_1}\sum_{\ell\geq 1}\frac{1}{1-w^\ell}\Big(k_2^{i_1}\Big[x_2^\ell - \Big(\frac{w}{x_2}\Big)^\ell\Big]$$

$$+ k_3^{i_1}\Big[(x_2x_3)^\ell - \Big(\frac{w}{x_2x_3}\Big)^\ell\Big]$$

$$+ k_4^{i_1}\Big[\Big(\frac{w}{x_5x_1}\Big)^\ell - (x_5x_1)^\ell\Big] + k_5^{i_1}\Big[\Big(\frac{w}{x_1}\Big) - x_1^\ell\Big]\Big)$$

$$+ \text{cyclic permutations}\Big\}_{5\ \text{terms}}\Big] \qquad (S.213)$$

where e.g.

$$K_4(2345) = \zeta_2^{i_2}\zeta_3^{i_3}\zeta_4^{i_4}\zeta_5^{i_5}\, k_2^{j_2}k_3^{j_3}k_4^{j_4}k_5^{j_5}\ \mathrm{tr}\left(R_o^{i_2j_2}R_o^{i_3j_3}R_o^{i_4j_4}R_o^{i_5j_5}\right) . \qquad (S.214)$$

As usual, for the nonorientable Moebius strip we replace ψ_{IJ}, w, x_1 by $\psi_{N,IJ}$, - w, - x_1 respectively.

This completes the M = 5 results. Higher point functions are computed similarly. We now make an asymptotic expansion for $q \to 0$

$$A_P = \int_0^1 \frac{dq}{q}\left[a_o + \frac{a_1}{\ell nq} + \frac{a_2}{(\ell nq)^2} + \cdots\right] . \tag{S.215}$$

The infra-red divergence ($q \to 0$) is from the a_o and a_1 terms only. From the physical picture of the dilaton pole, presumably all $a_k = 0$ for $k \geq 1$. We should note the (ℓnq) factors are $(\ell nq)^5$ and $(\ell nq)^{-M-1}$ from the momentum integrand and Jacobian respectively. For M = 4 it is trivially meromorphic and the annulus and Moebius strip diagrams cancelled. Recall that we put $q = q'^4$ to relate the q variables and used the change

$$\int_0^2 \prod_{I=1}^{M-1} d\nu'_I = 2^{M-1} \int_0^1 \prod_{I=1}^{M-1} d\nu_I \tag{S.216}$$

together with

$$\frac{dq}{q} = 4\,\frac{dq'}{q'} \tag{S.217}$$

to obtain (for M = 4) 4 times 8 = 32. But how can the 2^{M-1} become independent of M, the number of external lines? The answer lies in the mode sum

$$F_p(\nu, w) = \sum_{n=1}^{\infty} \frac{c^n - (w/c)^n}{1 - w^n} \quad \text{and } c = w^\nu \tag{S.218}$$

$$= \sum_{n=1}^{\infty} \sum_{m=0}^{\infty} \left(e^{n\nu \ln w} - e^{n(1-\nu)\ln w}\right) e^{mn \ln \nu} \tag{S.219}$$

$$= \sum_{m=0}^{\infty} \left(\frac{e^{(\nu+m)\ln w}}{1 - e^{(\nu+m)\ln w}} - \frac{e^{(1-\nu+m)\ln w}}{1 - e^{(1-\nu+m)\ln w}}\right) \tag{S.220}$$

$$= \frac{1}{(-\ln w)} \left(A_o + A_1(-\ln w) + A_2(-\ln w)^2 + \cdots\right) . \tag{S.221}$$

Now we find easily that

$$A_o = \frac{1}{\nu} + 2\nu \sum_{m=1}^{\infty} \frac{1}{\nu^2 - m^2} = \pi \cot \pi\nu \quad . \tag{S.222}$$

The nonorientable Moebius strip contains

$$F_N(\nu, w) = \sum_{\ell=1}^{\infty} \frac{1}{1 - (-w)^\ell} \left[c^\ell - (-w/c)^\ell\right] \tag{S.223}$$

$$\sim -\left(\frac{1}{2}\right) \frac{\ln q}{2\pi} \cot\left(\frac{\pi\nu}{2}\right)\left[1 + 0\left(\frac{1}{\ln q}\right)\right] \tag{S.224}$$

as $q \to 0$. Note that

$$F_p(\nu, w) \sim -\frac{\ln q}{2\pi} \cot \pi\nu \left[1 + 0\left(\frac{1}{\ln q}\right)\right] \tag{S.225}$$

the extra (1/2) in Eq. (S.224) is the essential point. When we generate the extra $(\ln q)^{M-4}$ from the mode sums, necessary to identify the coefficient a_o in Eq. (S.215), we also generate a relative factor 2^{4-M} in the Moebius strip relative to the annulus. Thus the terms may be combined as

$$\frac{1}{2} \int_0^1 \frac{d\lambda}{\lambda} \left[NF(\lambda) - \frac{4 \cdot 2^{M-1}}{2^{M-4}} F(-\lambda)\right] = 16 \int_{-1}^{+1} \frac{d\lambda}{\lambda} F(\lambda) \tag{S.226}$$

and the principal part prescription generalizes to all M for the gauge group O(32).

In the above analysis we have tacitly assumed that the integrand is meromorphic. In particular, we took $a_k = 0$ $(k \geq 1)$ in Eq. (S.215). If $a_1 \neq 0$ it gives an unacceptable non-leading divergence which violates unitarity and factorization. To show $a_1 = 0$ in light-cone gauge is nontrivial[61] but has been done explicitly for the pentagon diagram. The one-loop finiteness of the parity-conserving piece in the O(32) open superstring has been discussed also in covariant gauge[62].

Let us now consider finiteness of the N = 2 chiral and nonchiral superstrings. Because these strings are oriented, there is only one amplitude (a sphere with ℓ handles) for each number ℓ of loops. For $\ell = 1$, we consider the torus diagram. The modes for the closed string sector are a superposition of "right movers" and "left movers". For example, the transverse coordinate $X^i(\sigma, \tau)$ is written

$$X^i(\sigma, \tau) = \frac{1}{2}\left(X^i(\sigma - \tau) + \tilde{X}^i(\sigma + \tau)\right) \tag{S.227}$$

$$X^i(\tau - \sigma) = \frac{1}{2}x^i + \frac{1}{2}p^i(\tau - \sigma) + \frac{i}{2}\sum_{n\neq 0}\frac{1}{n}\partial^i_n e^{-2in(\tau-\sigma)} \tag{S.228}$$

$$X^i(\tau + \sigma) = \frac{1}{2}x^i + \frac{1}{2}p^i(\tau + \sigma) + \frac{i}{2}\sum_{n\neq 0}\frac{1}{n}\tilde{\alpha}^i_n e^{-2in(\tau+\sigma)} \quad . \tag{S.229}$$

The corresponding commutation relations are

$$[\alpha^i_n, \alpha^j_m] = [\tilde{\alpha}^i_n, \tilde{\alpha}^j_m] = \delta^{ij}\delta_{m+n,o}n \tag{S.230}$$

$$[\alpha^i_n, \tilde{\alpha}^j_m] = 0 \tag{S.231}$$

$$[x^i, p^j] = i\delta^{ij} \quad . \tag{S.232}$$

The fermionic coordinates for the N = 2 closed superstring are similarly superpositions of two open-like sectors

$$S^a(\tau - \sigma) = \sum_{n=-\infty}^{\infty} S_n^a \, e^{-2in(\tau-\sigma)} \tag{S.233}$$

$$\gamma^+ S_n = \frac{1}{2}(1 - \gamma_{11}) S_n = 0 \tag{S.234}$$

$$\{S_m^a, \bar{S}_n^b\} = [\gamma^+ \frac{1}{2}(1 - \gamma^{11})]^{ab} \delta_{m+n,o} \tag{S.235}$$

and

$$\tilde{S}^a(\tau + \sigma) = \sum_{n=-\infty}^{\infty} \tilde{S}_n^a \, e^{-2in(\tau+\sigma)} \tag{S.236}$$

$$\gamma^+ \tilde{S}_n = \frac{1}{2}(1 \mp \gamma^{11}) S_n = 0 \tag{S.237}$$

$$\{\tilde{S}_m^a, \bar{\tilde{S}}_n^b\} = [\gamma^+ \frac{1}{2}(1 \mp \gamma^{11})]^{ab} \delta_{m+n,o} \tag{S.238}$$

where the upper and lower signs in the left-mover sector correspond to chiral and nonchiral N = 2 closed superstrings respectively.

The states of such a closed superstring are direct products of the Fock states of the right- and left-movers. The vertex for emission of a closed string massless boson is the product of two factors of the form Eq. (S.132) above, that is

$$g\zeta^{i\tilde{i}}\tilde{\zeta}^{\tilde{i}}(P^i + k^j R^{ij}) \, V_o \left(\frac{k}{2}\right)(\tilde{P}^{\tilde{i}} + k^{\tilde{j}} \tilde{R}^{\tilde{i}\tilde{j}}) \, \tilde{V}_o \left(\frac{k}{2}\right) \tag{S.239}$$

where e.g. symmetrizing the indices $i\tilde{i}$ gives the vertex for emitting the graviton state. The closed string propagator is

$$\int d^2z\ |z|^{1/4}\ p^{2-2}\ z^{N}\ \bar{z}^{\tilde{N}} \tag{S.240}$$

$$N = \sum_{n=1}^{\infty} (\alpha^i_{-n}\alpha^i_n + \frac{n}{2}\bar{S}_{-n}\gamma^- S_n) \tag{S.241}$$

$$\tilde{N} = \sum_{n=1}^{\infty} (\tilde{\alpha}^i_{-n}\ \tilde{\alpha}^i_n + \frac{n}{2}\bar{\tilde{S}}_{-n}\gamma^-\tilde{S}_n)\ . \tag{S.242}$$

From the structure of Eqs. (S.239) and (S.240) it is clear that for tree diagrams the closed superstring involves a product of two factors, one for the right-movers and left-movers. The same is true for the one-loop diagram up to the loop momentum integral: note that the external momenta are shared equally between left-movers and right-movers.

Since one needs eight each of R_o and $\tilde{R}_o$ to achieve a nonvanishing supertrace, there is a superstring nonrenormalization theorem that the 2- and 3-point one-loop diagrams vanish. Let us then look at M = 4 external ground states. The resulting amplitude is

$$A = K(1,2,3,4) \int \prod_{I=1}^{4} d^2z_I\ |w|^{-2} \left(\frac{-4\pi}{\ell n|w|}\right)^5 \prod_{I<J} (\chi_{IJ})^{\frac{k_I\cdot k_J}{2}} \tag{S.243}$$

where the kinematic factor is

$$K(1,2,3,4) = \zeta_1^{i_1}\tilde{\zeta}_1^{\tilde{i}_1}\zeta_2^{i_2}\zeta_2^{\tilde{i}_2}\zeta_3^{i_3}\zeta_3^{\tilde{i}_3}\zeta_4^{i_4}\zeta_4^{\tilde{i}_4}\ k_1^{j_1}k_1^{\tilde{j}_1}k_2^{j_2}k_2^{\tilde{j}_2}k_3^{j_3}k_3^{\tilde{j}_3}k_4^{j_4}k_4^{\tilde{j}_4}$$
$$\mathrm{Tr}\left(R_o^{i_1j_1}R_o^{i_2j_2}R_o^{i_3j_3}R_o^{i_4j_4}\right)\mathrm{Tr}\left(\tilde{R}_o^{\tilde{i}_1\tilde{j}_1}\tilde{R}_o^{\tilde{i}_2\tilde{j}_2}\tilde{R}_o^{\tilde{i}_3\tilde{j}_3}\tilde{R}_o^{\tilde{i}_4\tilde{j}_4}\right) \tag{S.244}$$

and

$$w = \prod_{I=1}^{4} z_I \tag{S.245}$$

$$\chi_{IJ} = \chi(C_{JI}, w) \tag{S.246}$$

$$C_{JI} = z_{I+1} z_{J+2} \cdots z_J \tag{S.247}$$

$$\chi(z,w) = \exp\left(\frac{\ell n^2 \left|\frac{z}{w}\right|}{2\ell n|w|}\right) \left| \frac{1 - z}{\sqrt{z}} \prod_{m=1}^{\infty} \frac{(1 - w^m z)(1 - w^m/z)}{(1 - w^m)^2} \right| . \tag{S.248}$$

It is useful to change to torus variables defined by

$$\nu_I = \frac{\ell n z_1 z_2 \cdots z_I}{2\pi i} \tag{S.249}$$

$$\tau = \frac{\ell n w}{2\pi i} \tag{S.250}$$

whereupon the amplitude becomes

$$A = K(1,2,3,4) \int d^2\tau\, (\mathrm{Im}\tau)^{-5} \prod_{I=1}^{3} d^2\nu_I \prod_{I<J} (\chi_{IJ})^{\frac{k_I \cdot k_J}{2}} \tag{S.251}$$

$$\chi_{IJ} = 2\pi \exp\left(- \frac{\pi(\mathrm{Im}\nu_{JI})^2}{\mathrm{Im}\tau}\right) \left|\frac{\theta_1(\nu_{JI}|\tau)}{\theta_1'(o|\tau)}\right| \tag{S.252}$$

with $\nu_{JI} = (\nu_J - \nu_I)$.

Now the three torus variables ν_I define the relative positions of the four external lines, the complex variable τ has a special role. It turns out that the integrand (including the measure) possesses an invariance under the modular group

$$\tau' = \frac{a\tau + b}{c\tau + d} \tag{S.253}$$

with a, b, c, d integers such that ad - bc = 1. This is the group SL(2, Z). Under Eq. (S.253) one easily checks that

$$d^2\tau \to |c\tau + d|^{-4}\, d^2\tau \tag{S.254}$$

$$\text{Im}\tau \to |c\tau + d|^{-2}\, \text{Im}\tau \tag{S.255}$$

and then finds that

$$(\text{Im}\tau)^{-3} \int \prod_{I=1}^{3} d^2\nu_I \prod_{I<J} (\chi_{IJ})^{\frac{1}{2} k_I \cdot k_J} \tag{S.256}$$

is invariant using $\sum_{I<J} k_I \cdot k_J = 0$.

The complex parameter τ characterises the conformal equivalence class of the torus. Note that for higher genus ($\ell \geq 2$ loops) there are $(3\ell-3)$ such parameters. At one loop the 2-torus can be described by factoring the complex plane (C) with a lattice Γ generated by two nonparallel complex numbers w_1, w_2; that is, $T_2 = C/\Gamma$. The same torus is generated by w_1', w_2' given by

$$\begin{pmatrix} w_1' \\ w_2' \end{pmatrix} = \begin{pmatrix} a & b \\ c & d \end{pmatrix} \begin{pmatrix} w_1 \\ w_2 \end{pmatrix} \tag{S.257}$$

where the transformation is an SL(2, Z) matrix. If we conformally map the complex plane by $z \to w_1^{-1} z$ so that $w_1 \to 1$ and $w_2 \to w_1^{-1} w_2 = \tau$ then the torus is characterized by τ. One can choose $\text{Im}\tau > 0$ by interchanging w_1, w_2 if necessary. The upper half τ- plane is divided into an infinite number of equivalent regions by the modular invariance of Eq. (S.253). The fundamental region (F) is taken as

$$\text{Im}\tau > 0,\ |\tau| > 1,\ |\text{Re}\tau| < \frac{1}{2} \tag{S.258}$$

so by integrating τ only over F we avoid the potential singularities at $\tau = 0$. This modular invariance was studied first by Shapiro[63] in 1972 for the bosonic string where there remained divergence difficulties associated with tachyons. In the case of the N = 2 superstring, the M = 4 torus is completely finite because of modular invariance.

Using the explicit pentagon results of Ref. 61, it can be checked that modular invariance persists for the M = 5 pentagon torus of the N = 2 superstrings. One finds an expression of the form[64]

$$A_5^{(1\ \mathrm{loop})} = \int d^2\tau \prod_{I=1}^{4} d\nu_I \left(\frac{1}{\mathrm{Im}\tau}\right)^5 \left(K(z)\overline{K}(\overline{z}) + \frac{C}{\mathrm{Im}\tau}\right) \prod_{I<J} (\chi_{IJ})^{\frac{k_I \cdot k_J}{2}} \tag{S.259}$$

where K(z) is the kinematic factor for the open string case. The additional piece $C(\mathrm{im}\tau)^{-1}$ is generated through the loop momentum integral by an extra loop momentum dependence. The modular group is generated by $\tau \to (\tau + 1)$ and $\tau \to -1/\tau$, the latter being accompanied for convenience by $\nu_I \to -\nu_I/\tau$. Under both such transformations the integral of Eq. (S.259) is invariant and hence the pentagon is finite.

Results for higher-loop diagrams in N = 2 closed superstrings are incomplete, although the relevant generalization of the modular group to surfaces of genus greater than one is known [e.g. Ref. 65].

Note that the closed strings of the N = 2 type are necessarily orientable; there is no complete symmetry between left- and right-movers although the eigenvalues of N and $\tilde{N}$ are equal. By contrast, the closed strings which would appear through unitarity in open superstring theories are non-orientable: not only do N and $\tilde{N}$ have equal eigenvalues but there is total left-right symmetry; this allows further worldsheet topologies (e.g. Klein bottle) at one loop. Even at one loop, the graphs involving external closed strings in the O(32) open superstring theory have not all been calculated; at least, the way finiteness is achieved is more complicated[59] than for the oriented closed superstring.

Finally, one may ask: why is it more likely that the superstring theories are finite to all orders than, say, the D = 4, N = 8 supergravity theory which is finite to two loops? The answer is that the latter theory is fully expected to have ultra-violet divergences by power counting. In the superstring, the ultra-violet problem has been removed by the nature of the string as an extended object. All that remains is the infra-red problem and this is plausibly taken care of by modular invariance and its generalization. Some preliminary indications [e.g. Ref. 66] suggest that finiteness for the closed superstring does hold for any number of loops, at least for vacuum graphs with no external lines. This is encouraging.

S.7 HETEROTIC STRING

It was noted above in section (S.5) that the algebra of gauge and gravitational chiral anomaly cancellation works equally[53,67] for O(32) or E(8) x E(8). Only for O(32) was there a pre-existing string theory. Both groups have a common dimension, 496, and common rank, 16. The latter number (16) is the difference between the critical dimensions 26 and 10 for the bosonic and supersymmetric strings. Although compactification on a 16-torus would, in general, give only the isometry $U(1)^{16}$, it was pointed out by Freund [Ref. 68, Reprint No. 3; see also Ref. 69] that the groups O(32) and E(8) x E(8) have root diagrams corresponding to the only two even self-dual euclidean lattices in 16 dimensions[70].

The construction of a string with gauge group E(8) x E(8) was achieved by Gross, Harvey, Martinec and Rohm [Ref. 71, Reprint No. 4; Refs. 72, 73]. This remarkable model--named the heterotic string as a "vigorous cross-breed" of the bosonic string and the superstring--is our present topic.

A key observation in formulating the heterotic string is the fact that, as mentioned in the preceding subsection, the right- and left-movers of a closed string can be treated independently--the connection being only that the total momentum and position of the two components be identical. In particular, the right- and left-movers can be treated asymmetrically provided each sector is internally consistent. Even in the interacting string theory the vertex operators are direct products of right and left vertices.

In the heterotic string the right-movers are those of an N = 1 superstring sector in D = 10 spacetime dimensions. In light-cone gauge there are eight transverse coordinates

$$X^i(\tau - \sigma) = \frac{1}{2} x^i + \frac{1}{2} p^i(\tau - \sigma) + \frac{i}{2} \sum_{n\neq 0} \frac{1}{n} \alpha_n^i e^{-2in(\tau-\sigma)} \qquad (S.260)$$

together with eight Majorana-Weyl fermionic coordinates

$$S^a(\tau - \sigma) = \sum_{n=-\alpha}^{\infty} S_n^a e^{-2in(\tau-\sigma)} \qquad (S.261)$$

satisfying the conditions of Eq. (S.111) above.

The left-movers, on the other hand, are taken to be eight transverse coordinates $\tilde{X}^i(\tau + \sigma)$ together with sixteen internal coordinates $\tilde{X}^I$ given by

$$\tilde{X}^I(\tau + \sigma) = x^I + p^I(\tau + \sigma) + \frac{i}{2} \sum_{n\neq 0} \frac{1}{n} \tilde{\alpha}_n^I e^{-2in(\tau+\sigma)} \qquad (S.262)$$

where

$$[\tilde{\alpha}_n^I, \alpha_m^J] = n\delta_{m+n,o}\delta^{IJ} \qquad (S.263)$$

$$[x^I, p^J] = \frac{1}{2} i\delta^{IJ} . \qquad (S.264)$$

The $\frac{1}{2}$ in Eq. (S.264) is characteristic of light-cone gauge and reflects the fact that $\tilde{X}^I$ depends only on $(\tau + \sigma)$.

Unless $p^I = 0$, we must arrange that X^I satisfies the appropriate periodic boundary conditions of the closed string by allowing x^I to parametrize a compact manifold T; the only known consistent choice is the 16 torus T_{16}. The momenta p^I then becomes quantized in units of 1/R where R is the radius of the manifold.

In order to make such a left-moving sector consistent--in particular for the one-loop diagram to be modular invariant (see below)--the torus must be chosen such that it is a product of sixteen circles each of radius $R = \sqrt{\alpha'} = 1/\sqrt{2}$ (we choose units $\alpha' = 1/2$) and with the identification

$$X^I \equiv X^I + \sqrt{2}\,\pi R \sum_{i=1}^{16} e_i^I n_i \, . \tag{S.265}$$

Here, e_i^I are sixteen basis vectors which generate an integer even self-dual euclidean lattice, i.e.

$$(e_i^I)^2 = 2 \tag{S.266}$$

$$g_{ij} = \sum_{I=1}^{16} e_i^I e_j^I = \text{integer} \tag{S.267}$$

$$\det g = 1 \tag{S.268}$$

$$n_i = \text{integer} \, . \tag{S.269}$$

The allowed internal momenta are

$$p^I = \sum_{i=1}^{16} n_i e_i^I \tag{S.270}$$

showing that the minimum$(p^I)^2$ is two.

Taking the number operators

$$N = \sum_m \alpha^i_m \alpha^i_{-m} + \frac{1}{2} \sum_m \bar{S}_{-m} \gamma^- S_m \tag{S.271}$$

$$\tilde{N} = \sum_m (\alpha^i_m \alpha^i_{-m} + \tilde{\alpha}^I_m \tilde{\alpha}^I_{-m}) \tag{S.272}$$

the square mass operator is

$$\frac{1}{2} \alpha' \, (\text{mass})^2 = N + (\tilde{N} - 1) + \frac{1}{2} \sum_I (p^I)^2 \; . \tag{S.273}$$

The physical states satisfy

$$N = \tilde{N} - 1 + \frac{1}{2} \sum_I (p^I)^2 \tag{S.274}$$

since $2\left(N - \tilde{N} + 1 - \frac{1}{2} \sum (p^I)^2\right)$ is the generator of shifts in σ and we must ensure reparametrization invariance (no distinguished points).

The impressive fact is that in 16 dimensions the only even self-dual lattices are[70] the root lattices of the groups E(8) x E(8) and O(32) [strictly speaking spin(32)/Z_2] and this fact underlies the freedom of chiral anomalies in the heterotic string.

The spectrum of states is formed by direct products of right and left Fock space states. The ground state of the right-movers is annihilated by α^i_n, S^i_n $(n > 0)$ and N and comprises eight transverse bosons $|i\rangle_R$ and eight fermions $|a\rangle_R$. The ground state of the left-movers, annihilated by $\tilde{\alpha}^i_n$, $\tilde{\alpha}^I_n$ $(n > 0)$ and $\tilde{N}$ with zero p^I cannot satisfy Eq. (S.274) so this gets rid of the tachyon by heterosis; the tachyon would be present in the bosonic sector alone but reparametrization invariance of the complete closed string excludes it.

The ground states of the heterotic string are massless and are the direct products of $|i\rangle_R$ and $|a\rangle_R$ with $\tilde{\alpha}^i_{-1}|0\rangle_L$, $\tilde{\alpha}^I_{-1}|0\rangle_L$ and $|p^I\rangle_L\left((p^I)^2\right)$. The products with $\tilde{\alpha}^i_{-1}|0\rangle_L$ form the supermultiplet of N = 1, D = 10 supergravity. The products with $\tilde{\alpha}^I_{-1}|0\rangle_L$ form the

Kaluza-Klein gauge bosons of a $U(1)^{16}$, the isometry group of the torus. But the products with $|p^I, (p^I)^2 = 2>_L$ form the remaining 480 gauge bosons filling out the 496 dimensional adjoint of E(8) x E(8) or spin $(32)/Z_2$. The fact that this works is related to the construction of irreducible representations of Kac-Moody algebras using vertex operators [74; for a review, see Ref. 75]; this work was inspired by the role of vertices in the old dual resonance models, so one has in that sense followed a circle from physics to mathematics and back to physics. The massive excitations of the heterotic string have arbitrarily large p^I and hence arbitrary representations of the gauge group; this distinguishes the O(32) heterotic string from the O(32) open superstring which uses the Chan-Paton prescription and hence limits the representations to only the singlet and 496-dimensional adjoint.

The separation of left- and right-movers is a geometrical one which allows interactions to be introduced consistently. This is nontrivial: for example, one can consider the free bosonic string in $d < 26$ dimensions by simply projecting out a subspace, but then this is not geometrical and interactions will not be lorentz invariant within the subspace.

Because the heterotic string is closed and orientable (obviously orientable because of the asymmetry between right- and left-movers) only <u>one</u> interaction is possible--the splitting of one string into two (the "pants" diagram). For simplicity, let us consider the emission of a massless ground-state from the heterotic string.

First, take the bosons within the graviton supermultiplet with polarization tensor $\rho^{\mu\nu}(k)$, $k^2 = 0$, and satisfying (in light-cone gauge)

$$\rho^{+\mu} = \rho^{\mu +} = k_\mu \rho^{\mu\nu} = k_\nu \rho^{\mu\nu} = 0 \ . \tag{S.275}$$

The symmetric traceless part describes the graviton, the trace is the dilaton, and the rest is the antisymmetric tensor field. The integrand (inside the σ integration) of the vertex factors into right- and left-mover parts, namely (we put $k^+ = 0$ to avoid boosting and $k^- = 0$ by using complex components)

$$V(k) = \frac{4g}{\pi} \rho_{\mu\nu}(k) \int_0^\pi d\sigma\, B^\mu \tilde{P}^\nu e^{ik\cdot X} \tag{S.276}$$

where

$$B^i = P^i + \frac{1}{2} k^j k R^{ij} \tag{S.277}$$

$$P^i = \frac{dX^i}{d(\tau - \sigma)} = \frac{1}{2} p^i + \sum_{n\neq 0} \alpha_n^i e^{-2in(\tau-\sigma)} \tag{S.278}$$

$$R^{ij} = \frac{1}{8} \bar{S}\gamma^{ij-}S \tag{S.279}$$

$$\tilde{P}^i = \frac{1}{2} p^i + \sum_{n\neq 0} \tilde{\alpha}_n^i e^{-2in(\tau+\sigma)} . \tag{S.280}$$

For the gauge boson emission vertices we must distinguish the neutral gauge bosons

$$|i\rangle_R \times \tilde{\alpha}_{-1}^I |0\rangle_L \qquad \text{(16 of these)}$$

and the charged gauge bosons

$$|i\rangle_R \times |K^I, (K^I)^2 = 2\rangle \qquad \text{(480 of these)} .$$

For the neutrals the vertex is

$$V^I = \frac{4g}{\pi} \rho_\mu^I(k) \int_0^\pi d\sigma\, B^\mu \tilde{P}^I e^{ik\cdot X} \tag{S.281}$$

with

$$k^2 = k^{\mu}\rho^I_{\mu} = \rho^{+I} = 0 \qquad (S.282)$$

$$\tilde{P}^I = \frac{dX^I}{d(\tau + \sigma)} = P^I + \sum_{n\neq 0} \tilde{\alpha}^I_n \, e^{-2in(\tau+\sigma)} \quad . \qquad (S.283)$$

For the chargeds the emission vertex is

$$V^{K_I} = \frac{4g}{\pi} \rho_{\mu}(k) \int_0^{\pi} d\sigma \, B^{\mu} : e^{2iK^I X^I(\tau+\sigma)} : C(K_I) \, e^{ik\cdot X} \quad . \qquad (S.284)$$

The normal ordered factor is

$$: e^{2iK^I X^I} : = \exp\Big[- K^I \sum_{-\infty}^{-1} \frac{\tilde{\alpha}^I_n}{n} e^{-2in(\tau+\sigma)}\Big]$$

$$\exp\big(2iK^I(x^I + P^I(\tau + \sigma))\big) \exp\Big[-K^I \sum_{1}^{\infty} \frac{\tilde{\alpha}^I_n}{n} e^{-2in(\tau+\sigma)}\Big]. \qquad (S.285)$$

The different factors here do not commute because $(K^I)^2 \neq 0$. Note that $2K^I$ is the translation operator in the internal space because of the light-cone quantization $[x^I, P^J] = \frac{1}{2} i\delta^{IJ}$. Finally in Eq. (S.284), $C(K_I)$ is an important cocycle operator occurring in the Frenkel-Kac construction[74] and satisfying

$$C(k)\, C(L) = \varepsilon(K, L)\, C(K + L) \qquad (S.286)$$

with

$$\varepsilon(K, L)\, \varepsilon(K + L, M) = \varepsilon(L, M)\, \varepsilon(K, L + M) \ . \qquad (S.287)$$

Mathematically, the point is that it is the operators P^I and

$$E(K^I) = \int \frac{dx}{2\pi iz} : \exp\big(2iK^I X^I(z)\big) : C(K^I) \qquad (S.288)$$

which satisfy the Lie algebra of the gauge group. In the physics of the heterotic string, the cocycle factors are essential for summing over the different permutations of external legs since the cocycle guarantees the commutation properties of the vertices.

Let us consider the evaluation of tree amplitudes in the heterotic string. The relevant propagator is

$$\Delta = \int_{|z|<1} d^2z \, |z|^{1/4 \, p^2-2} \, z^N \bar{z}^{\tilde{N}-1+\sum_I 1/2 (P^I)^2} \, . \qquad (S.289)$$

The N-point function is then

$$A(12...N) = \langle N|\tilde{V}_{N-1}(k_{N-1}) \, \Delta \, \tilde{V}_{N-2}(K_{N-2}) \cdots \tilde{V}_2(k_2)|1\rangle \qquad (S.290)$$

where the $\tilde{V}$ are the vertices discussed above. Summing over the τ-orderings corresponds to extending the z integrals, for a given ordering, to the entire z-plane; for this to be true, the left-mover cocycle factors are necessary to compensate the interchange of vertex orderings.

The simplest computation is for the charged gauge bosons since they involve essentially the "tachyonic" vertex in the left-moving sector; the amplitudes for the other 16 (neutral) gauge bosons follow by gauge invariance. It is amusing to calculate the 4-point function with amplitude[73]

A(1234) =

$$g^2 \int \frac{d^2z}{4\pi} \langle \rho_4 K_4|\rho_3 B(K_3) \, e^{ik_3 \cdot X} :e^{2iK_3^I X^I}:$$

$$C(K_3) \, z^N \bar{z}^{\tilde{N}-1+1/2 (P^I)^2} |z|^{1/4 \, p^i-2} \rho_2 B(k_2) \, e^{ik_2 \cdot X} :e^{2iK_2^I X^I}:$$

$$C(K_2)|\rho, \, K_1\rangle \qquad (S.291)$$

where $(K_a)^2 = 2$ and

$$(k_a^i)^2 = \sum_{a=1}^{4} k_a^i = \sum_{a=1}^{4} K_a^I = \rho_a^i K_a^i = 0 \ . \tag{S.292}$$

Defining external and internal Mandelstam variables

$$s = (K_1 + K_2)^2 \qquad S = (K_1 + K_2)^2$$

$$t = (K_2 + K_3)^2 \qquad T = (K_2 + K_3)^2$$

$$u = (K_1 + K_3)^2 \qquad U = (K_1 + K_3)^2$$

one has $s + t + u = 0$, $S + T + U = 8$ and the allowed values of S, T, U are 0, 2, 4, 6, 8. One finds the result

$A(1234) =$

$$g^2 \; K_{1234} \frac{\Gamma(-1 - \frac{1}{8}s + \frac{1}{2}S)\Gamma(-1 - \frac{1}{8}t + \frac{1}{2}T)\Gamma(-1 - \frac{1}{8}u + \frac{1}{2}U)}{\Gamma(1 + \frac{1}{8}s)\Gamma(1 + \frac{1}{8}t)\Gamma(1 + \frac{1}{8}u)} \ . \tag{S.293}$$

Here, for example, the factor $\Gamma(-1 - \frac{1}{8}s + \frac{1}{2}S)$ has a massless pole ($s = 0$) corresponding to $S = 0$ (graviton, neutral gauge bosons) and to $S = 2$ (charged gauge bosons). Clearly, the gauge sector is incorporated into the heterotic string more aesthetically than it is in the open superstring.

This is a good point to emphasize that the connection between the gauge coupling g and the Newton coupling $\kappa = (8\pi G)^{1/2}$ is different in the heterotic string from the open superstring. Recall that the dimensions of g and κ are M^{-3} and M^{-4} respectively in 10 dimensions (the gravity force falls like R^{-8}). In the heterotic tree we have $g^2 \sim G \sim \kappa^2$ and hence $(\alpha')^{1/2} g = \kappa$ since α' has dimension M^{-2}. In the open superstring loop we have $g^4 \sim G \sim \kappa^2$ and hence $g^2 = \alpha'\kappa$.

We may proceed to compute the one-loop heterotic torus for four charged gauge bosons[73,76]. As we shall see, it is the modular invariance--freedom from reparametrization anomalies--of the torus diagram that requires the choice of gauge group corresponding to those that will be anomaly free at the hexagon level. We have

$$A(1234) = \mathrm{Tr}[\Delta\, V(4)\, \Delta\, V(3)\, \Delta\, V(2)\, \Delta\, V(1)] \,. \tag{S.294}$$

The right-mover factors contribute just like the open superstring:

$$K_{1234} \qquad \text{(from } S_o\text{)} \tag{S.295a}$$

$$f(\omega)^8 \qquad \text{(from } S_n\text{)} \tag{S.295b}$$

$$f(w)^{-8} \prod_{I<J} \prod_{m=1}^{\infty} \left[\frac{(1 - w^{m-1} C_{JI})(1 - w^m/C_{JI})}{(1 - w^m)^2}\right]^{\frac{1}{4} K_I \cdot K_J} \qquad \text{(from bosons).} \tag{S.295c}$$

The left-movers give factors

$$f(\bar{w})^{-24} \prod_{I<J} \prod_{m=1}^{\infty} \left[\frac{(1 - \bar{w}^{m-1} \bar{C}_{JI})(1 - \bar{w}^m/\bar{C}_{JI})}{(1 - w^m)^2}\right]^{\frac{1}{4} K_I \cdot K_J + K_I \cdot K_J} \,. \tag{S.296}$$

The ten-dimensional loop momentum integral is

$$\int d^{10}\ell \prod |z_I|^{\frac{1}{4} P_I^2} = \left(-\frac{2\pi}{\ln|w|}\right)^5 \prod_{I<J} \left[|C_{JI}|^{-\frac{1}{2}} \exp\left(\frac{\ln^2|C_{JI}|}{2\ln|w|}\right)\right]^{\frac{1}{4} K_I \cdot K_J} \,. \tag{S.297}$$

There is also the completely new sum over values of the lattice loop momentum L on the lattice Λ giving

$$\sum_{L \text{ in } \Lambda} \prod_{I=1}^{4} \bar{z}_I^{\rho_I^2} = \prod_{I<J} \left[\frac{1}{(\bar{C}_{JI})^{\frac{1}{2}}} \exp\left(\frac{\ln^2 C_{JI}}{2\ln \bar{w}}\right)\right]^{K_I \cdot K_J} \tilde{L} \tag{S.298}$$

$$\tilde{L} = \sum_{L \text{ in } \Lambda} \exp\left[\frac{1}{2} \ln \bar{w} \left(L - \sum_{I=1}^{4} \frac{\ln \bar{z}_I}{\ln \bar{w}} Q_I\right)^2\right] . \quad (S.299)$$

Collecting together all factors gives

$$A = K \int \Pi d^2 z \frac{1}{|w|^2} \left(-\frac{4\pi}{\ln|w|}\right)^5 \prod_{I<J} (\chi_{JI})^{\frac{1}{2} k_I \cdot k_J}$$

$$\frac{1}{w} f(\bar{w})^{-24} \prod_{I<J} (\psi_{JI})^{K_I K_J} \tilde{L} . \quad (S.300)$$

The integration over the phases of the z_i gives the physical state condition $N = \tilde{N} - 1 + \frac{1}{2} \sum_I (P^I)^2$. When we transform to torus variables, as in Eqs. (S.249) and (S.250), we may check the invariance of the integral under the changes $\nu_I \to (\nu_I + \tau)$ and $\nu_I \to (\nu_I + 1)$ allowing to restrict as ususal to $0 < \text{Im}\nu_I < \text{Im}\,\tau$ and $|\text{Re}\,\nu_I| < \frac{1}{2}$. Also, the integrand is invariant under the modular group Eq. (S.253) allowing the τ integral to be restricted to the fundamental region Eq. (S.258). Of all rank 16 groups it is essential to choose E(8) x E(8) or spin $(32)/Z_2$ in order that the expression $\tilde{L}$ in Eq. (S.299) transforms simply under the modular group. Roughly speaking, the left-movers depend only on $(\sigma + \tau)$ and hence one must treat equivalently the coefficients of σ which are winding numbers on the internal torus and the coefficients of τ which are the allowed lattice momenta; this is what forces the lattice to be self-dual.

The one-loop finiteness for M = 4 external legs can be extended to the M = 5 pentagon[64]; thus, the heterotic string is on a footing with the other superstrings as far as consistency and finiteness is concerned.

Some variations on the theme of the heterotic string have been offered[77-80]. Narain[78] points out that using a Lorentzian lattice $\Gamma_{26-d,10-d}$ instead of the Euclidean lattice Γ_{16} allows one to algebraically compactify to e.g. d = 4 with an enlarged gauge group such as E(8) x E(7) x E(6) x SU(2) but with no chiral fermions. The Narain model has acceptable one-loop properties[78,64].

Finally, a new twist on the heterotic string has been recently provided by Alvarez-Gaume et al.[79] and independently by Dixon and Harvey[80]. By modifying the fermion number projection operator in an apparently trivial way they arrive at quite different-looking physics. The gauge group is O(16) x O(16), there is no tachyon and no ten-dimensional supersymmetry. The Weyl fermions are in the representation of O(16) x O(16) as follows

$$(128, 1)_L + (1, 128)_L + (16, 16)_R \tag{S.301}$$

which is fully anomaly-free with no gravitino. There is a dilaton tadpole infinity, however, indicating that the vacuum of flat d = 10 Minkowski space is unstable. No stable vacuum is known for this theory, but it does seem impossible for the stable vacuum to have ten dimensional spacetime supersymmetry because the numbers of bosons and fermions are unequal at each level.

S.8 SUPERSTRING PHENOMENOLOGY

The phenomenological requirement that there are chiral fermions on four-dimensional Minkowski spacetime M_4 dictates that the initial superstring in ten dimensions must have Yang-Mills gauge fields. This then eliminates the two closed superstrings with N = 2 supersymmetry (types IIA and IIB) which have no such gauge symmetry.

This leaves two superstrings with O(32) gauge group and one with E(8) x E(8)' gauge group. So let us begin by arguing[81] that the prospects for the latter group are the much brighter. The most attractive grand unified theory (GUT) groups are SU(5), SO(10), and E(6). SO(32) does not contain E(6). Obviously SO(32) contains SO(10) but we need the spinor 16 of SO(10) to describe families of quarks and leptons. Since the 32 of SO(32) is real, the only possible embedding which involves the SO(10) spinor is $32 = 16 + 16^*$; but the chiral fermions are in the adjoint of SO(32) which is the antisymmetric part of 32 x 32, is completely vector like and contains no spinors. Hence an SO(10) GUT is impossible starting from SO(32). Of course, we may embed SU(N), including SU(5)[82], according to

$$32 = N + N^* + (32 - 2N)1 \qquad (S.302)$$

and then the 496 contains the $(N^2 - 1)$-dimensional adjoint of SU(N), the combinations $[2] + [2]^*$, $[1] + [1]^*$ and singlets. (Here [k] means the antisymmetric k-rank tensor.) The surviving chiral fermions on M_4 must be anomaly free and hence must be

$$f\left([2] + (N - 4)[1]^*\right) \qquad (S.303)$$

where f is the family number. This is a trivial replication of families; nevertheless, it may be worth pursuing further since the open superstring may well be finite. If O(32) were the only allowed gauge group we would be happy to work with it.

But E(8) x E(8)' is even much more promising. It provides an embarrassment of riches: the role of the second E(8)' is still unclear. Just one E(8) contains the famous nesting of subgroups

$$E(8) \supset E(7) \supset E(6) \supset SO(10) \supset SU(5) \qquad (S.304)$$

so all the most popular GUT groups are possible. There are interesting maximal subalgebras such as SU(3) x E(6) under which the adjoint 248 of chiral fermions decomposes as

$$248 = (1, 78) + (8, 1) + (3, 27) + (\overline{3}, \overline{27}) \ . \qquad (S.305)$$

This gives rise to the possibility, as we shall see, of an E(6) GUT with 27-plet families. It is worth mentioning one other maximal subalgebra SU(9) of E(8) under which

$$248 = 80 + 84 + 84^* \ . \qquad (S.306)$$

This could give rise to SU(8) grand unification with chiral fermions

$$[56 + 8(\overline{28}) + 27(8)] \qquad (S.307)$$

with multiples of six families[81].

To be more specific about the phenomenology on M_4 it is necessary to obtain more information about the compact manifold K in

$$M_{10} \to M_4 \otimes K \ . \qquad (S.308)$$

The study of the manifold K involves geometry and topology and has been pursued mainly by Witten and collaborators[83-90]. The final word on K is not yet given, and whether what follows on compactification is physics or only very interesting mathematical physics in unclear. Several authors[91-102] have investigated the superstring phenomenology based on specific choices for K in Eq. (S.308).

We follow the analysis of Candelas, Horowitz, Strominger and Witten[83]. The zero-mass sector of the heterotic string couple according to N = 1 supergravity in d = 10 coupled to N = 1 super Yang-Mills with gauge group either 0(32) or E(8) x E(8). The supergravity fields are (g_{MN}, ψ_M, B_{MN}, λ, ϕ) and the Yang-Mills fields are (F^a_{MN}, χ^a). Since the background fields on M_4 are required to be maximally symmetric, the background values for fermion fields vanish. We assume the compact manifold K in (S.308) is such that N = 1 supersymmetry is unbroken on M_4. This seems to be a useful assumption in restricting the choice of K. The vacuum should be invariant under a supersymmetry transformation generated by a parameter ε. The bose fields are automatically invariant while for the fermi fields the variations are [Greek indices are on M_4, latin indices on K)

$$\delta\psi_\mu = \nabla_\mu \varepsilon + \frac{1}{32}\sqrt{2}\, e^{2\phi}(\gamma_\mu \gamma_5 \otimes H)\varepsilon \tag{S.309}$$

$$\delta\lambda = \sqrt{2}\,(\gamma^m \nabla_m \phi)\varepsilon + \frac{1}{8} e^{2\phi} H\varepsilon \tag{S.310}$$

$$\delta\psi_m = \nabla_m \varepsilon + \frac{1}{32}\sqrt{2}\, e^{2\phi}(\gamma_m H - 12 H_m)\varepsilon \tag{S.311}$$

$$\delta\chi^a = -\frac{1}{4} e^{\phi} F^a_{mn} \gamma^{mn} \varepsilon \tag{S.312}$$

where H_{pqr} is the field strength derived from B_{pq} according to Eq. (S.76) above and $H = H_{pqr}\gamma^{pqr}$, $H_m = H_{mqr}\gamma^{qr}$. The idea now is to find the geometrical restrictions imposed on the compact manifold K by requiring $\delta\psi_\mu = \delta\lambda = \delta\psi_m = \delta\chi^a = 0$ for some nonzero ε.

Start with $\delta\psi_\mu = 0$ from Eq. (S.309). By acting again on ε with ∇_ν and forming the commutator one finds

$$[\nabla_\mu, \nabla_\nu]\varepsilon - \frac{1}{(16)^2} e^{4\phi}(\gamma_{\mu\nu} \otimes H^2)\varepsilon = 0 \; . \tag{S.313}$$

Let M_4 be a maximally symmetric space, so that [see e.g. Chapter 13 of Ref. 103]

$$R_{\mu\nu\rho\sigma} = \kappa(g_{\mu\rho} g_{\nu\sigma} - g_{\mu\sigma} g_{\nu\rho}) \tag{S.314}$$

where $\kappa > 0$ for de Sitter space, $\kappa < 0$ for anti-de Sitter space and $\kappa = 0$ for flat Minkowski space. Then

$$[\nabla_\mu, \nabla_\nu]\varepsilon = \frac{1}{4} R_{\mu\nu\rho\sigma}\gamma^{\rho\sigma}\varepsilon = \frac{1}{2}\kappa\,\gamma_{\mu\nu}\varepsilon \; . \tag{S.315}$$

It follows from Eqs. (S.313) and (S.315) that

$$H^2\varepsilon = (16)^2\, e^{-4\phi}\,\frac{1}{2}\kappa\varepsilon \; . \tag{S.316}$$

That is, ε is an eigenspinor of H^2. Since H is antihermitian, its eigenvalues are pure imaginary and therefore $\kappa < 0$.

From Eq. (S.310), $\delta\lambda = 0$ requires that

$$(\sqrt{2}\ \gamma^m \nabla_m \phi + \frac{1}{8}\ e^{2\phi}\ H)^2 \varepsilon = 0 \tag{S.317}$$

and, using Eq. (S.316), this gives

$$\frac{1}{8}\ \sqrt{2}\ e^{2\phi}\{\gamma^m \nabla_m \phi, H\}\varepsilon = -\ 2(\nabla_m \phi \nabla^m \phi + \kappa)\varepsilon\ . \tag{S.318}$$

The operator on the left side of Eq. (S.318) is antihermitian and hence must have imaginary eigenvalues, but the coefficient on the right side is real and hence must vanish:

$$\nabla_m \phi \nabla^m \phi = -\ \kappa\ . \tag{S.319}$$

The dilaton field ϕ depends only on the internal coordinate in the compact manifold K. Since ϕ must have a maximum somewhere, we conclude that

$$\nabla_m \phi = 0 \tag{S.320}$$

$$\kappa = 0 \tag{S.321}$$

$$H\varepsilon = 0\ . \tag{S.322}$$

In particular, Eq. (S.321) tells us that M_4 is flat Minkowski space. This is nontrivial because unbroken supersymmetry is possible for $\kappa < 0$. The cosmological constant Λ_4 is determined to be zero at the compactification scale. Whether Λ_4 remains zero after supersymmetry is broken is an unsolved problem.

The vanishing of $\delta\psi_m$, Eq. (S.311), gives

$$(\nabla_m - \beta H_m)\varepsilon = 0\ . \tag{S.323}$$

Because of Eq. (S.322), Eq. (S.323) implies

$$\gamma^m \nabla_m \epsilon = 0 . \tag{S.324}$$

The supersymmetry parameter ϵ is real and is chiral satisfying

$$\Gamma^{11} \epsilon = (\gamma^5 \otimes \Gamma^7)\epsilon = - \epsilon . \tag{S.325}$$

This means that ϵ may be expanded as

$$\epsilon = \sum_{i=1}^{4} \xi_i \otimes \eta_i \tag{S.326}$$

where the η_i are four real c-number spinors in the internal space (ξ_i are real anticommuting-number spinors in M_4). Since these η_i may not be eigenspinors of Γ^7, the existence of ϵ satisfying the relevant equations implies at least two η satisfying the same equations. Since the η satisfy Eq. (S.323), it is covariantly constant (with torsion $-4\beta H_{pqr}$) and hence may be normalized $\eta^T \eta = 1$. From such a normalized 8-component real η the almost complex structure is

$$J_m^{\ n} = - i\eta^T \gamma_m^{\ n} \Gamma^7 \eta \tag{S.327}$$

which satisfies

$$J_m^{\ n} J_n^{\ p} = - \delta_m^{\ p} \tag{S.328}$$

a generalization of $(i)^2 = - 1$. For $J_m^{\ n}$ to be integrable, yielding a complex manifold, the Nijenhuis tensor [see e.g. Chapter 5 of Ref. 104]

$$N_{mn}^{\ \ p} = J_m^{\ q} J_{[n\ ;q]}^{\ \ p} - J_n^{\ q} J_{[m\ ;q]}^{\ \ p} \tag{S.329}$$

must vanish. Using Eqs. (S.323) and (S.327) gives

$$\nabla_p J_{mn} = - 8\beta H_{sp[m} J_{n]}^{\ \ s} \tag{S.330}$$

and hence the Nijenhuis tensor is proportional to

$$\eta^T \{H, \gamma_{mn}{}^p\} \eta \tag{S.331}$$

which vanishes. Thus K is a complex manifold. It can be shown that Eq. (S.330) implies

$$H_{mnp} = -\frac{1}{16\beta} (g)^{1/2} \varepsilon_{mnprst} \nabla^r J^{st} . \tag{S.332}$$

Now H must further satisfy

$$dH = \text{tr } R^2 - \frac{1}{30} \text{tr } F^2 \tag{S.333}$$

with a background F which $\delta\chi^a = 0$ from Eq. (S.312) requires

$$F^a_{mn} \gamma^{mn} \eta = 0 . \tag{S.334}$$

A nonzero solution for H_{mnp} satisfying Eqs. (S.332)-(S.334) may be possible but certainly the simplest solution at this point is $H_{mnp} = 0$. This then gives the covariantly constant spinor

$$\nabla_n \eta = 0 . \tag{S.335}$$

This implies by commutation of covariant derivatives

$$[\nabla_m, \nabla_n] \eta = \frac{1}{4} R_{mnpq} \gamma^{pq} \eta = 0 . \tag{S.336}$$

Now act with γ^n and use cyclicity of the Riemann-Christoffel tensor

$$R_{mnpq} + R_{mpqn} + R_{mqnp} = 0 \tag{S.337}$$

and Dirac algebra to conclude that

$$R_{mn} = 0 . \tag{S.338}$$

Also, Eqs. (S.327) and (S.335) give the covariantly constant complex structure

$$\nabla^m J_{pq} = 0 \,. \tag{S.339}$$

Eqs. (S.338) and (S.339) imply that the manifold K is both Ricci flat and Kahler.

In general, a compact manifold does not allow a Ricci flat Kahler metric because of topological obstructions which are conveniently described by the holonomy group of K. The spin connection on K would, in general, be an $O(6) \sim SU(4)$ gauge field. The real covariantly constant spinor η is in the $(4 + \bar{4})$ representation of SU(4) and the existence of such a nonzero η means the holonomy is SU(3)--just as the VEV of a Higgs scalar in such a representation would break SU(4) to SU(3).

To see what this means, first consider a six-dimensional manifold with holonomy $U(3) \simeq SU(3) \times U(1)$. Such a manifold is Kahler and its line element can be expressed in terms of the real scalar function $K(z, \bar{z})$ by

$$ds^2 = \frac{\partial^2 K}{\partial z^i \partial \bar{z}^j} dz^i d\bar{z}^j \tag{S.340}$$

where we have written the infinitesimal coordinate as a $dx^m = a_i dz^i + a_{\bar{j}} d\bar{z}^j$ $(1 \leqslant m \leqslant 6,\ 1 \leqslant i, j \leqslant 3)$ with dz^i, $d\bar{z}^j$ transforming as 3, $\bar{3}$ under U(3).

Whether K admits a Kahler metric with SU(3) holonomy depends on whether a Kahler potential can be chosen such that the U(1) part of the spin connection vanishes. Let F_{mn} be the corresponding U(1) field strength then if S is some closed two-surface in K the quantity

$$I(S) = \int_S dS^{mn} F_{mn} \tag{S.341}$$

obeys a (Dirac) quantization condition and hence cannot vary under smooth changes of $K(z, \bar{z})$. Thus if $I(S) \neq 0$ for any S, K has no Kahler metric with $F_{mn} = 0$. This is a topological obstruction.

If $I(S) = 0$ for all S, K has vanishing first Chern class. Calabi conjectured[105] in 1957 and Yau proved[106] in 1977 that a six-dimensional Kahler manifold with vanishing first Chern class admits a Kahler metric of SU(3) holonomy. Hence the name Calabi-Yau (CY) manifold.

Some simple examples of CY manifolds can be provided by subspaces of complex projective spaces CP^N, the space of $(N + 1)$ complex variables z^i with the identification $z_i \equiv \lambda z_i$ for all nonzero complex λ, and minus the origin.

Let us take k homogeneous polynomials of degree $d_1, \ldots d_k$ in CP^{k+3}; the subspace given by the simultaneous vanishing of all k polynomials is a six-dimensional Kahler manifold. The full Chern class $c = (1 + c_1 + c_2 + c_3)$ is generated by [a physicist's derivation is given in Ref. 86]

$$c = \frac{(1 + J)^{k+4}}{(1 + d_1 J) \ldots (1 + d_k J)} \tag{S.342}$$

where J is a normalized 2-form ($J^n = 0$, $n \geqslant 4$). The term in J^i is c^i and hence $c^1 = 0$ if and only if

$$\sum_{i=1}^{k} d_i = (k + 4) . \tag{S.343}$$

This gives five possibilities with all $d_i \geqslant 2$ [$d_i = 1$ simply reduces CP^N to CP^{N-1}]

(i) One quintic in CP^4

(ii) One quartic and one quadratic in CP^5

(iii) Two cubics in CP^5

(iv) A cubic and two quadratics in CP^6

(v) Four quadratics in CP^7.

These five CY manifolds are merely examples; some others are described in Ref. 107.

In order to proceed to phenomenology, let us first discuss the number of families on M_4 for the case $E(8) \rightarrow SU(3) \times E(6)$ using for the adjoint

$$248_L = (8, 1)_L + (1, 78)_L + (3, 27)_L + (\overline{3}, \overline{27})_L \ . \qquad (S.344)$$

The number of families is $|N|$ where $N = (n^L_{27} - n^L_{\overline{27}})$ and from Eq. (S.344) this is equal to the difference in the number of zero modes transforming like 3_L and $\overline{3}_L$ under SU(3) on K. This index of the Dirac operator on K is simply connected to the Euler characterisitc $\chi(K)$, a fundamental topological invariant of the manifold. In fact,

$$\chi(K) = \text{index } (4) - \text{index } (\overline{4}) \qquad (S.345)$$

$$= \text{index } (3 + 1) - \text{index } (\overline{3} + 1) \qquad (S.346)$$

$$= 2 \text{ index } (3) \qquad (S.347)$$

so that the number of families is

$$|N| = \frac{1}{2} |\chi(K)| \ . \qquad (S.348)$$

We may compute $\chi(K)$ for the five examples of CY manifolds given above. From Eq. (S.342) we compute the third Chern class as $c_3 = \alpha_3 J^3$ then use the fact that the integral of J^3 over K is the product $\prod_{i=1}^{k} d_i = D$. Thus $\chi = = \alpha_3 D$. The results are respectively

(i) $d_1 = D = 5$; $\alpha_3 = -40$; $\chi = -200$

(ii) $d_1 = 4$, $d_2 = 2$, $D = 8$; $\alpha_3 = -22$; $\chi = -176$

(iii) $d_1 = d_2 = 3$, $D = 9$; $\alpha_3 = -16$; $\chi = -144$

(iv) $d_1 = 3$, $d_2 = d_3 = 2$, $D = 12$; $\alpha_3 = -12$; $\chi = -144$

(v) $d_1 = d_2 = d_3 = d_4 = 2$, $D = 16$; $\alpha_3 = -8$; $\chi = -128$

The numbers of families are respectively 100, 88, 72, 72 and 64. Other connected Calabi-Yau manifolds have a similarly large Euler characteristic, giving unacceptably many familes by Eq. (S.348).

The Euler characteristic can sometimes be drastically reduced if there is a discrete symmetry (G) acting freely, with no fixed points, on K. Then the multiply-connected manifold K/G has $\chi(K/G) = \chi(K)/n(G)$ where n(G) is the number of elements of G. For example, the manifold (i) can be shown[83] to admit $G = Z_5 \times Z_5$ with n(G) = 25 so that $\chi(K/G) = -8$ and there are only 4 families. In this way, examples of CY manifolds with various small numbers of families, including three[107], have been constructed.

Making the manifold K multiply-connected has another attractive advantage for breaking of the gauge symmetry in M_4. A gauge field A_m^a (m is an index in K) may be nontrivial even though $F_{mn}^a = 0$ by arranging a Wilson line γ which is noncontractible and such that

$$U = \exp \left(\int_\gamma A_m \, dx^m \right) \qquad (S.349)$$

be unequal to one. The gauge group is then broken to the subgroup which commutes with U. Such a mechanism is similar to having a Higgs field in the adjoint representation[108]. If our discrete group G is abelian, the symmetry breaking is rank preserving; if G is nonabelian the rank can be reduced.

Let us consider the simple scenario where we begin with the E(8) x E(8)' heterotic string. We choose a CY manifold giving three families $3(27)_L$ and break E(8) x E(8)' to E(6) x E(8)'. Wilson lines can break E(6) to $SU(3) \times SU(2) \times U(1)^p$ but p = 2 or 3 (that is, overall rank is at least five) for the following reason[84]. Consider

$$E(6) \rightarrow SU(3)_C \times SU(3)_L \times SU(3)_R \qquad (S.350)$$

$$27_L = (1, \bar{3}, 3)_L + (3, 3, 1)_L + (\bar{3}, 1, \bar{3})_L \, . \qquad (S.351)$$

The $SU(3)_R$ singlets are the nine states (α = 1, 2, 3 for color)

$$\binom{u_\alpha}{d_\alpha}_L \quad d'_{\alpha L} \, .$$

The $SU(3)_L$ singlets are the nine states

$$\bar{u}_{\alpha L} \quad \bar{d}_{\alpha L} \quad \bar{d}'_{\alpha L} \, .$$

Consequently the weak hypercharge $Y = 2(Q - T_3)$ is given in the basis (S.350) by

$$Y = (0) + \begin{pmatrix} 1/3 & & \\ & 1/3 & \\ & & 2/3 \end{pmatrix} + \begin{pmatrix} -2/3 & & \\ & -2/3 & \\ & & 4/3 \end{pmatrix} . \qquad \text{(S.352)}$$

The U(1) charge Y' orthogonal to Y remains unbroken since we cannot break a U(1) by Wilson lines, and hence we are left with rank at least five. This has led to phenomenological discussions of new neutral currents.[86,96,98,100] Also, with 27-plets of E(6) one may look for additional chiral fermions belonging, for example, to the $(\bar{5} + \bar{5})$ of the SU(5) subgroup of E(6) since $10 + \bar{5} + (5 + \bar{5}) + 2(1)$.[101] Actually, more general analysis[89,102] shows that there exist candidate vacua where the gauge group in M_4 may be only rank four with no extra neutral currents and no extra chiral fermions; thus such new states, "suggested" by superstrings, are not required.

The Yukawa couplings on M_4 can be expressed[86,109] as integrals over K of the relevant zero modes. Remarkably, such integrals can be evaluated in terms of topology alone, being for example proportional to the intersection number of three four-dimensional surfaces within the six-dimensinal manifold K[109].

At the empirical level, if N = 1 supersymmetry remains unbroken down to an energy of order on the electroweak scale, dimension-four terms in the effective lagrangian may lead to fast proton decay[84]. In supersymmetric GUTs such terms were always eliminated by imposition of a symmetry; in superstrings, there is no such freedom and it is not obvious that the appropriate symmetry is already present. A related problem is that neutrino masses could be too large if low-energy supersymmetry is preserved. The mechanism of supersymmetry breaking in superstrings is unknown; one suggestion is in Ref. 85. Particularly urgent is to understand how the cosmological constant (see Eq. (S.321) above) remains zero after supersymmetry breaking. More generally, a question is: why is there compactificatin at all to $M_4 \times K$? What is unstable about flat M_{10}?

The consistency of a given background manifold for the superstring can be usefully examined by the study of two-dimensional sigma models. The conformal invariance of a supersymmetric sigma model with target manifold K is necessary for consistency of the superstring compactification[110,111]. The equations that the renormalization group β-function vanishes in the second quantized sigma model coincide with the classical first quantized equations of motion for the massless states of the superstring. In particular, the CY manifold K requires conformal invariance and hence ultra-violet finiteness of an N = 2 supersymmetric sigma model. The N = 4 case is known to be finite[112,113]. A "proof" of N = 2 finiteness had been given[114]. Grisaru, van de Ven and Zanon[115] have recently shown, however, that the N = 2 model is not finite: there are 4-loop counterterms. The low-energy analysis of the superstring S-matrix confirms this result[90,116]. Perhaps the answer lies in adding torsion [e.g. Ref. 117]. Or perhaps the suggestion of Ref. 88 is more relevant: the manifold K should be an "orbifold" or a smooth manifold, such as a torus (on which the superstring certainly exists) factored by a discrete group with fixed points. The resultant singularities must be "patched" to form a manifold.

In any case, the new constraints on the six-dimensional compact manifold K should be welcomed, since they may lead to some testable predictions.

S.9 SUMMARY

The superstring theory is not only able to describe quantum gravity but requires that gravity exist because, even starting with open superstrings containing only massless states of spin one and lower, the massless spin two graviton appears inevitably as a closed string state since an open string can become a closed string by joining its two ends together. The remarkable thing about this quantum gravity model is that it was discovered not by studying Einstein's theory of general relativity but by identifying the graviton within a framework initially invented for strong interaction phenomenology.

The five standard superstrings are the O(32) open superstring, the two N = 2 closed superstrings and the two heterotic strings with O(32) and E(8) x E(8). As far as finiteness is concerned, results at one loop are very encouraging. These theories have no ultraviolet difficulties and the good infrared properties seem likely to persist to all orders.

The attempts at superstring phenomenology are still in their infancy. The nature of the six-dimensional compactified manifold is still under investigation and more formal progress is essential before linking superstrings to particle physics.

REFERENCES

1. Y. Nambu, in Preludes in Theoretical Physics, ed. A. De Shalit, North Holland (1966).
H. Fritzsch and M. Gell-Mann, in Proceedings of the Sixteenth International Conference on High Energy Physics, Chicago (1972) vol. 2, p. 135;
S. Weinberg, Phys. Rev. Lett. 31, 494 (1973);
D. J. Gross and F. Wilczek, Phys. Rev. D8, 3633 (1973);
H. Fritzsch, M. Gell-Mann and H. Leutwyler, Phys. Lett. 47B, 365 (1973).
2. D. J. Gross and F. Wilczek, Phys. Rev. Lett. 26, 1343 (1973);
H. D. Politzer, Phys. Rev. Lett. 26, 1346 (1973).
3. K. G. Wilson, Phys. Rev. D10, 2445 (1974).
4. M. Creutz, Phys. Rev. Lett. 43, 553 (1979).
5. T. Yoneya, Nuov. Cim. Lett. 8, 951 (1973).
6. T. Yoneya, Prog. Theor. Phys. 51, 1907 (1974).
7. J. Scherk and J. H. Schwarz, Nucl. Phys. B81, 118 (1974).
8. J. Scherk and J. H. Schwarz, Phys. Lett. 52B, 347 (1974).
9. J. Scherk and J. H. Schwarz, Phys. Lett. 57B, 463 (1975).
10. J. A. Shapiro, Phys. Lett. 33B, 361 (1970).
11. M. A. Virasoro, Phys. Rev. 177, 2309 (1969).
12. E. Del Guidice and P. Di Vecchia, Nuov. Cim. 5A, 90 (1971).
13. M. Yoshimura, Phys. Lett. 34B, 79 (1971).
14. S. N. Gupta, Proc. Phys. Soc. 63A, 46 (1950); ibid. 64A, 850 (1951); ibid. 65A, 161 (1952); Rev. Mod. Phys. 29, 334 (1957).
15. S. Weinberg, Phys. Rev. 135, B1048 (1964); ibid. 138, B988 (1965).
16. P. Ramond, Phys. Rev. D3, 2415 (1971).
17. A. Neveu and J. H. Schwarz, Nucl. Phys. B31, 86 (1971).
18. J. L. Gervais and B. Sakita, Nucl. Phys. B34, 632 (1971).

19. D. B. Fairlie and D. Martin, Nuov. Cim. 18A, 373 (1973); ibid. 21A, 647 (1974).
20. C. Montonen, Nuov. Cim. 19A, 69 (1974).
21. P. H. Frampton, Nuov. Cim. Lett. 8, 525 (1973); Phys. Rev. D9, 487 (1974).
22. P. Frampton in Laws of Hadronic Matter, ed. A. Zichichi, Academic Press (1975) p. 112-136.
23. F. Gliozzi, J. Scherk and D. Olive, Nucl. Phys. B122, 23 (1977).
24. M. B. Green and J. H. Schwarz, Nucl. Phys. B181, 502 (1981).
25. Y. A. Golfand and E. P. Likhtman, JETP Lett. 13, 323 (1971).
26. D. Volkov and V. P. Akulov, Phys. Lett. 46B, 109 (1973).
27. J. Wess and B. Zumino, Nucl. Phys. B70, 39 (1974).
28. B. Zumino in Renormalization and Invariance in Quantum Field Theory, ed. E. Caianello, Plenum Press (1974) p. 367.
29. A. Salam and J. Strathdee, Nucl. Phys. B76, 477 (1974).
30. P. C. West, Proceedings of the 1983 Shelter Island Conference, MIT Press (1985) p. 127.
31. P. Fayet and S. Ferrara, Phys. Reps. 32C, 249 (1977); H. P. Nilles, Phys. Reps. 110C, 1 (1984).
32. J. Wess and J. Bagger, Supersymmetry and Supergravity, Princeton University Press (1983).
33. S. J. Gates, M. T. Grisaru, M. Rocek and W. Siegel, Superspace, Benjamin/Cummings Frontiers in Physics (1983).
34. D. Z. Freedman, P. Van Nieuwenhuizen and S. Ferrara, Phys. Rev. D13, 3214 (1976).
35. P. Van Nieuwenhuizen, Phys. Reps. 68, 189 (1981).
36. E. Bergshoeff, M. De Roo, B. De Wit and P. Van Nieuwenhuizen, Nucl. Phys. B195, 97 (1982).
37. G. F. Chapline and N. S. Manton, Phys. Lett. 120B, 105 (1983).
38. E. T. Whittaker and G. N. Watson, A Course of Modern Analysis, 4th edition, Cambridge University Press (1962) p. 470.

39. M. B. Green and J. H. Schwarz, Phys. Lett. 109B, 444 (1982).
40. J. H. Schwarz, in Proceedings of the Johns Hopkins Workshop on Current Problems in Particle Theory, Florence (June 1982) p. 233.
41. Z. Horvath, L. Palla, E. Cremmer and J. Scherk, Nucl. Phys. B127, (1977);
L. Palla, in Proceedings of the 19th International Conference on High Energy Physics, Tokyo (1978) p. 629.
42. P. H. Frampton, Phys. Lett. 122B, 351 (1983);
P. H. Frampton and T. W. Kephart, Phys. Rev. Lett. 50, 1343 (1983) [Reprint No. 1]; ibid. 50, 1347 (1983); Phys. Rev. D28, 1010 (1983).
43. P. H. Frampton and T. W. Kephart, Phys. Lett. 131B, 80 (1983).
44. P. H. Frampton, Phys. Lett. 140B, 313 (1984).
45. L. Caneschi, G. Farrar and A. Schwimmer, Phys. Lett. 138B, 386 (1984).
46. P. H. Frampton and K. Yamamoto, Phys. Rev. Lett. 52, 2016 (1984).
47. P. H. Frampton and K. Yamamoto, Nucl. Phys. 254B, 349 (1985).
48. S. Randjbar-Daemi, A. Salam and J. Strathdee, Phys. Lett. 124B, 345 (1983);
P. H. Frampton, P. J. Moxhay and K. Yamamoto, Phys. Lett. 144B, 354 (1984).
49. S. Randjbar-Daemi, A. Salam and J. Strathdee, Nucl. Phys. B214, 491 (1983).
50. S. Randjbar-Daemi, A. Salam and J. Strathdee, Phys. Lett. 132B, 386 (1984); Nucl. Phys. B242, 447 (1984).
51. H. Georgi, Nucl. Phys. B156, 126 (1979).
52. P. H. Frampton, Phys. Lett. 89B, 352 (1980).
53. M. B. Green and J. H. Schwarz, Phys. Lett. 149B, 117 (1984) [Reprint No. 2].

54. L. Alvarez-Gaume and E. Witten, Nucl. Phys. BH234, 269 (1984).
55. P. H. Frampton and K. Yamamoto, Phys. Lett. 156B, 345 (1985).
56. J. H. Schwarz, Phys. Reps. 89, 223 (1982).
57. Such a chapter had been sketched out for "Dual Resonance Models" in 1974 but was omitted for reasons of space.
58. D. Amati, C. Bouchiat and J. L. Gervais, Nuov. Cim. Lett. 2, 399 (1969).
59. M. B. Green and J. H. Schwarz, Phys. Lett. 151B, 21 (1985).
60. P. H. Frampton, P. Moxhay and Y. J. Ng, Phys. Rev. Lett. 55, 2107 (1985).
61. P. H. Frampton, P. Moxhay and Y. J. Ng, Nucl. Phys. B (1986, to appear).
62. L. Clavelli, Phys. Rev. D33, 1098 (1986).
63. J. A. Shapiro, Phys. Rev. D5, 1945 (1972).
64. P. H. Frampton, Y. Kikuchi and Y. J. Ng, Phys. Lett. B (1986, to appear).
65. G. Gilbert, University of Texas preprint UTTG-23-85 (1985); S. Mandelstam, UC Berkeley preprint (1985).
66. R. Catenacci, M. Cornalba, M. Martellini and C. Reina, Pavia-Milan preprint (1986).
67. J. Thierry-Mieg, Phys. Lett. 156B, 199 (1985).
68. P. G. O. Freund, Phys. Lett. 151B, 387 (1985) [Reprint No. 3].
69. A. Casher, F. Englert, H. Nicolai and A. Taormina, Phys. Lett. 162B, 121 (1985).
70. J.-P. Serre, Cours d'Arithmétique, Presses Universitaires de France (1970) pages 95 and 177.
71. D. J. Gross, J. A. Harvey, E. Martinec and R. Rohm, Phys. Rev. Lett. 54, 502 (1985) [Reprint No. 4].
72. D. J. Gross, J. A. Harvey, E. Martinec and R. Rohm, Nucl. Phys. B256, 253 (1985).

73. D. J. Gross, J. A. Harvey, E. Martinec and R. Rohm, Nucl. Phys. B267, 75 (1986).
74. I. B. Frenkel and V. G. Kac, Inv. Math. 62, 23 (1980); G. Segal, Comm. Math. Phys. 80, 301 (1981).
75. See the contributions by Louise Dolan and Peter Goddard in: Vertex Operators in Mathematical Physics, eds. J. Lepowsky, S. Mandelstam and I. M. Singer, Springer-Verlag (1985).
76. S. Yahikozawa, Phys. Lett. 166B, 135 (1986).
77. R. Bluhm and L. Dolan, Phys. Lett. 169B, 347 (1986).
78. K. S. Narain, Phys. Lett. 169B, 41 (1986).
79. L. Alvarez-Gaume, P. Ginsparg, G. Moore and C. Vafa, Harvard University preprint HUTP-86/A013 (1986).
80. L. J. Dixon and J. A. Harvey, Princeton University preprint (1986).
81. P. H. Frampton, H. van Dam and K. Yamamoto, Phys. Rev. Lett. 54, 1114 (1985).
82. E. Witten, Phys. Lett. 149B, 351 (1984).
83. P. Candelas, G. T. Horowitz, A. Strominger and E. Witten, Nucl. Phys. B258, 46 (1985).
84. E. Witten, Nucl. Phys. B258, 75 (1985).
85. M. Dine, R. Rohm, N. Seiberg and E. Witten, Phys. Lett. 156B, 55 (1985).
86. A. Strominger and E. Witten, Comm. Math. Phys. 101, 341 (1985).
87. X.-G. Wen and E. Witten, Nucl. Phys. B261, 651 (1985).
88. L. Dixon, J. A. Harvey, C. Vafa and E. Witten, Nucl. Phys. B261, 678 (1985) and Princeton University preprint (1986).
89. E. Witten, Nucl. Phys. B268, 79 (1986).
90. D. J. Gross and E. Witten, Princeton University preprint (1986).
91. J. D. Breit, B. A. Ovrut and G. C. Segre, Phys. Lett. 158B, 33 (1985).

92. M. Dine, V. Kaplunovsky, M. Mangano, C. Nappi and N. Seiberg, Nucl. Phys. B259, (1985).

93. K. Pilch and A. Schellekens, Nucl. Phys. B259, 637 (1985).

94. M. Dine and N. Seiberg, Phys. Rev. Lett. 55, 366 (1985).

95. V. S. Kaplunovsky, Phys. Rev. Lett. 55, 1036 (1985).

96. S. M. Barr, Seattle preprint (1985).

97. T. Hubsch, H. Nishino and J. C. Pati, Phys. Lett. 163B, 111 (1985).

98. R. Holman and D. B. Reiss, Phys. Lett. 166B, 305 (1986).

99. G. Lazarides, C. Panagiotakopoulos and Q. Shafi, Phys. Rev. Lett. 56, 432 (1986).

100. P. Binetruy, S. Dawson, I. Hinchliffe and M. Sher, LBL preprint (October 1985).

101. V. Barger, N. Deshpande, R. J. N. Phillips and K. Whisnant, Phys. Rev. D33, 1912 (1986).

102. H. W. Braden, P. H. Frampton, T. W. Kephart and A. K. Kshirsagar, UNC-Chapel Hill preprint IFP-268-UNC (1986).

103. S. Weinberg, Gravitation and Cosmology, Wiley (1972).

104. Y. Choquet-Bruhat, C. DeWitt-Morette and M. Dillard-Bleick, Analysis, Manifolds and Physics, North Holland (Revised Edition, 1982).

105. E. Calabi, in Algebraic Geometry and Topology: A symposium in honor of S. Lefschetz, Princeton University Press (1957) p. 78.

106. S.-T. Yau, Proc. Nat. Acad. Sci. 74, 1798 (1977).

107. S. T. Yau, in Symposium on Anomalies, Geometry, Topology, eds. W. A. Bardeen and A. R. White, World Scientific (1985) p. 395.

108. Y. Hosotani, Phys. Lett. 129B, 193 (1983).

109. A. Strominger, Phys. Rev. Lett. 55, 2547 (1985).

110. C. G. Callan, D. Friedan, E. J. Martinec and M. J. Perry, Nucl. Phys. B262, 593 (1985).

111. A. Sen, Phys. Rev. Lett. 55, 1846 (1985).

112. L. Alvarez-Gaume and P. Ginsparg, Comm. Math. Phys. 102, 311 (1985).
113. C. M. Hull, Nucl. Phys. B260, 182 (1985).
114. L. Alvarez-Gaume, S. Coleman and P. Ginsparg, Comm. Math. Phys. 103, 423 (1986).
115. M. T. Grisaru, A. E. M. van de Ven, and D. Zanon, Harvard University preprints (1986).
116. M. D. Freeman and C. N. Pope, Imperial College preprint (1986).
117. A. Strominger, Santa Barbara preprint NSF-ITP-86-16 (1986).

VOLUME 50, NUMBER 18 PHYSICAL REVIEW LETTERS 2 MAY 1983

Explicit Evaluation of Anomalies in Higher Dimensions

Paul H. Frampton and Thomas W. Kephart
Institute of Field Physics, Department of Physics and Astronomy, University of North Carolina, Chapel Hill, North Carolina 27514

(Received 17 February 1983)

The one-loop Kaluza-Klein anomaly is evaluated explicitly for gauge theories in six, eight, and ten dimensions. The result is well defined and unique, despite the nonrenormalizability of the theory.

PACS numbers: 11.10.Gh, 11.15.Ex, 11.30.Rd

Strings have played a role in many branches of physics from polyatomic molecules to vortices in type-II superconductors. Recently the space-time analog of vortices has entered cosmology (cosmic strings) in the theory of fluctuations and galaxy formation. It is highly probable that strings may soon play a role in particle physics. There has long been the curiosity that the $D=10$ string theory may be finite and recent advances in the superstring have demonstrated the absence of tachyons, etc. Since $D=10$ dimensions is *required* for the string, a successful dimensional compactification of this theory could lead to a viable quantum gravity.

It is generally believed that all elementary-particle interactions *except* gravity are describable by non-Abelian gauge theories. Such gauge theories share with quantum electrodynamics the property of perturbative renormalizability and hence complete calculability for small couplings. This domain of applicability covers the electroweak forces for any normal energy (less than a few gigaelectronvolts) and the strong forces at high energy (more than a few gigaelectronvolts).

Historically the realization that gauge theories are renormalizable, even with the gauge group spontaneously broken, focused attention on the subject of chiral anomalies since the latter provided the only known consistency condition required for gauge theories. The anomalies had been studied much earlier and appeared then as only an arcane curiosity, but in gauge theories they become of central importance: The chiral fermions *must* be such that the anomaly cancels, otherwise the gauge theory is inconsistent and hence meaningless.

Now we wish to consider[1] a gauge theory in more than four dimensions, generally in an even number of dimensions such as 6, 8, 10, ... Such a theory is *a priori* nonrenormalizable and thus inconsistent and meaningless. The reasons for considering this inconsistent theory are two in number: First, contrary to expectations, we shall find that the one-loop anomaly in such a theory is well defined and unique, and independent of the severe ultraviolet divergences associated with nonrenormalizability. Second, this kind of nonrenormalizable gauge theory can arise as the limit of small Regge slope of a string model. We have in mind, for example, the ten-dimensional supersymmetric string.[2] Here there is the hope that the string theory is completely

FIG. 1. Triangle diagram for $D = 4$.

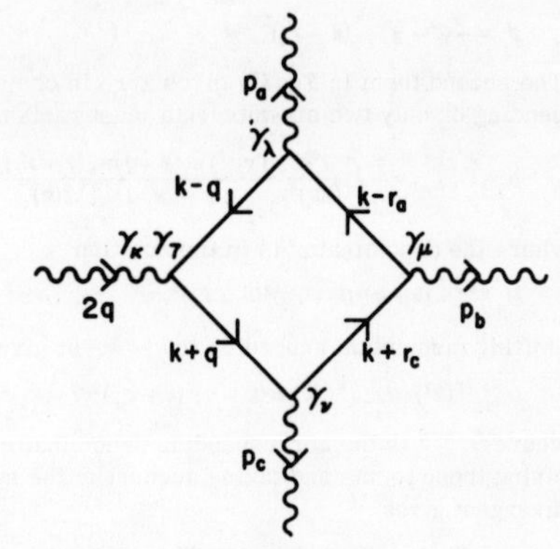

FIG. 2. Box diagram for $D = 6$.

VOLUME 50, NUMBER 18 PHYSICAL REVIEW LETTERS 2 MAY 1983

finite and hence all the ultraviolet divergences associated with the nonrenormalizability of the low-energy gauge theory are successfully cut off by the very-high-mass Regge recurrences and daughters.

In discussing the string theory, we must emphasize that although the ultraviolet divergences may be modified by the very-high-mass particles in the theory, the anomalies are *not* so modified. This is because, as has been shown elsewhere,[3] the anomaly receives contributions only from zero-mass fermions. This is why the anomalies in nonrenormalizable Kaluza-Klein gauge theory are interesting: First, they are well defined despite nonrenormalizability, and second, they are germane to finite higher-dimensional string theory.

In what follows, we shall first consider anomalies only in Abelian theories. The generalization to non-Abelian theory will be discussed later; it involves no new complication.

Let us first briefly recall the case of the triangle anomaly[4] for $D=4$. The relevant three-point function is written

$$V_{\kappa\lambda\mu}(p_1p_2)=S\Gamma_{\kappa\lambda\mu}(p_1p_2)+(a/8\pi^2)\epsilon_{\kappa\lambda\mu\alpha}(p_{2\alpha}-p_{1\alpha}), \tag{1}$$

where $\Gamma_{\kappa\lambda\mu}$ is the Feynman integral for Fig. 1 $[q=\frac{1}{2}(p_1+p_2),\ r=\frac{1}{2}(p_1-p_2)]$,

$$\Gamma_{\kappa\lambda\mu}(p_1p_2)=\int\frac{d^4k}{(2\pi)^4}\frac{\mathrm{Tr}[\gamma_\lambda(\not k-\not q)\gamma_\kappa\gamma_5(\not k+\not q)\gamma_\mu(\not k+\not r)]}{(k-q)^2(k+q)^2(k+r)^2}, \tag{2}$$

and $S\Gamma_{\kappa\lambda\mu}$ implies Bose symmetrization of $(p_{1\lambda},p_{2\mu})$. Setting $p_{1\lambda}V_{\kappa\lambda\mu}=0$ requires $a=+1$ in Eq. (1), whereupon

$$2q_\kappa V_{\kappa\lambda\mu}=2^3X_3\epsilon_{\lambda\mu\alpha\beta}p_{1\alpha}p_{2\beta} \tag{3}$$

with

$$X_3=+1/2^4\pi^2. \tag{4}$$

The factor 2^3 in Eq. (3) takes account of the three $(1+\gamma_5)/2$ projection operators needed in Eq. (2).

These steps can be carried out for $D>4$. For $D=6$, we write the four-point vertex

$$V_{\kappa\lambda\mu\nu}(p_ap_bp_c)=S\Gamma_{\kappa\lambda\mu\nu}(p_ap_bp_c)+(a/2^4\pi^3)\epsilon_{\kappa\lambda\mu\nu\beta}(p_{b\alpha}p_{c\beta}+p_{c\alpha}p_{a\beta}+p_{a\alpha}p_{b\beta}). \tag{5}$$

Here the Feynman amplitude $\Gamma_{\kappa\lambda\mu\nu}(p_ap_bp_c)$ for Fig. 2 is given by

$$\Gamma_{\kappa\lambda\mu\nu}(p_ap_bp_c)=\int\frac{d^6k}{(2\pi)^6}\frac{\mathrm{Tr}[\gamma_\lambda(\not k-\not q)\gamma_\kappa\gamma_7(\not k+\not q)\gamma_\nu(\not k+\not r_c)\gamma_\mu(\not k-\not r_a)]}{(k-q)^2(k+q)^2(k+r_c)^2(k-r_a)^2}. \tag{6}$$

Here $q=\frac{1}{2}(p_a+p_b+p_c)$ and $r_a=(q-p_a)$, etc. To evaluate $p_{a\lambda}\Gamma_{\kappa\lambda\mu\nu}$ one uses

$$\not p_a=-(\not k-\not q)+(\not k-\not r_a). \tag{7}$$

The second term in Eq. (7) gives zero in conjunction with Eq. (6) since a third-rank pseudotensor depending on only two six-momenta must vanish. The first term in (6) gives

$$p_{a\lambda}\Gamma_{\kappa\lambda\mu\nu}=-\int\frac{d^6k}{(2\pi)^6}\frac{\mathrm{Tr}[\gamma_\kappa\gamma_7(\not k+\not q)\gamma_\nu(\not k+\not r_c)\gamma_\mu(\not k-\not r_a)]}{D_+{}^{abc}(k)}, \tag{8}$$

where the denominator is in the notation

$$D_\pm{}^{abc}=(k\pm q)^2(k+r_c)^2(k-r_a)^2. \tag{9}$$

Shifting momentum according to $k'=k-p_a$ gives

$$D_+{}^{abc}(k')=D_-{}^{bca}(k)=(k-q)^2(k+r_a)^2(k-r_b)^2, \tag{10}$$

where $D_-{}^{bca}$ is the corresponding denominator occurring in the contraction $p_{a\lambda}\Gamma_{\kappa\mu\nu\lambda}(p_bp_cp_a)$. Combining these terms and taking account of the surface term arising from the integration being linearly divergent gives

$$p_{a\lambda}(\Gamma_{\kappa\lambda\mu\nu}+\Gamma_{\kappa\mu\nu\lambda})=-(48\pi^3)^{-1}\epsilon_{\kappa\mu\nu\alpha\beta\gamma}p_{a\gamma}p_{b\beta}p_{c\gamma}. \tag{11}$$

VOLUME 50, NUMBER 18 PHYSICAL REVIEW LETTERS 2 MAY 1983

Noting that this is totally Bose symmetric and that $p_{a\lambda}\Gamma_{\kappa\mu\lambda\nu}=p_{a\lambda}\Gamma_{\kappa\nu\lambda\mu}=0$ gives

$$p_{a\lambda}S\Gamma_{\kappa\lambda\mu\nu}=-(24\pi^3)^{-1}\epsilon_{\kappa\mu\nu\alpha\beta\gamma}p_{a\alpha}p_{b\beta}p_{c\gamma}. \tag{12}$$

Thus, in Eq. (5), we need $a=+1$ to ensure $p_{a\lambda}V_{\kappa\lambda\mu\nu}=0$. To evaluate $2q_\kappa\Gamma_{\kappa\lambda\mu\nu}$ we use a similar procedure because the same denominators occur. For example,

$$2q_\kappa\Gamma_{\kappa\lambda\mu\nu}\simeq\frac{8i}{(2\pi)^6}\int d^6k\,\epsilon_{\lambda\mu\nu\alpha\beta\gamma}\left[\frac{-p_{b\beta}p_{c\gamma}}{D_+{}^{abc}(k)}+\frac{p_{a\beta}p_{b\gamma}}{D_-{}^{abc}(k)}\right] \tag{13}$$

allows us to use again $D_+{}^{abc}(k')=D_-{}^{bca}(k)$, with $k'=k-p_a$, and hence combine with another term in $2q_\kappa S\Gamma_{\kappa\lambda\mu\nu}$. The overall result is

$$2q_\kappa S\Gamma_{\kappa\lambda\mu\nu}=-(8\pi^3)^{-1}\epsilon_{\lambda\mu\nu\alpha\beta\gamma}p_{a\alpha}p_{b\beta}p_{c\gamma}, \tag{14}$$

and finally the square anomaly for $D=6$, normalized by

$$2q_\kappa V_{\kappa\lambda\mu\nu}(p_ap_bp_c)=2^4X_4\epsilon_{\lambda\mu\nu\alpha\beta\gamma}p_{a\alpha}p_{b\beta}p_{c\gamma}, \tag{15}$$

is

$$X_4=-2^{-6}\pi^{-3}. \tag{16}$$

Of relevance to the $D=10$ string[1,2] is the hexagon anomaly which we have calculated to the one-loop level. In an obvious generalization of the notation used above we have

$$V_{\kappa\lambda\mu\nu\rho\sigma}(p_ap_bp_cp_dp_e)=S\Gamma_{\kappa\lambda\mu\nu\rho\sigma}(p_ap_bp_cp_dp_e)+(160\pi^5)^{-1}\epsilon_{\kappa\lambda\mu\nu\rho\sigma\alpha\beta\gamma\delta}(p_{b\alpha}p_{c\beta}p_{d\gamma}p_{e\delta}+\ldots). \tag{17}$$

In $\Gamma_{\kappa\lambda\mu\nu\rho\sigma}(p_ap_bp_cp_dp_e)$, contraction with, say, $p_{a\lambda}$ gives rise to denominators of the form [$q=\frac{1}{2}(p_a+p_b+p_c+p_d+p_e)$, $r_a=(q-p_a)$, etc., $s_{ab}=(q-p_a-p_b)$, etc.]

$$D_\pm{}^{abcde}(k)=(k\pm q)^2(k+r_e)^2(k+s_{de})^2(k-s_{ab})^2(k-r_a)^2 \tag{18}$$

and, with $k'=k-p_a$, one finds $D_+{}^{abcde}(k')=D_-{}^{bcdea}(k)$. The other pairs of denominators obtained by permuting $(p_{b\mu},p_{c\nu},p_{d\rho},p_{e\sigma})$ all lead to additive contributions and one has finally

$$p_{a\lambda}S\Gamma_{\kappa\lambda\mu\nu\rho\sigma}=-(160\pi^5)^{-1}\epsilon_{\kappa\mu\nu\rho\sigma\alpha\beta\gamma\delta\epsilon}p_{b\alpha}p_{c\beta}p_{d\gamma}p_{e\delta}p_{f\epsilon}. \tag{19}$$

Similarly, all the $2\times5!=240$ terms in $(2q)_\kappa S\Gamma_{\kappa\lambda\mu\nu\rho\sigma}$ combine to give equal-sign surface terms when we permute the five external momenta. This sum gives for the hexagon anomaly

$$2q_\kappa V_{\kappa\lambda\mu\nu\rho\sigma}=2^6X_6\epsilon_{\lambda\mu\nu\rho\sigma\alpha\beta\gamma\delta\epsilon}p_{a\alpha}p_{b\beta}p_{c\gamma}p_{d\delta}p_{e\epsilon} \tag{20}$$

with

$$X_6=-2^{-10}\pi^{-5}. \tag{21}$$

The general case $D=2n$ can be shown[5] to give

$$X_{n+1}=(-1)^n2^{-2n}\pi^{-n}. \tag{22}$$

The generalization to a non-Abelian theory is straightforward since it has been shown[6] that the group-theoretic piece of the anomaly is an overall multiplicative factor so that, for example, in $D=10$

$$2q_\kappa V^{ABCDEF}_{\kappa\lambda\mu\nu\rho\sigma}(p_ap_bp_cp_dp_e)=-(16\pi^5)^{-1}S\,\mathrm{Tr}(\Lambda^A\Lambda^B\Lambda^C\Lambda^D\Lambda^E\Lambda^F)\epsilon_{\lambda\mu\nu\rho\sigma\alpha\beta\gamma\delta\epsilon}p_{a\alpha}p_{b\beta}p_{c\gamma}p_{d\delta}p_{e\epsilon}, \tag{23}$$

where $S\,\mathrm{Tr}$ denotes the averaged totally symmetric trace over the group generators Λ^A, Λ^B,... in the appropriate basis.

We find it remarkable that these one-loop diagrams can be calculated completely to find unique anomalies in the higher-dimensional theories despite their nonrenormalizability. Such anomalies must be absent (cancelled) in a consistent higher-dimensional string theory.

We have only summarized the calculations which are actually rather lengthy and will be published in detail elsewhere.[5] The crucial point is that the anomalies are dictated by the homotopy group of mappings of the gauge group on the $(D-1)$-sphere in Euclidean space, and this is independent of the severe ultraviolet divergences. We are throughout regarding the higher space-time dimensions as physical, and the corresponding degrees of freedom as dynamical ones which must satisfy

VOLUME 50, NUMBER 18 PHYSICAL REVIEW LETTERS 2 MAY 1983

canonical commutation relations in a quantum theory.

This work was supported in part by the U. S. Department of Energy under Contract No. DE-AS05-79ER-10448.

[1]P. H. Frampton, Phys. Lett. 122B, 351 (1983).

[2]L. Brink, M. B. Green, and J. H. Schwarz, Nucl. Phys. B198, 474 (1982).

[3]P. H. Frampton, J. Preskill, and H. van Dam, to be published.

[4]S. Adler, Phys. Rev. 177, 2426 (1969); J. S. Bell and R. Jackiw, Nuovo Cimento 60, 47 (1969).

[5]P. H. Frampton and T. W. Kephart, University of North Carolina–Chapel Hill Report No. IFP-193-UNC, 1983 (to be published).

[6]P. H. Frampton and T. W. Kephart, following Letter [Phys. Rev. Lett. 50, 1347 (1983)].

Volume 149B, number 1,2,3 PHYSICS LETTERS 13 December 1984

ANOMALY CANCELLATIONS IN SUPERSYMMETRIC $D = 10$ GAUGE THEORY AND SUPERSTRING THEORY ☆

Michael B. GREEN
Queen Mary College, University of London, London E1 4NS, UK
and California Institute of Technology, Pasadena, CA 91125, USA

and

John H. SCHWARZ
California Institute of Technology, Pasadena, CA 91125, USA

Received 10 September 1984

Supersymmetric ten-dimensional Yang–Mills theory coupled to $N = 1$, $D = 10$ supergravity has gauge and gravitational anomalies that can be partially cancelled by the addition of suitable local interactions. The remaining pieces of all the anomalies cancel if the gauge group is SO(32) or $E_8 \times E_8$. These cancellations are automatically incorporated in the type I superstring theory based on SO(32). A superstring theory for $E_8 \times E_8$ has not yet been constructed.

It is not easy to obtain an effective four-dimensional theory containing chiral fermions by spontaneous compactification of a theory in $D > 4$ dimensions. It may be necessary that the D-dimensional theory itself contains chiral fermions and elementary Yang–Mills gauge fields [1]. But such theories are, in general, plagued by anomalies in the divergences of Yang–Mills gauge currents [2] and gravitational (local Lorentz) gauge currents [3]. Certain superstring theories are promising candidates for ten-dimensional quantum theories containing gravitation. (For reviews see ref. [4].) An essential requirement for their consistency is the absence of gauge and gravitational anomalies.

An interesting example of an anomaly-free chiral theory in higher dimensions is IIB $D = 10$ supergravity [5], which can be viewed as the low-energy approximation to the corresponding superstring theory (based on oriented closed strings). Gravitational anomalies have been shown to cancel in IIB supergravity [3], and probably do so in IIB superstring theory, as well. Since these theories contain no elementary Yang–Mills gauge fields, they are trivially free from Yang–Mills gauge anomalies. Unfortunately, the absence of such gauge fields may also exclude compactifications leading to a chiral theory in four dimensions [‡1].

In this letter we wish to report the discovery of other anomaly-free chiral theories in ten dimensions. One of these is type I superstring theory (which contains unoriented open and closed strings) with SO(32) as the gauge group. The mechanisms responsible for anomaly cancellations are explained below in more conventional field-theoretic terms by describing the low-energy effective field theory, which consists of $N = 1$, $D = 10$ supergravity coupled to supersymmetric Yang–Mills theory, even though any truncation of the complete string theory is probably not fully consistent. At the level of the effective field theory, $E_8 \times E_8$ is shown to be another possible anomaly-free theory.

The superstring results are summarized at the end of this paper, but details of the analysis are left for a future publication. The coupled $D = 10$ super Yang–

☆ Work supported in part by the US Department of Energy under Contract No. DEAC-03-81ER400050 and by the Fleischmann Foundation.

‡1 An interesting suggestion for circumventing this problem has been made recently [6].

Volume 149B, number 1,2,3 PHYSICS LETTERS 13 December 1984

Mills plus supergravity theory in its minimal form [7] contains gauge and gravitational anomalies. In order to understand the cancellations, the minimal lagrangian must be supplemented by additional higher-derivative local interactions that arise in the low-energy expansion of the superstring theory. Some of these terms are noninvariant under local Lorentz and Yang–Mills gauge transformations, with variations arranged to cancel those arising from fermion loops. They are analogous to "Wess–Zumino terms" that have been introduced in other contexts. Although such terms can be added to the action for any gauge group, for SO(32) or $E_8 \times E_8$ they can be chosen to completely cancel the anomalies. From the point of view of the effective field theory this appears to involve arbitrary choices of coefficients. However, these are precisely the coefficients that are contained in the SO(32) superstring theory.

The action for the interacting theory contains the bosonic terms

$$S_0 = \int d^{10}x\, e[-(1/2\kappa^2)R - (1/\kappa^2)\varphi^{-2}\partial_\mu\varphi\partial^\mu\varphi - (1/4g^2)\varphi^{-1}F^a_{\mu\nu}F^{\mu\nu a} - (3\kappa^2/2g^4)\varphi^{-2}H_{\mu\nu\rho}H^{\mu\nu\rho}]. \qquad (1)$$

This appears to contain two coupling constants, the Yang–Mills coupling g and the gravitational coupling κ, with dimensions of (length)^3 and (length)^4, respectively. (The ratio κ/g^2 is proportional to the string tension.) However, g^2 is not an arbitrary parameter of the action since it can be absorbed in the definition of φ. Thus there are no arbitrary dimensionless parameters. (We are grateful to E. Witten for pointing this out to us.) In eq. (1) e is the determinant of the zehnbein e^m_μ, and φ is a scalar field. The Yang–Mills field strength is defined (in the language of forms) by

$$F \equiv \tfrac{1}{2}F_{\mu\nu}dx^\mu \wedge dx^\nu = dA + A^2 . \qquad (2)$$

We usually omit $\wedge$ signs for simplicity, so that $A^2 \equiv A \wedge A$. The one-form potential A is a matrix representation of the gauge algebra

$$A = A^a_\mu \lambda^a dx^\mu . \qquad (3)$$

We only discuss SO(N), USp(N) and $E_8 \times E_8$ explicitly, but other groups can be analyzed in an analogous manner. It is important for the analysis that the λ matrices are antihermitean matrices in the *fundamental* representation of the algebra (the N of SO(N) or USp(N)), even though the massless fermions of the Yang–Mills sector belong to the adjoint representation. The trace of a matrix M is denoted by tr M when M is in the fundamental representation and by Tr M when M is in the adjoint representation. (Of course, these are identical for E_8.) The field strength H associated with the two-form potential $B = B_{\mu\nu}dx^\mu \wedge dx^\nu$ of the supergravity multiplet is defined for the case of SO(N) or USp(N) by

$$H = dB + \omega^0_{3L} - \omega^0_{3L} . \qquad (4)$$

In the case of $E_8 \times E_8$ this equation is replaced by

$$H = dB + \tfrac{1}{30}\omega^0_{3Y_1} + \tfrac{1}{30}\omega^0_{3Y_2} - \omega^0_{3L} . \qquad (4')$$

In eq. (4) ω^0_{3Y} is the Yang–Mills Chern–Simons three-form

$$\omega^0_{3Y} = \mathrm{tr}(AF - \tfrac{1}{3}A^3) , \qquad (5)$$

which satisfies

$$d\omega^0_{3Y} = \mathrm{tr}\, F^2 . \qquad (6)$$

In eq. (4′) $\omega^0_{3Y_1}$ and $\omega^0_{3Y_2}$ are Yang–Mills Chern–Simons forms (given by eq. (5)) for each of the two E_8 factors. The Yang–Mills Chern–Simons terms in eq. (4) would have a coefficient of $\frac{1}{30}$ (as in eq. (4′)) for the special case of SO(32) if it were defined in terms of matrices in the adjoint representation. The Lorentz group Chern–Simons form is given in identical fashion by

$$\omega^0_{3L} = \mathrm{tr}(\omega R - \tfrac{1}{3}\omega^3) \qquad (7)$$

and satisfies

$$d\omega^0_{3L} = \mathrm{tr}\, R^2 . \qquad (8)$$

Here $\omega = \omega_\mu dx^\mu$, the local Lorentz connection, is a 10×10 matrix of the tangent-space algebra SO(9,1), and $R = d\omega + \omega^2$ is the Lorentz-curvature two-form. The last term in eq. (4) is a higher-derivative term arising in the low-energy expansion of the superstring and therefore does not appear in ref. [7]. Its inclusion leads to a very similar treatment of the Yang–Mills local Lorentz symmetries.

Under infinitesimal local Yang–Mills transformations (with parameter Λ) and local Lorentz transformations (with parameter Θ), the fields transform as follows

$$\delta A = d\Lambda + [A, \Lambda] , \quad \delta\omega = d\Theta + [\omega, \Theta] , \qquad (9a,b)$$

Volume 149B, number 1,2,3 PHYSICS LETTERS 13 December 1984

$$\delta B = -\operatorname{tr}(A\mathrm{d}\Lambda) + \operatorname{tr}(\omega \mathrm{d}\Theta)\,,$$

$$\text{for } SO(N) \quad \text{or } USp(N)\,, \tag{9c}$$

$$\delta B = -\tfrac{1}{30}\operatorname{tr}_1(A_1 \mathrm{d}\Lambda_1) - \tfrac{1}{30}\operatorname{tr}_2(A_2\mathrm{d}\Lambda_2) + \operatorname{tr}(\omega\mathrm{d}\Theta)\,,$$

$$\text{for } E_8 \times E_8\,, \tag{9c'}$$

$$\delta F = [F, \Lambda]\,, \quad \delta R = [R, \Theta]\,, \quad \delta H = 0\,. \tag{10a,b,c}$$

The subscripts in eq. (9c′) refer to the two E_8 factors. It follows from eq. (10) that the terms in eq. (1) are invariant. It is crucial that the potential B transforms under both kinds of transformations and that the field strength H is invariant.

Cancellation of Yang–Mills gauge anomalies. Anomalies in ten-dimensional gauge theories arise from loop diagrams with six or more external lines and chiral fields circulating around the loop. The anomalies corresponding to diagrams with more than six external lines are uniquely determined by the hexagon diagrams. Their form can be deduced by implementing the Wess–Zumino consistency conditions [8]. The analysis of ref. [9] relates the anomalies to a formal expression in twelve dimensions, namely the gauge-invariant twelve-form Ω_{12}, where

$$\Omega_{4n} = \operatorname{tr} F^{2n}\,. \tag{11}$$

The cases of SO(N) and USp(N) are considered first. In these cases Ω_{4n} can be reexpressed in terms of fundamental-representation matrices

$$\Omega_4 = (N + 2l)\operatorname{tr} F^2\,, \tag{12a}$$

$$\Omega_8 = (N + 8l)\operatorname{tr} F^4 + 3(\operatorname{tr} F^2)^2\,, \tag{12b}$$

$$\Omega_{12} = (N + 32l)\operatorname{tr} F^6 + 15\operatorname{tr} F^2 \operatorname{tr} F^4\,, \tag{12c}$$

where

$$l = -1 \quad \text{for } SO(N)\,, \tag{13a}$$

$$l = +1 \quad \text{for } USp(N)\,. \tag{13b}$$

The first term in eq. (12c) corresponds to anomalies that cannot be cancelled by the addition of local interactions to the effective action. However, it is absent if the gauge group is chosen to be SO(32). The second term in eq. (12c) can be expressed as

$$\operatorname{tr} F^2 \operatorname{tr} F^4 = \mathrm{d}(\omega^0_{3\mathrm{Y}} \operatorname{tr} F^4) = \mathrm{d}(\omega^0_{7\mathrm{Y}} \operatorname{tr} F^2)\,, \tag{14}$$

where $\omega^0_{2n+1\,\mathrm{Y}}$ is defined (up to an exact form) by

$$\mathrm{d}\omega^0_{2n+1\,\mathrm{Y}} = \operatorname{tr} F^{n+1}\,. \tag{15}$$

Explicit expressions for $\omega^0_{2n+1\,\mathrm{Y}}$ are given in ref. [9]. Similarly, applying an infinitesimal gauge transformation, one defines $\omega^1_{2n\,\mathrm{Y}}$ (up to an exact form) by

$$\delta\omega^0_{2n+1\,\mathrm{Y}} = -\mathrm{d}\omega^1_{2n\,\mathrm{Y}}\,, \tag{16}$$

In the present case, the consistent anomaly corresponding to the second term in eq. (12c) is

$$G = c\int(\tfrac{1}{3}\omega^1_{2\mathrm{Y}}\operatorname{tr} F^4 + \tfrac{2}{3}\omega^1_{6\mathrm{Y}}\operatorname{tr} F^2)\,, \tag{17}$$

where the ratio of the two terms is uniquely determined by Bose symmetry. The numerical constant. c can be deduced from refs. [3,9]. This anomaly can be cancelled by adding to the effective action

$$S_1 = c\int(B\operatorname{tr} F^4 + \tfrac{2}{3}\omega^0_{3\mathrm{Y}}\omega^0_{7\mathrm{Y}})\,. \tag{18}$$

The nontrivial Λ transformation of B in eq. (9c) is the key to the cancellation between δS_1 and G. This amounts to a complete cancellation of the Yang–Mills anomaly for the gauge group SO(32).

The extra terms in the action contribute anomalous pieces to tree amplitudes. For example, there is a tree diagram with six external gauge fields (fig. 1) in which the $B \operatorname{tr} F^4$ vertex is connected to the $\omega^0_{3\mathrm{Y}}\, \mathrm{d}B$ vertex (contained in the H^2 term of eq. (1)) by a B propagator. This gives rise to an anomalous piece of the six-particle amplitude, which together with the second term in eq. (18) cancels the hexagon anomaly.

For the group E_8, eq. (12) is replaced by

$$\Omega_4 = \operatorname{tr} F^2 \tag{12a'}$$

$$\Omega_8 = \tfrac{1}{100}(\operatorname{tr} F^2)^2 \tag{12b'}$$

$$\Omega_{12} = \tfrac{1}{7200}(\operatorname{tr} F^2)^3\,. \tag{12c'}$$

It is important in comparing the analysis of the mixed

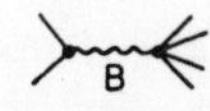

Fig. 1. Six-particle tree amplitude that contributes to anomalies.

Volume 149B, number 1,2,3 PHYSICS LETTERS 13 December 1984

anomalies for SO(32) and $E_8 \times E_8$ that for both these gauge groups

$$\Omega_{12} = \tfrac{1}{48}\Omega_4[\Omega_8 - \tfrac{1}{300}(\Omega_4)^2] . \tag{19}$$

It follows from eq. (12c′) by the same reasoning as before that the consistent anomaly for $E_8 \times E_8$ is given by

$$G' = \frac{c}{108\,000}\int [\omega^1_{2Y_1}(\mathrm{tr}_1 F^2)^2 + \omega^1_{2Y_2}(\mathrm{tr}_2 F^2)^2] . \tag{17'}$$

This can be cancelled by adding

$$S'_1 = \frac{c}{108\,000}\int \{30B[(\mathrm{tr}_1 F^2)^2 + (\mathrm{tr}_2 F^2)^2 - \mathrm{tr}_1 F^2\,\mathrm{tr}_2 F^2] - \omega^0_{3Y_1}\omega^0_{3Y_2}(\mathrm{tr}_1 F^2 - \mathrm{tr}_2 F^2)\} . \tag{18'}$$

Cancellation of gravitational anomalies. Gravitational anomalies can be viewed either as a breakdown of general coordinate invariance or local Lorentz invariance [10,11]. They can be shifted from the one to the other by adding a suitable extra term to the effective action [11]. This term is zero when either anomaly is absent. We choose to study local Lorentz anomalies, rather than general-coordinate ones, since they bear the greater similarity to the Yang–Mills gauge ones. As described above for the Yang–Mills case, Lorentz anomaly structures arising from hexagon (and higher) diagrams with external gravitons only can be associated with the formal twelve-form [3]

$$-[\tfrac{1}{32} + (n-496)/13824]\,(\mathrm{tr}\,R^2)^3 - [\tfrac{1}{8} + (n-496)/5760]\,\mathrm{tr}\,R^2\,\mathrm{tr}\,R^4 - [(n-496)/7560]\,\mathrm{tr}\,R^6. \tag{20}$$

In this expression we have included the contributions of one left-handed spin 3/2 gravitino and one right-handed spin 1/2 field from the supergravity sector and n left-handed spin 1/2 fields from the matter sector, which only depends on the dimension of the gauge group.

The last term in eq. (20) corresponds to an anomaly of the form $\int\omega^1_{10L}$, which cannot be cancelled by adding local terms to the action. Therefore 496 left-handed spin 1/2 fields are needed in the matter sector in order that it vanish. Remarkably, since the dimension of the adjoint representation of SO(32) or $E_8 \times E_8$ is 496, the cancellation occurs for either of these gauge groups. The anomalies associated with the first two terms of eq, (20) can be cancelled (putting $n = 496$) by adding to the effective action for SO(32) or $E_8 \times E_8$

$$S_2 = -c\int [\tfrac{1}{32}B(\mathrm{tr}\,R^2)^2 + \tfrac{1}{8}B\,\mathrm{tr}\,R^4 + \tfrac{1}{12}\omega^0_{3L}\omega^0_{7L}] . \tag{21}$$

Thus, not only does the cancellation of the Yang–Mills anomaly require coupling to the supergravity multiplet, but also the cancellation of the gravitational anomalies requires the contributions of Yang–Mills supermultiplets.

Cancellation of mixed anomalies. The extra terms in the action, which are needed to cancel the pure Yang–Mills and local Lorentz anomalies, also contribute to mixed anomalies. (Consider, for example, the Lorentz variation of the B field in eq. (18)). The relative normalization of all the anomalies can be deduced from eq. (63) of ref. [3]. In particular, the twelve-form that characterizes the mixed anomalies for SO(32) is

$$-\mathrm{tr}\,R^2[\mathrm{tr}\,F^4 + \tfrac{1}{8}(\mathrm{tr}\,F^2)^2] + \mathrm{tr}\,F^2[\tfrac{1}{8}\mathrm{tr}\,R^4 + \tfrac{5}{32}(\mathrm{tr}\,R^2)^2] . \tag{22}$$

The twelve-forms in (14), (20) and (22) are given with the correct relative normalization. The mixed anomalies can be cancelled by adding the local terms

$$S_3 = c\int (\tfrac{1}{8}B\,\mathrm{tr}\,R^2\,\mathrm{tr}\,F^2 + \tfrac{1}{48}\omega^0_{3L}\omega^0_{3Y}\,\mathrm{tr}\,R^2 - \tfrac{1}{24}\omega^0_{3Y}\omega^0_{3L}\,\mathrm{tr}\,F^2 - \tfrac{2}{3}\omega^0_{3L}\omega^0_{7Y} + \tfrac{1}{12}\omega^0_{3Y}\omega^0_{7L}) . \tag{23}$$

The cancellation requires that $N + 8l = 24$ and $N + 2l = 30$ in eq. (12). Thus the cancellation of mixed anomalies gives two independent confirmations of the consistency of choosing the gauge group SO(32). In summary, modifying the action of ref. [7] by including the last term in eq. (4) and adding $S_1 + S_2 + S_3$ to the action gives complete cancellation of all anomalies at one-loop level for the gauge group SO(32).

Given the cancellation of mixed anomalies in SO(32) the possibility of their cancellation in $E_8 \times E_8$ follows from eq. (19). Explicitly, for $E_8 \times E_8$ one obtains

Volume 149B, number 1,2,3 PHYSICS LETTERS 13 December 1984

$$S_3' = \frac{c}{7200} \int (30\, B \,\mathrm{tr}\, R^2 \,\mathrm{tr}_1 F^2 + 5\omega_{3L}^0 \omega_{3Y_1}^0 \,\mathrm{tr}\, R^2 + 20\omega_{3Y_1}^0 \omega_{7L}^0 + \omega_{3Y_1}^0 \omega_{3L}^0 \,\mathrm{tr}_1 F^2 + \omega_{3L}^0 \omega_{3Y_1}^0 \,\mathrm{tr}_2 F^2) + (1 \leftrightarrow 2). \tag{23'}$$

Thus, as before, the addition of $S_1' + S_2 + S_3'$ to the action cancels all the one-loop anomalies for $E_8 \times E_8$. It may be interesting to look for analogous anomaly-free theories in other dimensions.

Superstring anomalies. The results described above were discovered by calculating the anomalies of superstring diagrams and studying the low-energy expansions of the formulae. Since type I superstrings appear otherwise to be consistent quantum theories, the possibility of anomalies is of crucial importance. Their occurrence in a covariant formalism implies that unphysical modes of gauge fields cannot be consistently decoupled, leading to a breakdown of unitarity in the quantum theory. In the light-cone gauge formalism the problem would be expressed as a breakdown of Lorentz invariance.

The groups SO(N), USp(N), and U(N) have been shown to be possible superstring gauge groups at the classical level [4,12]. The U(N) case was shown in ref. [4] to give inconsistencies at the quantum level. In view of the preceding results, it seems very likely to us that $E_8 \times E_8$ is also possible for superstrings. Since such a superstring theory has not yet been formulated, the following discussion is restricted to the cases SO(N) and USp(N).

We have explicitly evaluated the anomalies of one-loop superstring hexagons with external massless Yang–Mills particles in a covariant formalism. The results are summarized here, but the details will be presented elsewhere. The one-loop diagrams come in three topologies, shown in fig. 2, called planar, nonorientable, and nonplanar. The planar and nonorientable diagrams have a group theory structure corresponding to the tr F^6 term in eq. (12) whereas the nonplanar diagrams have the group-theoretic structure of the tr F^2 tr F^4 term. The nonplanar diagrams turn out to be anomaly-free. This means that the superstring theory necessarily contains terms such as in eq. (18) with the requisite coefficients.

The planar and nonorientable superstring diagrams do have anomalies. They are evaluated by associating polarization vectors ζ_r^μ to five of the external lines (r = 2, 3, 4, 5, 6)

$$k_r \cdot \zeta_r = k_r^2 = 0\,, \tag{24}$$

and the momentum vector k_1^μ in place of ζ_1^μ for line no. 1. Just as in ordinary field-theory calculations, various infinities must be carefully regulated in order to get the correct nonzero result. The result has the form

$$G = (\mathrm{const.})\,(N + 32l)\,\mathrm{tr}(\lambda_1\lambda_2\lambda_3\lambda_4\lambda_5\lambda_6) \times \epsilon_{\mu_1 \ldots \mu_5 \nu_1 \ldots \nu_5} \zeta_1^{\mu_1} \ldots \zeta_5^{\mu_5} k_1^{\nu_1} \ldots k_5^{\nu_5} \times \int_0^1 \left(\prod_{i=1}^{5} \vartheta(\nu_{i+1} - \nu_i)\,\mathrm{d}\nu_i \right) \times \prod_{1 \le i < j \le 8} [\sin \pi(\nu_j - \nu_i)]^{2\alpha' k_i \cdot k_j} + \text{permutations}\,, \tag{25}$$

where $\nu_6 = 1$. The coefficient N comes from the planar diagram and the coefficient $32l$ from the nonorientable one. As before, a perfect cancellation occurs for SO(32). Eq. (25) is different from the anomaly of the low-energy effective field theory, because (contrary to common belief) massive fermions also contribute to the anomaly. It reduces to the usual anomaly at low energies ($\alpha' k_i \cdot k_j \ll 1$). Anomalies involving excited external states can be deduced by factorizing on poles of eq. (25).

The field-theoretic discussion in the earlier part of this paper only displayed certain higher-derivative terms, occuring in the low-energy expansion of the

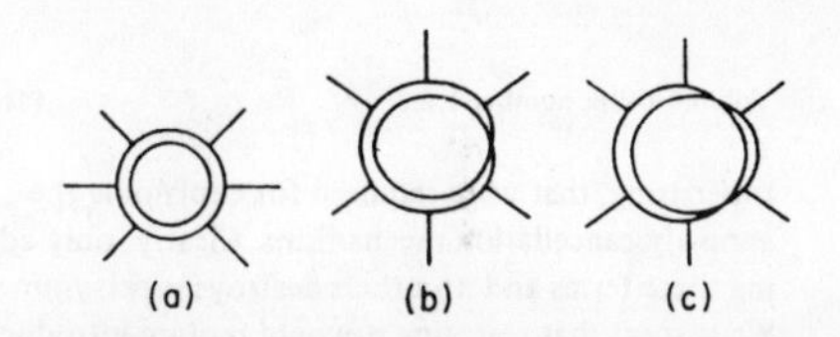

Fig. 2. (a) The planar hexagon string diagram. (b) The nonorientable hexagon string diagram. (c) The nonplanar hexagon string diagram.

superstring, that were required for explaining the anomaly cancellation mechanisms. Clearly, only adding these terms and no others destroys supersymmetry. We suspect that restoring it would require introducing the entire infinite expansion of the superstring effective action. We have not investigated anomalous divergences of the supersymmetry current, but since the gravitino and the graviton are in the same supermultiplet it seems likely that the cancellation of gravitational anomalies should also ensure the cancellation of supersymmetry anomalies in the superstring theory.

In conclusion, it appears very likely that SO(32) superstring theory is a consistent quantum theory. More analysis of anomaly cancellation in various diagrams involving closed strings still needs to be carried out. Also, the possibility of anomalies associated with "big" gauge transformations, not connected to the identity, remains to be investigated. The anomaly cancellations in the case of the $E_8 \times E_8$ effective theory suggest that it is also the low-energy expansion of a superstring theory, which has not yet been formulated. The fact that anomaly-free type I superstring theories are chiral theories containing elementary gauge fields makes them appealing candidates for a fundamental theory.

We wish to acknowledge helpful and stimulating discussions with W. Bardeen, D. Friedan, S. Shenker, P. West, B. Zumino, and other participants at the Aspen Center for Physics. In an earlier version of this paper it was noted that the gravitational anomalies cancel for $E_8 \times E_8$, but the cancellation of the other anomalies in that theory was overlooked. We are grateful to G. Chapline and R. Slansky for pointing out to us that E_8 has no independent fourth- and sixth-order Casimir invariants. L. Dixon, J. Harvey and E. Witten and, independently, J. Thierry-Mieg have also observed that all the Yang–Mills and mixed anomalies can be cancelled for the case of $E_8 \times E_8$.

References

[1] E. Witten, Fermion numbers in Kaluza–Klein theory, Princeton University preprint.
[2] P.H. Frampton and T.W. Kephart, Phys. Rev. Lett. 50 (1983) 1343, 1347;
P.K. Townsend and G. Sierra, Nucl. Phys. B222 (1983) 493;
B. Zumino, Y.S. Wu and A. Zee, Nucl. Phys. B239 (1984) 477;
L. Baulieu, Nucl. Phys. B241 (1984) 557.
[3] L. Alvarez-Gaumé and E. Witten, Nucl. Phys. B234 (1983) 269.
[4] J.H. Schwarz, Phys. Rep. 89 (1982) 223; Caltech preprint CALT-68-1137;
M.B. Green, Surv. High Energy Phys. 3 (1983) 127.
[5] M.B. Green and J.H. Schwarz, Phys. Lett. 122B (1983) 143;
J.H. Schwarz and P.C. West, Phys. Lett. 126B (1983) 301;
J.H. Schwarz, Nucl. Phys. B226 (1983) 269;
P.Howe and P.C. West, Nucl. Phys. B238 (1984) 181.
[6] M. Gell-Mann and B. Zwiebach, Phys. Lett. 147B (1984) 111.
[7] E. Bergshoeff, M. de Roo, B. de Wit and P. van Nieuwenhuizen, Nucl. Phys. B195 (1982) 97;
G.F. Chapline and N.S. Manton, Phys. Lett. 120B (1983) 105.
[8] J. Wess and B. Zumino, Phys. Lett. 37B (1971) 95.
[9] B. Zumino, Y.S. Wu and A. Zee, Nucl. Phys. B239 (1984) 477.
[10] Z. Zumino, Les Houches lectures (1983);
R. Stora, Cargese lectures (1983).
[11] W.A. Bardeen and B. Zumino, Berkeley preprint LBL-17639;
L. Alvarez-Gaumé and P. Ginsparg, Harvard University preprint HUTP-84/AO16.
[12] N. Marcus and A. Sagnotti, Phys. Lett. 119B (1982) 97.,

Volume 151B, number 5,6 PHYSICS LETTERS 21 February 1985

SUPERSTRINGS FROM 26 DIMENSIONS?

Peter G.O. FREUND[1]
Institute for Theoretical Physics, University of California, Santa Barbara, CA 93106, USA

Received 13 November 1984

The finite type I superstring theories of Green and Schwarz (SO(32) and (?)$E_8 \times E_8$) in ten dimensions are viewed as special dimensional reductions on 16-tori from the non-supersymmetric Veneziano–Nambu–Goto strings in 26 dimensions. Fermions appear as solitons of the two-dimensional string field theory. Various problems of such an approach are pointed out and possible solutions outlined.

Green and Schwarz [1] have discovered that the $N = 1$ supersymmetric theory of open and closed strings with SO(32) Chan–Paton rules [2] is anomaly-free and very likely finite. The freedom of anomalies, though not the finiteness, persists in the Scherk limit [3] of point-like strings ($\alpha' \to 0$). In this limit the theory reduces to a coupled $N = 1$ supersymmetric SO(32) Yang–Mills plus supergravity system with some additional terms [1]. Were one, in this Scherk limit, to change the gauge group G, anomaly cancellation would occur also for $G = E_8 \times E_8$ (which has the same rank and dimension as SO(32)), but for no other compact semi-simple group [3]. The existence of superstrings with $E_8 \times E_8$ symmetry is therefore expected. Alas, no generalization of the Chan–Paton rules to exceptional groups has, so far, been found, and this theory has not yet been constructed. Both the SO(32) and the $E_8 \times E_8$ superstrings (should the latter exist) would allow chiral fermions upon dimensional reduction to four dimensions [4]. They both look phenomenologically promising: one of them could be a "theory of the world", which unifies all interactions (gravity included) and all forms of matter in a consistent manner. Yet, two theories of the world are one theory too many. If one of these two theories had an as yet undiscovered flaw not shared by the other, this, of necessity, would decide the uniqueness problem. Here we contemplate the possibility that both the SO(32) and $E_8 \times E_8$ theories are equally problem-free, and are but two realizations of one and the same underlying theory. Specifically we propose that this primordial theory involves $d > 10$ space–time dimensions, and that the SO(32) and $E_8 \times E_8$ superstrings in ten dimensions are but different compactifications of this primordial theory. I shall argue that $d = 26$ and that the primordial theory is the old *non-supersymmetric* Veneziano–Nambu–Goto (VNG) string [5].

Right away this raises a number of questions:

(a) How does the compactification of the VNG string produce the Green–Schwarz groups SO(32) and $E_8 \times E_8$?

(b) How does one account for the appearance of fermions and supersymmetry in the resulting ten-dimensional theory?

(c) How do the ten-dimensional theories free themselves from the tachyon and from other diseases of the VNG string?

(d) How do the Chan–Paton rules in ten dimensions emerge?

In an attempt to justify this picture, questions (a)–(d) will be addressed below, though no complete answers will be provided.

A string theory is a conformally invariant two-dimensional σ-model. As such, it admits the infinite Virasoro algebra of two-dimensional conformal transformations, and is uniquely characterized by the Kac–Moody algebra of its "currents". Any two conformally invariant two-dimensional quantum field theories that share the same "current algebra" are equivalent [6].

Consider, on the one hand, the two-dimensional

[1] On leave from the Department of Physics and The Enrico Fermi Institute, The University of Chicago, Chicago, IL 60637, USA.

0370-2693/85/$ 03.30 © Elsevier Science Publishers B.V. (North-Holland Physics Publishing Division)

Volume 151B, number 5,6 PHYSICS LETTERS 21 February 1985

σ-model corresponding to the rank-26 Lie group SO(44) × $\mathbf{R}^4$ along with a Wess–Zumino term with quantized coefficient $N = 1$, at the value of the coupling constant at which the β-function vanishes and conformal invariance is restored. On the other hand, consider the 26-dimensional VNG string, dimensionally reduced to Minkowski four space–time on a 22-torus $\bar{\mathrm{T}}^{22}$ with all 22 radii equal. Based on results of Frenkel and of Goddard and Olive [7], Witten [8] has shown these two, on the face of it such very different theories, to be equivalent in the above sense. Notice that the rank of the group SO(44) is the same as the dimension of $\bar{\mathrm{T}}^{22}$. This torus is the factorization $\mathbf{R}^{22}/\Lambda_{22}$ of euclidean 22-space $\mathbf{R}^{22}$ by the 22-dimensional hypercubic lattice Λ_{22}. The crucial feature is, as will be recalled below, that setting the lattice spacing equal to one (unimodular lattice), Λ_{22} contains precisely 924 vectors of length squared equal to 2 and that these vectors present the root diagram of the Lie algebra $\mathcal{SO}(44)$. In the same way, the $(26-d)$-torus $\bar{\mathrm{T}}^{26-d}$ with all radii equal is $\mathbf{R}^{26-d}/\Lambda^{26-d}$ (Λ_{26-d} being the $(26-d)$-dimensional unimodular hypercubic lattice). Again the vectors of length squared equal to 2 in the Λ_{26-d} span the root diagram of $\mathcal{SO}(52-2d)$, the rank of which is $26-d$, the common dimension of Λ_{26-d} and of $\bar{\mathrm{T}}^{26-d}$.

Now let us specialize to the case where we reduce from 26 down to $d = 10$ dimensions. The lattice $\Lambda_{26-10} = \Lambda_{16}$ describes the quantization of 16 momentum components when the string theory is compactified on $\bar{\mathrm{T}}^{16}$. At every vector $k \in \Lambda_{16}$ we can define a Nambu vertex operator $U(k,z)$ [5,7]. Consider now the operators $U(k_i,z)$ $i = 1, ..., 480$ where k_i are the 480 vectors of length squared equal to 2 ($k_i^2 = 2$) of the unimodular lattice Λ_{16}. From these $U(k_i,z)$ by suitable contour integration over the complex variable z, and some further manipulations, we recover [7] 480 operators which, together with the components of the momentum operator in the 16 lattice directions, span the Lie algebra $g_{\Lambda_{16}} = \mathcal{SO}(32)$. Taking higher moments in the z-integration yields the Kac–Moody algebra $\tilde{g}_{\Lambda_{16}}$. Its representation theory ensures the presence of 496 massless vector particles in the spectrum, i.e., of a gauged SO(32) symmetry. It now appears as if we had arrived at a bosonic string theory in 10 large and 16 compactified dimensions in which a gauged SO(32) symmetry is present. But repeating the just-mentioned construction, this time choosing for k_i the unit vectors of Λ_{16}, one can reconstruct [7] the *fermionic* oscillators of the Neveu–Schwarz and Ramond model as well. Fermions are thus obtained out of bosons, but then a two-dimensional system is involved here, so that, as in the sine-Gordon and other two-dimensional systems, fermions can arise in an originally bosonic theory [9]. These fermions are solitons and they fall in a supersymmetric pattern (for the Ramond-like excitations, their spinorial character, as far as ten-dimensional space–time is concerned, is not clear, though). So a kind of dynamical supersymmetry appears. We believe this ten-dimensional supersymmetric theory contains the SO(32) Green–Schwarz superstring.

Before continuing on this point, let us now consider the $E_8 \times E_8$ theory. Already in the SO(32) case, we considered the SO(32) σ-model with $N = 1$ Wess–Zumino terms at the zero of its β-function. To get the proper central extension in the Virasoro algebra, the rank of the group had to be 16. The whole world manifold $\mathbf{R}^{10} \times$ SO(32) is then a 506-dimensional group manifold (Minkowski 10-space $\mathbf{R}^{10}$ is, after all, an abelian group itself) of total rank 26. What in the older literature is referred to as the critical dimension having to be 26, can now be rephrased as the critical *rank* equalling 26. (This raises the question as to whether 26 or 506 is the "true" dimensionality of space–time in this case.) In that sense, any other candidate group must also have rank 26 which is the case for $E_8 \times E_8 \times \mathbf{R}^{10}$. There exists a well-known even unimodular integral lattice $\Gamma_8 + \Gamma_8$, obtained by iterating the 16-dimensional root diagram of $E_8 \times E_8$ [10,11]. $\Gamma_8 + \Gamma_8$ and another lattice Γ_{16}, whose vectors of length squared span the root diagram of $\mathcal{SO}$ (32), are the only two 16-dimensional euclidean, even unimodular, integral lattices up to isomorphism [10–12]. The $\mathcal{SO}$ (32) root diagram, as we have seen, also sits in Λ_{16} an odd lattice which contains unit vectors. By contrast, there exists no 16-dimensional odd integral euclidean unimodular lattice containing unit vectors such that its length squared two vectors span $E_8 \times E_8$ [13]. The construction of the Neveu–Schwarz–Ramond sector in the $E_8 \times E_8$ case is therefore still open. Maybe dropping unimodularity will help.

With the Wess–Zumino parameter set at $N = 1$ the rank of the group G (other than Minkowski space) had to equal $26 - d$, the dimension of the torus $\bar{\mathrm{T}}^{26-d}$.

Volume 151B, number 5,6 PHYSICS LETTERS 21 February 1985

For Wess–Zumino integers $N \neq 1$, this is replaced by the condition [14]

$$r = (26 - d)\left(\frac{d_G}{Nd_G + (N-1)(d-26)}\right)$$

with d_G the dimension of the group G (for $N = 1$ this reduces to $r = 26 - d$, as it should). For $d = 10$, the case considered here, this allows the further groups G_2, SU(3), SU(4) ~ SO(6), and SO(8) for $N = -48$, $N = -6$, $N = -64$, and $N = +8$, respectively. (Similar considerations for $d = 4$ have been made earlier by Friedan [15], see also ref. [4].)

We now return to the problem of recovering the SO(32) superstring theory from the compactified VNG string and to the questions (b)–(d) posed in the introduction.

First of all, the VNG string has a tachyon [5] which means that one is expanding around an unstable vacuum, which, as pointed out by Gross and Witten [16], is not even a classical solution. The tachyon cannot possibly appear in the superstring, on account of ten-dimensional supersymmetry. But in a superstring theory, there are two kinds of supersymmetry, the two-dimensional conformal supersymmetry on the string's world sheet, and the Poincaré supersymmetry of the ten-dimensional host space. In the original Neveu–Schwarz–Ramond formulation the latter supersymmetry was not enforced and tachyons were present. They were then eliminated via a projection onto an even "*G*-parity" sector [17]. Coming from 26 dimensions, whatever supersymmetry is achieved via the Frenkel–Goddard–Olive construction is also of a world-sheet type, and a further projection is called for onto a ten-dimensional Poincaré supersymmetric sector.

Second, the original 26-dimensional theory is not finite or even consistent [16]. Toroidal compactifications can be contemplated toward final dimensions other than 10. But only in the critical ten-dimensional case is there a chance of obtaining a non-anomalous super-Virasoro algebra. Maybe this drives the inconsistent [16] 26-dimensional theory down to ten dimensions. The original 26-dimensional string has no fermions and no self-dual antisymmetric tensor fields. Then it cannot have gauge or gravitational anomalies. This may be the reason why the reduced superstring theories are anomaly-free themselves.

Now to the Chan–Paton rules: These rules were inspired by the two-component duality exhibited by hadronic reactions [18]. In string theory they occupy a rather peculiar place. One produces a detailed dynamics for a string devoid of internal attributes and then appends these Chan–Paton rules as an afterthought, not at the lagrangian level, but when calculating *S*-matrix elements. What is the dynamics behind these rules? One can require the internal attributes to be carried at the ends of the string. In the SO(32) case, the ends carry the fundamental representation 32, so that the open strings carry the representations contained in the product 32 × 32. The rules vouch for the consistency of this picture. A priori, one could have expected strings in higher representations of SO(32) as well. But the rules assure the consistency of having intermediate one-particle states in all channels only in those representations contained in the product 32 × 32. For $E_8 \times E_8$ such a truncation does not seem possible, as suggested also by the σ-model picture. This raises the general question as to what algebraic combination laws for the internal attributes of strings are compatible with duality. At the simplest level, is there a set A of $E_8 \times E_8$ representations including (1,1) + (1,248) + (248,1) which is "closed" under crossing, in that particle poles in representations from the set A in the s-channel involve, by duality, poles again only in representations from the set A in the t- and u-channels for all combinations of external particles. One may also ask whether there are string models, which, while producing the same $\alpha' \to 0$ Scherk limit as the SO(32) superstring, involve for finite α' not just the representations contained in the product 32 × 32 but also higher ones. In other words, are the Chan–Paton rules the unique prescription even in the SO(32) case? A general theory of internal attribute algebraic rules is needed (with usual Chan–Paton rules only O(n) and Sp(n) groups are allowed; $E_8 \times E_8$ is ruled out [19]).

In conclusion then, the rank 16 of the groups SO(32) and $E_8 \times E_8$, along with Witten's σ-model considerations, suggest that one derive these superstrings from a 26-dimensional string theory. The VNG string is an obvious candidate, and the more obvious difficulties (lack of fermions, supersymmetry, ...) can be solved taking advantage of the remarkable properties of two-dimensional field theories (fermionic solitons, Kac–Moody algebras, ...). Ultimately, only a detailed derivation of the ten-dimensional superstring from the 26-dimensional VNG string can vindicate this admittedly quite unorthodox point of view. Still, one cannot but find it ironical that a theory originally meant to describe the hadronic S-matrix and then abandoned, should now, whether in ten- or 26-dimensional guise, make a comeback as a field theory "of the world".

I wish to thank Professor J.R. Schrieffer and Professor R. Sugar for their kind hospitality at the Institute for Theoretical Physics and Professor D. Friedan, Professor J. Schwarz and Professor E. Witten for their patient criticism and comments. This material was based upon research supported in part by the National Science Foundation under Grant Nos. PHY83-01221 and PHY77-27084, supplemented by funds from the National Aeronautics and Space Administration.

References

[1] M.B. Green and J. Schwarz, Phys. Lett. 149B (1984) 117; 151B (1985) 21.
[2] J.E. Paton and H.M. Chan, Nucl. Phys. B10 (1969) 516; T. Matsuoka, K. Ninomiya and S. Sawada, Prog. Theor. Phys. 42 (1969) 56; H. Harari, Phys. Rev. Lett. 22 (1969) 562; J. Rosner, Phys. Rev. Lett. 22 (1969) 689.
[3] J. Scherk, Nucl. Phys. B31 (1971) 222.
[4] E. Witten, Nucl. Phys. B, to be published; S. Randjibar Daemi, A Salam and J. Strathdee, Phys. Lett. 132B (1983) 56.
[5] See, e.g., C. Rebbi, Phsys. Rep. 12C (1974) 1.
[6] E. Witten, Commun. Math. Phys. 92 (1984) 455.
[7] I.B. Frenkel, preprint; P. Goddard and D. Olive, preprint.
[8] E. Witten, unpublished; see also P.G.O. Freund, P. Oh and J.T. Wheeler, Nucl. Phys. B, to be published.
[9] S. Coleman, Phys. Rev. D11 (1975) 1088; S. Mandelstam, Phys. Rev. D11 (1975) 3026.
[10] E. Witt, Abh. Math. Sem. Univ. Hamburg 14 (1941) 323.
[11] J. Milnor, Proc. Natl. Acad. Sci. USA 51 (1964) 542.
[12] J.-P. Serre, A course in arithmetic (Springer, Berlin, 1973) pp. 55, 110.
[13] M. Kneser, Arch. Math. 8 (1957) 241.
[14] A. Rocha-Caridi, private communication to D. Friedan.
[15] D. Friedan, private communication.
[16] D. Gross and E. Witten, private communication.
[17] J. Schwarz, Phys. Rep. 8C (1973) 269.
[18] R. Dolen, D. Horn and C. Schmid, Phys. Rev. Lett. 19 (1967) 402; P.G.O. Freund, Phys. Rev. Lett. 20 (1968) 235; H. Harari, Phys. Rev. Lett. 20 (1968) 1395.
[19] N. Marcus and A. Sagnotti, Phys. Lett. 119B (1982) 97.

VOLUME 54, NUMBER 6 PHYSICAL REVIEW LETTERS 11 FEBRUARY 1985

Heterotic String

David J. Gross, Jeffrey A. Harvey, Emil Martinec, and Ryan Rohm
Joseph Henry Laboratories, Princeton University, Princeton, New Jersey 08544
(Received 21 November 1984)

A new type of superstring theory is constructed as a chiral combination of the closed $D=26$ bosonic and $D=10$ fermionic strings. The theory is supersymmetric, Lorentz invariant, and free of tachyons. Consistency requires the gauge group to be $\mathrm{Spin}(32)/Z_2$ or $E_8\times E_8$.

PACS numbers: 11.30.Pb, 11.30.Ly, 12.10.En

Recent interest in superstring unified field theories has been sparked by the discovery of Green and Schwarz[1] that nonorientable (type I) open and closed superstrings[2] with $N=1$ supersymmetry are finite and free of anomalies if the gauge group is SO(32). Previously the only consistent, anomaly-free[3] superstring theory was that of orientable (type II) $N=2$ supersymmetric closed strings. The new theory has the advantage of already containing a large (and unique) gauge group. It is much easier to contemplate this theory producing the low-energy gauge group, as well as families of chiral massless fermions, upon compactification of the original ten dimensions. Witten has discussed some of the phenomenology of this theory and has shown that it is easy to imagine compactifications that yield an SU(5) theory with any number of standard fermionic generations.[4]

The anomaly cancellation mechanism of Green and Schwarz is based on group theoretical properties of SO(32) which are shared by only one other semisimple Lie group, namely $E_8\times E_8$. Such a group, however, cannot appear in the standard form of open-string theory, in which gauge groups are introduced by attaching quantum numbers to the ends of the string and Chan-Paton factors[5] to the scattering amplitudes. This procedure yields only the gauge groups $\mathrm{SO}(N)$ and $\mathrm{Sp}(2N)$.[6] The correspondence between the low-energy limit of existing supersymmetric string theories with anomaly-free, $D=10$, supergravity field theories suggests the existence of a new kind of string theory whose low-energy limit would have an $E_8\times E_8$ gauge group. Eschewing the Chan-Paton route to gauge groups for open strings, one might try to obtain $E_8\times E_8$ by compactifying a higher-dimensional closed-string theory. An important clue to how such a theory might arise is provided by the work of Frenkel and Kac.[7]

In this Letter we shall outline the construction of a new kind of closed-string theory, whose low-energy limit is $D=10$, $N=1$ supergravity coupled to supersymmetric Yang-Mills theory with gauge group $\mathrm{Spin}(32)/Z_2$ *or* $E_8\times E_8$. This theory is constructed as a hybrid of the $D=10$ fermionic string and the $D=26$ bosonic string, which preserves the appealing features of both. We show that the orientable, closed heterotic[8] string has an $N=1$ supersymmetric spectrum of states of positive metric, is free of tachyons and is Lorentz invariant. The requirement that gravitational and gauge anomalies be absent necessitates the compactification of the extra sixteen bosonic coordinates of the heterotic string on a maximal torus of determined radius, in a way that produces gauge groups $\mathrm{Spin}(32)/Z_2$ or $E_8\times E_8$. We further argue that the heterotic loop diagrams are free of all infinities—thus yielding new consistent candidates for a unified field theory.

The construction of the heterotic string is based on the observation that the states of the first quantized type-II closed strings, fermionic or bosonic, are essentially direct products of left- and right-moving modes. The physical degrees of freedom of the bosonic string are the 24 transverse coordinates $X^i(\tau-\sigma)$ and $\tilde{X}^i(\tau+\sigma)$ which describe right- (left-) moving two-dimensional free fields, with periodic boundary conditions on the circle $0\leqslant\sigma\leqslant\pi$. The fermionic string contains eight transverse coordinates as well as eight right- and left-moving two-dimensional real fermions, $S^a(\tau-\sigma)$ and $\tilde{S}^a(\tau+\sigma)$ $(a=1,\ldots,8)$ which are Majorana-Weyl ten-dimensional light-cone spinors.[1] The right- and left-handed components of the string are tied together by the constraint that the total momentum and position of each component be identical. Thus the bosonic coordinates are given by the operators (we choose units in which the slope parameter is $\alpha'=\frac{1}{2}$)

$$X^i(\tau-\sigma)=\tfrac{1}{2}x^i+\tfrac{1}{2}p^i(\tau-\sigma)+\frac{i}{2}\sum_{n\neq0}\frac{1}{n}[\alpha_n^i e^{-2in(\tau-\sigma)}],$$
$$\tilde{X}^i(\tau+\sigma)=\tfrac{1}{2}x^i+\tfrac{1}{2}p^i(\tau+\sigma)+\frac{i}{2}\sum_{n\neq0}\frac{1}{n}[\tilde{\alpha}_n^i e^{-2in(\tau+\sigma)}], \tag{1}$$

where

$$[\alpha_n^i,\alpha_m^j]=[\tilde{\alpha}_n^i,\tilde{\alpha}_m^j]=\delta^{ij}\delta_{m+n,0}n,$$

$$[\tilde{\alpha}_n^i,\alpha_m^j]=0,\quad [x^i,p^j]=i\delta^{ij},$$

whereas the fermionic coordinates are

$$S^a(\tau-\sigma)=\sum_{n=-\infty}^{+\infty}S_n^a e^{-2in(\tau-\sigma)},$$

where

$$\gamma^+S_n=hS_n=0,\quad \{S_m^a\bar{S}_n^b\}=(\gamma^+h)^{ab}\delta_{m+n,0} \tag{2}$$

$[h=\frac{1}{2}(1\pm\gamma_{11})]$, with a similar expression for the left movers $\tilde{S}^a(\tau+\sigma)$.

The states of the first quantized, type-II, closed strings are direct products of the Fock space states of the right and left movers. In addition, the basic one-particle operators of the theory (e.g., the super Poincaré generators) are direct sums of operators in the right- and left-moving sectors separately, and the vertex operators that describe the splitting and joining of the strings are direct products of left and right vertices.[2,9] Therefore one can, in principle, construct a consistent string theory in which the left- and right-handed components are treated asymmetrically—as long as each sector is internally consistent. We therefore construct the heterotic string to consist of fermionic-string right movers: eight transverse coordinates X^i $(i=1,\ldots,8)$, eight Majorana-Weyl fermionic coordinates S^a; and of bosonic string left movers: eight transverse coordinates $\tilde{X}^i$ and sixteen internal coordinates $\tilde{X}^I$, $I=1,\ldots,16$.

$$\tilde{X}^I(\tau+\sigma)=x^I+p^I(\tau+\sigma)+\frac{i}{2}\sum_{n\neq 0}\frac{1}{n}\tilde{\alpha}_n^I e^{-2in(\tau+\sigma)}, \tag{3}$$

where

$$[\tilde{\alpha}_n^I,\tilde{\alpha}_m^J]=n\delta_{m+n,0}\delta^{IJ},\quad [x^I,p^J]=\tfrac{1}{2}i\delta^{IJ}.$$

(The factor of $\frac{1}{2}$, familiar in light-cone quantization, arises since X^I is a function of $\tau+\sigma$ alone.)

If X^I is to satisfy the periodic boundary conditions of the closed string, either $p^I=0$ or else x^I parametrizes some compact space T. The simplest (and at present the only known consistent) choice for T is a sixteen-dimensional torus. Compactification of a conventional string coordinate on a nonsimply connected manifold introduces several new features.[10] There now exist classically stable topological configurations where a string coordinate, $X(\sigma)=x+2\alpha'p\tau+2NR\sigma+\ldots$, winds N times around the manifold as σ runs from 0 to π. In addition, of course, the momenta, p^I, are quantized in units of the inverse radius, $1/R$, of the manifold.

In order to achieve a consistent string theory involving only left-moving coordinates X^I to cancel anomalies and to preserve the geometrical structure of string interactions, we are forced to compactify on a special torus. T must be a "maximal" torus—a product of circles of equal radii, $R=(\alpha')^{1/2}=1/\sqrt{2}$, on which points are identified according to

$$X^I\equiv X^I+\sqrt{2}\pi R\sum_{i=1} e_i^I n_i, \tag{4}$$

where e_i^I are sixteen basis vectors [normalized so that $(e_i^I)^2=2$] which generate an integer, even, self-dual lattice. This means that $g_{ij}=\sum_{I=1}^{16}e_i^Ie_j^I$ is integer valued and $\det g=1$. In this case the allowed values of the momenta are

$$p^I=\sum_{i=1}^{16}n_ie_i^I\quad (n_i=\text{integer}), \tag{5}$$

and must equal the winding numbers N^I. [Note that the minimal value of $(p^I)^2$ is two.] There exist only two lattices of this type. One is Γ_{16} and coincides with the lattice of weights of $\mathrm{Spin}(32)/Z_2$, the other is a direct product, $\Gamma_8\times\Gamma_8'$, where Γ_8 is the lattice of weights of E_8.

To complete the construction of the heterotic string we express $X^{\pm}=(1/\sqrt{2})(X^0\pm X^9)$ in terms of the physical degrees of freedom, as usual in light-cone gauge,[11] $X^+=x^++p^+\tau$ and

$$X^-(\sigma,\tau)=x^-+p^-\tau+\frac{i}{2}\sum\frac{1}{n}(\alpha_n^-e^{-2in(\tau-\sigma)}+\tilde{\alpha}_n^-e^{2in(\tau+\sigma)}). \tag{6}$$

In the heterotic string α_n^- is constructed as in the fermionic string

$$\alpha_n^-=\frac{1}{p^+}\sum_m\alpha_m^i\alpha_{n-m}^i+\frac{1}{2p^+}\sum_m\left(m-\frac{n}{2}\right):\bar{S}_{n-m}\gamma^-S_m: \tag{7}$$

whereas $\tilde{\alpha}_n^-$ is constructed as in the bosonic string

$$\tilde{\alpha}_n^-=\frac{1}{p^+}\sum_m(\tilde{\alpha}_m^i\tilde{\alpha}_{n-m}^i+\tilde{\alpha}_m^I\tilde{\alpha}_{n-m}^I). \tag{8}$$

The mass operator is

$$\tfrac{1}{2}\alpha'(\text{mass})^2=N+(\tilde{N}-1)+\tfrac{1}{2}\sum_I(p^I)^2, \tag{9}$$

where $N(\tilde{N})$ are the number operators for the right (left) movers (i.e., $N=p^+\alpha_0^-$, $\tilde{N}=p^+{:}\tilde{\alpha}_0^-{:}$). The subtraction of 1 from $\tilde{N}$ arises from the normal ordering of the bosonic number operator $\tilde{N}$; it is also required in order to maintain Lorentz invariance. Finally, we must constrain the physical states so that

$$N=\tilde{N}-1+\tfrac{1}{2}\textstyle\sum_I(p^I)^2. \tag{10}$$

This implies that the unitary operator $U(\Delta)=\exp 2i\Delta[N-\tilde{N}+1-\frac{1}{2}\sum_I(p^I)^2]$, which shifts σ (in X^i, $\tilde{X}^i$, $\tilde{X}^I$) by Δ, is the identity operator on physical states, thereby ensuring that there is no distinguished point on the closed string.

The physical states of the heterotic string are direct products of Fock space states ($|\,\rangle_R\times|\,\rangle_L$) of the right-moving fermionic string and the left-moving bosonic string, subject to the constraint (10). The right-handed ground state is annihilated by α_n^i, S_n^i $(n>0)$, and N, and forms an irreducible representation of the zero mode oscillators S_0^a, containing eight transverse bosonic states $|i\rangle_R$ and eight fermionic states $|a\rangle_R$. The left-handed ground state, annihilated by $\tilde{\alpha}_n^i$, $\tilde{\alpha}_n^I$ $(n>0)$, and $\tilde{N}$, with zero p^I is removed from the physical Hilbert space by the constraint. Therefore the heterotic string is free of tachyons! The lowest mass states are direct products of $|i\rangle_R$ or $|a\rangle_R$ with $\tilde{\alpha}_{-1}^i|0\rangle_L$, $\tilde{\alpha}_{-1}^I|0\rangle_L$, or $|p^I\rangle_L$ [with $(p^I)^2=2$] and are all massless. The states $|i \text{ or } a\rangle_R\times\tilde{\alpha}_{-1}^j|0\rangle_L$ form the irreducible $N=1$, $D=10$ supergravity multiplet. The states $|i \text{ or } a\rangle_R\times\alpha_{-1}^I|0\rangle_L$ and $|i \text{ or } a\rangle_R\times|p^I\rangle_L$ [with $(p^I)^2=2$] form an irreducible $N=1$, $D=10$, super Yang-Mills multiplet of G [either SO(32) or $E_8\times E_8$]—the Lie group whose maximal torus coincides with T. These consist of sixteen neutral vector mesons (plus their supersymmetric partners) with $p^I=0$, which are the ordinary Kaluza-Klein gauge bosons arising from the U(1)[16] isometry of T. The additional 480 charged vectors with $(p^I)^2=2$, which complete the adjoint representation of G, are special to closed-string theory. Given the work of Frenkel and Kac, who have used string vertices to construct representations of Kac-Moody-Lie algebras, it is not surprising that the standard geometrical string interactions will yield a heterotic string whose states form representations of G with G-invariant interactions (note that massive excited states can have arbitrarily large values of the charges p^I, corresponding to arbitrary representations of G).

Lorentz invariance in $D=10$ is easily established since the generators, $J^{\mu\nu}=\int_0^\pi d\sigma[\bar{X}^\mu P^\nu-\bar{X}^\nu P^\mu]+K^{\mu\nu}$ [where $\bar{X}^\mu=X^\mu+\tilde{X}^\mu$, $P^\mu=(d/d\tau)\bar{X}^\mu$, $K^{ij}=i/8\times\sum\bar{S}_{-n}\gamma^{ij-}S_n$, etc.] act separately on the right and left movers. Since the right movers (left movers) are those of the fermionic (bosonic) string in its critical dimension of 10(26), the would-be anomalies cancel. In addition the hybrid string contains one supersymmetry, generated by

$$Q^a=i(p^+)^{1/2}(\gamma+S_0)^a+2i\frac{1}{(p^+)^{1/2}}\sum(\gamma_iS_{-n})^a\alpha_n^i, \tag{11}$$

which acts on right movers alone, and satisfies (as in the fermionic string)

$$\{Q^a,Q^b\}=-2(n\gamma^\mu P)^{ab}. \tag{12}$$

The interactions of the heterotic string, which correspond geometrically to the splitting and joining of closed strings, can be constructed as direct products of the vertex operators for the fermionic and bosonic strings. We have constructed the vertices for the emission of massless closed strings and have shown that they are Lorentz and G invariant.[12] We have also examined the one-loop diagrams and have seen that these are consistent with unitarity (the restriction to self-dual lattices comes from this requirement) and are finite.[12] While we have not calculated the string hexagon diagrams, the equivalence of the heterotic low-energy field theory to the anomaly-free $D=10$ $N=1$ theories convinces us that they will be anomaly free. Thus we have established the existence of two new consistent closed-string theories, which naturally lead, by a string Kaluza-Klein mechanism, to the gauge symmetries of SO(32) or $E_8\times E_8$. These theories differ in essential ways from open-string gauge theories (for example the gauge coupling $g^2=\kappa^2/\alpha'$ as opposed to $g^2=\kappa\alpha'$ for open strings).

Fermionization of two-dimensional field theories often simplifies their structure, as in the case of the nonlinear sigma model with a Wess-Zumino term.[13] We have also constructed a version of the heterotic string in which the internal coordinates are fermions [in this case, $E_8\times E_8$ is realized on its SO(16) $\otimes$ SO(16) subgroup]. Details of this construction will be presented in Ref. 12.

Additional possibilities for use of the above mechanism to generate new string theories are severely limited. Attempts to compactify the $D=26$, bosonic string to $D=10$ on a sixteen-dimensional torus[14] are doomed, since they would produce a gauge group of $G\times G$, would not contain $D=10$ fermions or supersymmetry, and would have tachyons. Compactification of the type-II closed fermionic string will produce no new gauge symmetries associated with winding strings.

Finally, we note that the heterotic $E_8\times E_8$ string is perhaps the most promising candidate for a unified field theory. One can easily contemplate physically interesting compactifications of this theory to four dimensions, including the possibility that the $E_8 \rightarrow E_8'$ symmetry is unbroken, thereby implying the existence of a "shadow world" consisting of E_8' matter which interacts with us (E_8 matter) only gravitationally.[12]

We would like to acknowledge conversations with M. Green, M. Peskin, and E. Witten. This work was supported in part by the National Science Foundation under Grant No. PHY-80-19754.

[1]M. B. Green and J. H. Schwarz, California Institute of Technology Reports No. CALT-68-1182, and No. CALT-68-1194 (unpublished).

[2]J. H. Schwarz, Phys. Rep. **89**, 223 (1982); M. B. Green, Surv. High Energy Phys. **3**, 127 (1983).

[3]L. Alvarez-Gaumé and E. Witten, Nucl. Phys. **B234**, 269 (1983).

[4]E. Witten, to be published.

[5]J. Paton and H. M. Chan, Nucl. Phys. **B10**, 519 (1969).

[6]J. H. Schwarz, in Proceedings of Johns Hopkins Workshop, Florence, 1982 (unpublished); M. Marcus and A. Sagnotti, Phys. Lett. **119B**, 97 (1982).

[7]I. B. Frenkel and V. G. Kac, Inv. Math. **62**, 23 (1980); G. Segal, Commun. Math. Phys. **80**, 301 (1981); P. Goddard and D. Olive, to be published.

[8]From the Greek "heterosis": increased vigor displayed by crossbred animals or plants.

[9]M. B. Green and J. H. Schwarz, Nucl. Phys. **243**, 475 (1984).

[10]E. Cremmer and J. Scherk, Nucl. Phys. **B103**, 399 (1976).

[11]P. Goddard, J. Goldstone, C. Rebbi, and C. Thorn, Nucl. Phys. **B56**, 109 (1973).

[12]D. Gross, J. Harvey, E. Martinec, and R. Rohm, to be published.

[13]E. Witten, Commun. Math. Phys. **92**, 455 (1984).

[14]P. Freund, to be published.

INDEX